Progress in Mathematical Physics
Volume 31

Editors-in-Chief

Anne Boutet de Monvel, *Université Paris VII Denis Diderot*
Gerald Kaiser, *The Virginia Center for Signals and Waves*

Editorial Board

D. Bao, *University of Houston*
C. Berenstein, *University of Maryland, College Park*
P. Blanchard, *Universität Bielefeld*
A.S. Fokas, *Imperial College of Science, Technology and Medicine*
C. Tracy, *University of California, Davis*
H. van den Berg, *Wageningen University*

L.R. Rakotomanana

A Geometric Approach to Thermomechanics of Dissipating Continua

Birkhäuser
Boston • Basel • Berlin

L.R. Rakotomanana
Université de Rennes 1
Institut Mathématique de Rennes
Campus de Beaulieu
Rennes Cedex 35042
France

Library of Congress Cataloging-in-Publication Data

Rakotomanana, Lalao.
　A geometric approach to thermomechanics of dissipating continua / Lalao Rakotomanana.
　　p. cm. – (Progress in mathematical physics ; v. 31)
　Includes bibliographical references and index.
　ISBN 0-8176-4283-8 (alk. paper) – ISBN 3-7643-4283-8 (alk. paper)
　　1. Thermodynamics. 2. Continuum mechanics. I. Title. II. Series.
　Series.

　QC311.2.R34 2003
　536'.7–dc21
　　　　　　　　　　　　　　　　　　　　　　　　　　　2003050229
　　　　　　　　　　　　　　　　　　　　　　　　　　　CIP

AMS Subject Classifications: 76-XX; 74-XX

Printed on acid-free paper.
©2004 Birkhäuser Boston

Birkhäuser

ISBN 0-8176-4283-8　　SPIN 10863565
ISBN 3-7643-4283-8

Reformatted from the author's files by TeXniques, Inc., Cambridge, MA.
Printed in the United States of America.

9 8 7 6 5 4 3 2 1

Birkhäuser　Boston • Basel • Berlin
A member of BertelsmannSpringer Science+Business Media GmbH

Preface

Across the centuries, the development and growth of mathematical concepts have been strongly stimulated by the needs of mechanics. Vector algebra was developed to describe the equilibrium of force systems and originated from Stevin's experiments (1548–1620). Vector analysis was then introduced to study velocity fields and force fields. Classical dynamics required the differential calculus developed by Newton (1687). Nevertheless, the concept of particle acceleration was the starting point for introducing a structured spacetime. Instantaneous velocity involved the set of particle positions in space. Vector algebra theory was not sufficient to compare the different velocities of a particle in the course of time. There was a need to (parallel) transport these velocities at a single point before any vector algebraic operation. The appropriate mathematical structure for this transport was the connection.[1]

The Euclidean connection derived from the metric tensor of the referential body was the only connection used in mechanics for over two centuries. Then, major steps in the evolution of spacetime concepts were made by Einstein in 1905 (special relativity) and 1915 (general relativity) by using Riemannian connection. Slightly later, nonrelativistic spacetime which includes the main features of general relativity

---

[1] It took about one and a half centuries for connection theory to be accepted as an independent theory in mathematics. Major steps for the connection concept are attributed to a series of findings: Riemann 1854, Christoffel 1869, Ricci 1888, Levi-Civita 1917, Weyl 1918, Cartan 1923, Eshermann 1950. See Appendix C for a brief history of affine connection.

theory, was proposed by Cartan (1923) by means of the affine connection, independent of the relativization of time since it did not involve large speeds of material points.

Besides the evolution of the spacetime concept, development of the geometry underlying continuum mechanics seemed to stagnate in spite of the success of the affine connection in other fields of physics. The essential difficulty of the finite deformation theory of material bodies is that the displacement is not infinitesimal and has to be considered at least with respect to two sets of base vectors (in fact, two local spacetimes) respectively for the undeformed state and for the deformed state. For elastic bodies, the problem does not arise since the use of homeomorphic mapping allows us to identify the topology of the two states (undeformed and deformed).

There is a wide ranging literature in continuum mechanics on the representation of the nonelastic response of a body to external loadings. In most works, the connection used for describing large continuum deformations remains the metric connection identified with the Christoffel symbols with a vanishing curvature. Thus, these continuum theories constitute more or less an extension of Cartesian tensor notation as a convenient shorthand for writing equations.

Originally, continuum theories faced the same old problem as Newtonian dynamics: the transport of tensor variables in the course of time (spatial connection) and the comparison of tensor variables from one material point to another (material connection). Based on the concept of material connection, Noll first introduced the notion of materially uniform simple body (1967) at the same time as Wang did (1967) in the field of continuum mechanics. Non-Riemanian geometry was introduced independently by Bilby (1955), Kondo (1955), and Kröner (1963, 1981) for modeling the mechanical response of dislocated and latticed solids. The main improvement in the simple material theory, polished up by different authors such as Edelen and Lagoudas (1988), Maugin (1993), Le and Stumpf (1996) to name but several, was the accounting for internal surfaces slipping within the matter (dislocations, some kinds of disclinations, vortices structure,...).

Noll's simple material allowed us to model singularity by introducing the affine connection as an independent variable in addition to the metric tensor. In the same way, the present work has been written with the intention of proposing a rational theory of continuum thermomechanics capable of capturing the distribution of scalar and vector discontinuity within the continuum. It has grown out of a lecture given in 1987–1988 to graduate students in mechanics at the University of Antsiranana, Madagascar. The title was chosen deliberately "A Geometric Approach to Thermomechanics of Dissipating Continua," since distribution of singularity and irreversible dissipation were indeed strongly related to the concept of affine connection, a joint concept between geometry and topology. Singularity is shown to be entirely captured by the torsion and curvature tensors of the affine connection while the rate of evolution of singularity induces a bulk irreversibility by increasing the production of internal entropy. This is due to the loss of affine equivalence between the continuum and the Euclidean referential body. The loss of affine equivalence may be interpreted, as shown all through this work, as an irreversible evolution of the continuum topology. The backbone of the theory relies on Noll's theory of simple material.

All the chapters of this book alternate mathematics and mechanics. I hope that, on the one hand, the elementary level of the mathematics presented here will not hurt pure mathematicians. On the other hand, the part of mechanics I limited only to general statements will remain self sustaining. My intention is for it to serve as a useful foundation for more specialized topics. Many important and fascinating topics have been entirely or almost entirely omitted. Among them I may mention the variational formulation, wave propagation (only a small part has been developed in this field), and stability. I doubt that it would be worthwhile to start a discussion on these topics without being able to pursue it at considerable length. Much more coverage and a lot of work would be required. Of course, the discussions that are included in the book have no claim to be complete accounts of the topics considered. Further reading on both solid and fluid mechanics is required.

I wish to thank to all my colleagues with whom I have worked at the University of Antsiranana (Strength of Material Laboratory), at the University of Lausanne (Hôpital Orthopédique), at the Federal Institute of Technology of Lausanne (Biomedical Engineering Laboratory, Applied Mechanics Laboratory), and at the University of Rennes 1 (Institute of Mathematics). Among them, I would like to mention namely J. Radofilao who taught me general relativity and differential geometry and A. Curnier who initiated me to the theory of large continuum deformations. Special thanks to N. Ramaniraka who has, with me, arrived at some important results in the last chapter devoted to the two models of microcracked solids.

Last but not least, I am greatly indebted to my wife Oly and my children Rindra, Herinarivo, and Haga, to whom this book is dedicated, for their understanding and moral support.

L.R. Rakotomanana
Préverenges, Switzerland September 1999
Thorigné-Fouillard, France May 2001

Contents

*A Geometric Approach
to Thermomechanics
of Dissipating Continua*

1
Introduction

The present study focuses on theoretical modeling of a "continuum with a singularity distribution" in the sense that the distribution of matter is assumed continuous but that discontinuity of scalar fields (density, temperature,...) and discontinuity of vectorial fields (micro-cracks, inter-granular de-cohesions, growth of adiabatic shear bands, e.g., [4]...) may appear in the continuum. These two types of discontinuity constitute the singularity type we deal with. For a single surface singularity, jump conditions across this surface (e.g., Green and Nagdhi [69]) were derived from the balance equations of mass, linear momentum, moment of momentum, energy, and entropy. The density of singularity is assumed sufficiently high to accept a continuous volume distribution of singularity.

Field singularity may appear in solid mechanics. On the one hand, modeling of microdefected solids with non-Riemannian geometry draws back to, e.g., Bilby et al. [13], Kondo [97], Kröner [101]. Marcinkowsky employed differential geometry to model internal boundaries in crystals (e.g., [126]). Tensor quantities such as distortion, torsion, lattice connection, curvature, the Burgers vector, dislocation density, and finally the Burgers circuit have been derived in [126] to capture the defects in crystals. By avoiding the crystalline structure assumption and within the framework of simple material, Noll, e.g., [148], Wang, e.g., [200], and Bloom [15] have defined nonhomogeneity (density of microdefects) of a continuum in terms of curvature and torsion of material connections. Zorawski [212] and Gairola, e.g., [62] have also used affine connection in studying large deformations of elastic continua in the presence of dislocations.

In the same way, dislocations and disclinations have been investigated in the framework of large deformation elasticity in [214], either in a 3-dimensional continuum or in elastic shells and membranes approximations. An alternative relativistic approach has been proposed by Baldacci et al. [3] by transferring nonhomogeneity of matter to ambient space. Minagawa [137] introduced a space of nonmetric connections to capture defects in a Cosserat continuum. When restricted to micropolar materials, he found that connections were of the metric type and defects reduced to dislocations and disclinations. Duan [43] combined non-Riemannian geometry and Noether's theorem to derive conservation laws of nonlinear continuum with dislocation. He employed a small rotation of local coordinate frame to obtain what he called the gauge angular momentum conservation law. In this sense, the theory he presented was an extension of the elasticity theory to a dislocation continuum theory.

Duan and Duan [44] proposed later two kinds of connections (affine connection, gauge connection) to describe plastic imperfections in materials. They showed that use of the gauge connection is equivalent to the use of the curvature tensor and is related to disclinations. More recently, Edelen and Lagoudas, e.g., [104] and Maugin, e.g., [132] have resorted to drawing an analogy between the geometry of defects and the structural equations of Cartan to model dislocations in solids. In the same way, Lagoudas and Edelen [104] introduced both spatial and material gauge connections. Then, they used these two connections for building two sets of Cartan's equations of structure (spatial and material) and showed that spatial torsion and curvature captured microcracks and microrotations. Furthermore, they found that material torsion and curvature are the variables associated to dislocations and disclinations.

In the same way, a model of continuum allowing a distribution of scalar and vectorial discontinuity was introduced in [163] where both torsion and curvature were not null. In a series of papers Le, Stumpf, and coworkers [109], [111] have developed the concept of crystal reference based on the metric tensor for modeling elastic strain and on the crystal connection for capturing translational dislocations. They have shown that integrability of the differential equations governing crystal reference location requires that crystal curvature vanishes. Recently, within the geometric framework of G-structure (symmetry-group-based theory), Epstein and Maugin [51] have shown that continuum large strain inelasticity resulted from the evolution of nonhomogeneity distribution. They proposed evolution laws analogous to the rate-independent plasticity law, with a threshold in stress, to connect the change of dislocation pattern and the Eshelby stress tensor. Within the same framework, the second-grade elasticity was used to capture field singularity as dislocations and disclinations [114].

On the other hand, the classic point of view consists in using metric manifolds as geometric support of the continuum and in introducing more and more complex constitutive laws, e.g., [36], [52], [58], [130]. Such a method was also used to model a defected continuum. Eshelby in, e.g., [54] reviewed fundamental aspects of dislocation theories based on Somigliana discontinuity of vector fields and Peach–Koehler force concerning dislocations [152]. For the sake of completeness, Eshelby (1975) has proposed as a dual variable an energy-momentum tensor such that the integral of its normal component over a closed surface gives the force on the defects and the nonhomogeneities within the surface [53]. Various topics in the theory of dislocations were reviewed in his paper along with physical arguments.

In the framework of internal variable thermodynamics, Marigo [128] developed a model enabling the study of the initiation and increase of the internal structure of the elastic continuum (nonholonomic deformation). He defined a convex damage threshold in terms of admissible strains. By assuming that microcracking was the fundamental mechanism of deformation, Audoin [2] formulated a continuum damage model based on thermodynamics with internal variables. A fourth-order tensor was introduced for capturing the anisotropic damage of elastic material. For elastic-viscoplastic material, recent studies, e.g., [172] relate positive strain hardening and continuum damage softening to model the creep fracture prediction by means of a nonlocal formulation of damage.

General principles involved in the mechanics of crystalline solids, particularly for inelastic solids with brittle microcracking, can be found in, e.g., [175]. It was suggested that a phenomenological theory of microcracking solely based on the internal variable theory of continuum thermodynamics seemed to be inadequate for a clear description of the phenomena that occur at a microscale. A need for constitutive theories allowing the capture of micro-structural changes during inelastic deformation appears. In this way, phenomenological approaches of microdefected material may be found in recent works that are based on the thermodynamics of internal variables, e.g., [77], [113], [131]. The choice of the internal variable remains actual, e.g., [79]. In the same way, Ashby and Dyson (1986) have derived damage-evolution law and creep-law to analyze the effects of micromechanisms of damage and the global law of creep. In their paper [1], they proposed a unified theory of continua and a micromechanical approach to study the creep fracture and to point out the dominant damage mechanism for the shape and tensile creep curve. However, their use for continua with singularity required further investigations. Indeed, recent theories on adiabatic shear bands, e.g., [4] confirmed high coupling effects of mechanical stress and temperature. It has been shown, e.g., [178] that adiabatic shear bands are due to plastic flow localization in the presence of material imperfection and defects rather than to load instability as shear strain concentration.

Stout in, e.g., [185] extended Griffith's theory in fracture mechanics and other dislocation continuum theories to elaborate a discontinuous model of the deformations and thermodynamics during brittle fracture. He described the crack density with a statistical mechanical approach and drew an analogy between crack and dislocation. By relaxing the adiabatic assumption, effects of the microstructural tensor and kinetics factors on shear band initiation have been considered in, e.g., [187]. More recently, Perzyna [153] conducted an analytical study to assess the influence of anisotropy on the occurrence of adiabatic shear bands. It has been shown that in an adiabatic process for elastic-plastic damaged solids, the discontinuity of the acceleration field is related to the acoustic tensor. Studies were led in this area to assess the heat propagation in anisotropic damaged solids, e.g., [203] in the framework of damaged continuum and irreversible thermodynamics.

Singularity of fields also occurs in fluid flow suffering a creation of internal slip surfaces. First, abrupt changes in pressure and density can occur across surfaces in gas flow. A shock wave is a surface in the flow region across which at least the velocity field, the density field, or the pressure field suffers a jump discontinuity. It is now well known that the entropy should increase across a shock surface, e.g., [179]. As a consequence, an important property of shock waves is that they introduce vorticity. Secondly, the new interest in the presence of coherent structures in turbulence has fostered the hope that the study of vortices—a velocity field singularities in some cases—will lead to models and a better understanding of turbulent fluid flow, e.g., [125], [174]. In some ways, the dynamics of vorticies is a natural paradigm for the field of chaotic motion and dynamical system theory. Despite the fact that the present work is a rather solid-oriented study, we present here basic geometric tools for capturing the creation of field singularity in fluid flow. In this sense, an attempt to relate some aspects of the creation of vorticies within the fluid will be addressed. The connection between solid dislocations and fluid vortex is straightforward. A fluid vortex is defined by drawing a closed loop and summing up the fluid velocities around the loop (the result, called circulation, is null if no vortex is enclosed; otherwise the sum gives a non-null number). For crystalline solid dislocation, consider an arrow of fixed length assigned to every pair of closest atoms and sum up these arrows over a closed loop. The sum usually amounts to zero if the crystal is perfect, but if the loop encloses a dislocation, it gives the Burgers vector **b**.

Despite a growing number of papers in the area of continua with singularity during the last decade, it seems that, in fact, the use of a Riemannian manifold affinely equivalent to a Euclidean referential body in elasticity, e.g., [71], [130], the elastic-plastic situation, e.g., [108], and fluid flows, e.g., [174], [179] as a geometric backbone may exclude a priori the possibility of modeling the continuous distribution of discontinuity of scalar and vectorial fields. Moreover, although most of the studies on continuum thermomechanics admit the irreversible aspects of field singularity and therefore admit at least implicitly the occurrence of a volume heat source, few have developed complete conservation laws compatible with singularity distribution, e.g., [148], [200]. This work is an attempt to combine two approaches: (a) by using a manifold with an affine connection as a geometric support of the continuum and (b) by using the notion of generalized standard material, e.g., [66] for modeling the thermomechanics of continuum with singularity. The present book shows the relation between the loss of affine equivalence between a referential Euclidean space and the continuum and the internal irreversible dissipation. The loss of affine equivalence means also a change in the topology of the continuum.

2
Geometry and Kinematics

2.1 Introduction to continuum motion

We fix once and for all on a referential solid body endowed with a metric tensor $\mathbf{g}$ and a positive volume form ω_0. The reference is assumed Euclidean, that is, characterized by the existence of a global Cartesian coordinate system covering the whole referential body. The vector space underlying the reference is denoted Σ.

2.1.1 Continuum deformation

A continuum B is a 3-dimensional manifold, the elements of which are called particles (or material points) M. The deformation of a continuum is an intuitive physical notion that is represented mathematically by a transformation of Euclidean space, associated to the referential solid body, onto itself. The time t describes the transformation and its range is supposed to span $-\infty$ to $+\infty$, $t = 0$ being the initial instant. Since we keep within the classical mechanics framework, we do not distinguish between Σ and R^3. The motion of a continuum B is defined by the placement of each constitutive point M at any time t:

$$\mathbf{OM}(t) \doteq \varphi(M, t) \qquad M \in B, \qquad -\infty < t < \infty. \tag{2.1}$$

A configuration of the continuum B is a homeomorphism $\varphi : B \longrightarrow R^3$. Classical continuum models are based on two continuity assumptions: the continuity of distri-

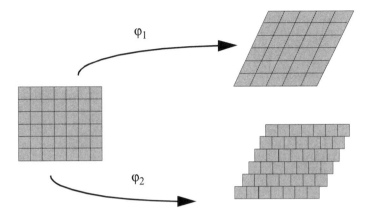

Figure 2.1. Solid deformation: Two classical examples of topological motion (φ_1: homeomorphism) and nontopological motion (φ_2: with internal slip surfaces distributed continuously in the body) although both of them are defined as macroscopic shear strain.

bution of matter and the continuity of transformation of matter, e.g., [125]. The most usual continuum theory is based on the so-called strongly continuous transformation for which $\varphi(M, t)$ is a homeomorphism. Such a transformation is called hereafter a strongly continuous transformation. When such is the case, the body's topology is not modified during the course of time. Although the deformation is completely determined by the transformation, the state of motion at a given point during the course of time may be described by the evolution of density $\rho(M, t)$ and the displacement $\mathbf{u}(M, t)$ or the velocity $\mathbf{v}(M, t)$ of the particle M at the time t.[1] The displacement of a point M between two instants t_0 and t is an element of the vector space Σ:

$$\mathbf{u}(M, t - t_0) \equiv \varphi(M, t) - \varphi(M, t_0). \tag{2.2}$$

The velocity of the particle M at the time t is the classical velocity with respect to the referential body, also belonging to the vector space $\sum$:

$$\mathbf{v}(M, t) \equiv \lim_{h \to \infty} \frac{1}{h}[\varphi(M, t + h) - \varphi(M, t)]. \tag{2.3}$$

There is no need to complicate the notion at this stage by defining either the concept of "material velocity" or of "spatial velocity." More precise definitions will be introduced

---

[1]A continuum is also assumed to be a measured space, thus endowed with a non-negative scalar measure called the mass distribution of the continuum $m(B) = \int_B \rho(M, t) dv$. The volume element dv indicates that the integral is defined in terms of Lebesgue measure in Euclidean space. Integration on a continuum, as a manifold, is introduced later.

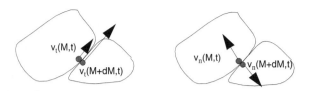

Figure 2.2. The first kind of singular surface used in continuum mechanics was the Helmholtz's slip surface where the inverse of the placement is discontinuous. Admissible discontinuity (tangential) and inadmissible discontinuity (normal) in continuum mechanics are sketched.

later. In this work, we assume that the continuum may contain singularity and that the transformation may not be a homeomorphism. The topology of the continuum B may then be different from one configuration to another. In nontopological motions, void nucleation and development of singularity as breakup, coalescence is allowed, e.g., [151]. Before going into details of the theoretical singularity characterization, two usual examples of field singularity in solids and in fluids are sketched in the two following sections.

2.1.2 Dislocations in solids

Physically, a microdefect is often associated with the creation of an internal slip surface within matter, e.g., [13]. Following a closed path around a defect, the displacement field has a jump. In a crystalline lattice, this brusque variation is equal to a period of the lattice b, e.g., [61], [62] (Burgers vector), $\|\mathbf{u}\|$ being the fictive translation required to close the dislocation:

$$\mathbf{b} = \int_C d\mathbf{u} \tag{2.4}$$

where C is a closed curve around a dislocation line which is a line of singularity of the displacement and the velocity fields. It is commonly admitted that dislocations are responsible for plastic deformation of crystalline solids. For a continuous distribution of defects, displacement and velocity are not single-valued functions of the position of the material point M. From the modeling point of view, a continuum B remains a set of material points continuously distributed in a certain region of ambient space. However, the occurrence of tangential discontinuity may appear with sufficiently high density in some continua. Such a microdefected continuous medium is a particular case of continuum with singularity.

For classification [61], consider the unit tangent vector to the dislocation line τ at point M. There are three classes of dislocations: (a) if $\tau \cdot \mathbf{b} = 0$, this is a screw dislocation; (b) if $\tau \wedge \mathbf{b} = 0$, we have an edge dislocation; (c) for any other $(\tau, \mathbf{b})$,

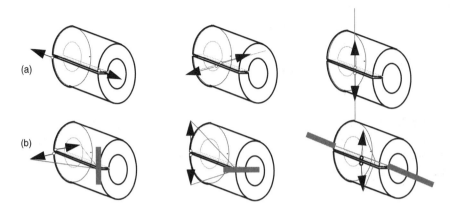

Figure 2.3. The Volterra process describes the creation of a surface singularity as shown by the figure. (a) Relative translations of two surfaces in a continuum. Screw and edge dislocations are represented. (b) Relative rotations of two surfaces in a continuum. They are called rotation dislocations or disclinations. Occurrences of these six dislocations are not possible in a continuum with singularity which does not permit a separation. Creation of voids is prohibited while internal slippage of surfaces is permitted.

the dislocation is mixed. Whatever the continuity order, the existence of a homeomorphism between material points and geometrical points at each time t allows us to transpose the geometrical properties of a Euclidean referential body to the continuum. From the dynamics point of view, each defect is not only characterized by the path and the speed of crack opening (see hereafter for the Volterra process). The crack tip itself transforms the (assumed) continuous solid by creating a cloud of microcracks and dislocations around itself. Of course, the increase of microcrack density ahead of the tip almost modifies the mechanical properties of the solid locally.

2.1.3 Vortices in fluids

Consider a fluid flow. Field singularity often appears when solid bodies are moving in the fluid, e.g., [199]. The fluid beneath the wing near the wing tips will end up spilling over the tip (Figure 2.4). By considering the circulation around a dotted loop somewhere behind the tip, the circulation around the closed curved (∂C) will be nonzero. There must be creation of the so-called vortex in the fluid as shown by using the celebrated Stokes theorem:

$$\Gamma = \int_{\partial C} \mathbf{v} \cdot d\mathbf{l} \neq 0 \quad \Longrightarrow \quad \int_{\partial C} \mathbf{v} \cdot d\mathbf{l} = \int_{C} \operatorname{rot} \mathbf{v} \cdot \mathbf{n} da \neq 0$$

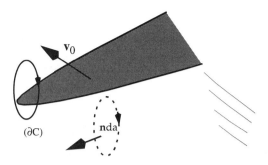

Figure 2.4. Consider a finite flat wing moving with a constant velocity in a fluid otherwise at rest, and suppose that the pressure on the lower surface of the wing is on average higher than on the upper surface in such a way that a net lift is experienced by the wing. A surface across which the velocity suffers a transversal discontinuity is called the vortex sheet.

in which we have defined the vorticity $\Omega \equiv \mathrm{rot}\,\mathbf{v}$, showing that the velocity circulation around a reducible curve ∂C (a curve that can be shrunk continuously to a point without going out of the fluid) is equal to half of the flux of vorticity field, the velocity rotation (or angular velocity) amounts to half of the vorticity $\frac{1}{2}\Omega = \frac{1}{2}\mathrm{rot}\,\mathbf{v}$. In regions of fluid where the vorticity $\Omega \equiv \mathrm{rot}\,\mathbf{v}$ does not vanish identically, we may draw curves parallel to Ω at each point of the curve. They are called vortex lines. When a vortex line is immediately surrounded by irrotational fluid, it is called a vortex filament or often just a vortex. Vortex requires that the fluid is ideal since viscosity diffuses vorticity.

To define the vorticity it is common to consider, around any fluid particle, the principal axes of the rate of strain (a triad of three mutually orthogonal straight lines) passing through the same fluid particle in such a way that if they move with the fluid flow then, after a short time dt, the angles between them remain right angles (the rates of shear strain are zero to the first order). Physically, the vorticity is a vector equal to twice the angular velocity of the previous principal triad relative to instantaneous position, e.g., [8], [67]. The concept of vorticity allows us to define irrotational flow and rotational flow. Irrotational flows are those fluid motions where $\Omega = 0$. Most fluid motions are rotational flow $\Omega \neq 0$. In (Figure 2.4) when the lift is present, the flow field as a whole cannot be completely free from vorticity. By analogy with dislocations in solid bodies, the circulation of a vector field can be described as a field singularity in fluid. The fluid flow is said to be rotational when the vorticity is nonzero. Turbulent flows are rotational, at least in certain regions of space. Vortex may be created by external causes such as rigid bodies moving in the fluid or created within the interior of the flow itself through some initial conditions as mixing. Additionally, shock waves in fluid introduce vorticity into an otherwise irrotational flow, probably due to the fact that energy is conserved across a shock surface while entropy is generated.

Any surface at which the normal velocity is continuous and the tangential velocity discontinuous is called a vortex sheet. By denoting $[\mathbf{v}] = \mathbf{v}^+ - \mathbf{v}^-$ the jump of the velocity across the interface separating the continuum B into two parts B^- and B^+, the vorticity is given by

$$\omega = \mathbf{n} \wedge [\mathbf{v}] \tag{2.5}$$

where $\mathbf{n}$ is the unit normal vector pointing into the domain B^+. As for solids, fields of velocity and vorticity with discontinuity are of considerable importance. Singularity must be consistent with the physical balance laws: mass, linear momentum, angular momentum, energy, and entropy. Anyway, the presence of singularity suggests an extension of the classical fluid model allowing one to capture scalar and vector field discontinuity, e.g., [163]. What we will develop in the next sections consists in defining the tensor variables to capture scalar and vector discontinuity in a continuum.

2.1.4 Singular surfaces

More generally, consider a regular surface S that is the common boundary of two disconnected regions B^- and B^+, with $B \equiv B^+ \cup B^-$. Let ψ be a function that is continuous, at each time t, in the interior of B^- and B^+ respectively and that approaches definite limit values $\psi^+ = \psi(M^+, t)$ and $\psi^- = \psi(M^-, t)$ as M^+ and M^- approaches a point M_0 on the surface S while remaining within B^- and B^+, respectively. The jump of the field $\psi(M, t)$ across S at M_0 is denoted as previously:

$$[\psi] \equiv \psi^+ - \psi^-. \tag{2.6}$$

Its value depends on the location of M_0. In this work, we are mainly interested in scalar-valued function $\psi(M, t)$ and vector-valued function $\mathbf{v}(M, t)$. If the jump $[\psi] \neq 0$, the surface S is said to be singular with respect to the field $\psi(M, t)$. In the case of a vector field, if the jump is orthogonal to S, $[\mathbf{v}] \parallel \mathbf{v}$ locally, we face a longitudinal discontinuity, and if the jump is tangent to S, $[\mathbf{v}] \perp \mathbf{n}$, a transversal discontinuity. From these variables, Emde defined the analogies of differential operators as follows [49]:

$$\text{Grad } \mathbf{v} \equiv \mathbf{n} \otimes [\mathbf{v}] \qquad \text{Div } \mathbf{v} \equiv \mathbf{n} \cdot [\mathbf{v}] \qquad \text{Rot } \mathbf{v} \equiv \mathbf{n} \wedge [\mathbf{v}]. \tag{2.7}$$

For either solids or fluids, the creation of dislocations (translation and rotation) and the creation of vorticity must respectively involve transformations that are not homeomorphisms (of class C^2). The transition can occur only in the presence of shocks (normal discontinuity, tangential continuity) or tangential discontinuity (normal continuity).

2.2 Geometry of continuum

We assume the continuum is modeled by a differentiable manifold B, e.g., [146], [163], [201] endowed with a metric tensor $\mathbf{g}$, which is an inner product defined at each tangent space. The (vector) tangent space of B (resp. its dual B^*) is denoted $T_M B$ (resp. $T_M B^*$).

2.2.1 Metric tensor and volume form

A Riemannian metric, assumed throughout the present work, on B is a 2-covariant tensor field on B that satisfies the following axioms at each point M (symmetric bilinear form):

1. $\mathbf{g}(M, t)(\mathbf{u}, \mathbf{v}) = \mathbf{g}(M, t)(\mathbf{v}, \mathbf{v})$, $\forall(\mathbf{u}, \mathbf{v}) \in T_M B$

2. $\mathbf{g}(M, t)(\mathbf{u}, \mathbf{u}) \geq 0$ where the equality holds only when $\mathbf{u} = 0$.

The metric tensor may be decomposed onto any vector base $(\mathbf{u}_1, \mathbf{u}_2, \mathbf{u}_3)$ and the components on this vector base form a matrix containing all the data on the length and the direction of the vector basis. Hereafter, the metric is called a material metric tensor when projected onto a material tangent vector.

Example. The matrix of the Euclidean metric associated to skew rectilinear coordinates:

$$\|\mathbf{u}_1\| = \|\mathbf{u}_2\| = \|\mathbf{u}_3\| = 1 \quad (\mathbf{u}_1, \mathbf{u}_2) = \gamma \quad (\mathbf{u}_2, \mathbf{u}_3) = \alpha \quad (\mathbf{u}_3, \mathbf{u}_1) = \beta$$

takes the form

$$(g_{ab}) = \begin{pmatrix} 1 & \cos\alpha & \cos\beta \\ \cos\gamma & 1 & \cos\alpha \\ \cos\beta & \cos\alpha & 1 \end{pmatrix}. \tag{2.8}$$

For the body orientation, we recall that a differentiable manifold B covered by a set B_i is orientable if, for any overlapping charts B_i and B_j, there exist local coordinates y_i^a for B_i and y_j^b for B_j such that the Jacobian $J = \frac{\partial y_i^a}{\partial y_j^b} > 0$. For orientable 3-dimensional continua, there exists (at least one) a 3-form denoted ω_0 that vanishes nowhere, called a volume form (or volume element). Details may be found in Appendix C. The volume form may be decomposed onto any base $(\mathbf{u}_1, \mathbf{u}_2, \mathbf{u}_3)$ and in the particular case where the volume form is associated to the chosen metric, the components are expressed by:

$$\omega_0(\mathbf{u}_a, \mathbf{u}_b, \mathbf{u}_c) = \begin{cases} \sqrt{g} & \text{if (a,b,c) cyclic sequence} \\ -\sqrt{g} & \text{if (a,b,c) anti-cyclic sequence} \\ 0 & \text{if (a,b,c) acyclic sequence} \end{cases} \tag{2.9}$$

where $\sqrt{g}^2 = \det[\mathbf{g}(\mathbf{u}_a, \mathbf{u}_b)]$. It should be noticed that the notion of orientation (volume form) need not be related to the shape and size of the continuum (metric notion). As will be shown in the next step, there is a need for an extra structure (affine connection) specifying how tensors are transported from one point of the body to its neighborhood.

2.2.2 Basics on affine connection

The gradient of a scalar field ψ, of class C^1 on a continuum B and denoted $\nabla \psi$, is a 1-form on B and defined by, e.g., [29]:

$$\lim_{\varepsilon \to 0} \frac{1}{\varepsilon} [\psi(M + \varepsilon \mathbf{u}) - \psi(M)] \equiv \nabla \psi(M)(\mathbf{u}) = \nabla_{\mathbf{u}} \psi(M). \tag{2.10}$$

More generally, the directional derivative along $\mathbf{u}$ is defined, e.g., [123] by $\nabla_{\mathbf{u}} \psi \equiv \mathbf{u}(\psi)$ with $\forall \mathbf{u} \in T_M B$. By extension, an affine connection ∇ on B is an operator associating two vector fields $\mathbf{u}$ and $\mathbf{v}$ to a third vector field $\nabla_{\mathbf{u}} \mathbf{v}$. The operator permits us to calculate the values of $\mathbf{v}$ at a neighboring point M' transported in M by $\mathbf{u}$, e.g., [29]:

$$\mathbf{v}(M', t) = \mathbf{v}(M, t) + \nabla_{\mathbf{u}} \mathbf{v}(M, t). \tag{2.11}$$

The vector $\nabla_{\mathbf{u}} \mathbf{v}$ is the covariant derivative of $\mathbf{v}$ along $\mathbf{u}$ with respect to ∇. The coefficients of the affine connection in a basis, say $(\mathbf{u}_1, \mathbf{u}_2, \mathbf{u}_3)$, denoted $\Gamma^c_{ab} = \mathbf{u}^c(\nabla_{\mathbf{u}_a} \mathbf{u}_b)$ generalize the Christoffel symbols of Riemannian geometry. The intrinsic definition of the gradient of a vector field $\mathbf{v}$ and that of a linear form ω are then deduced in, e.g., [159]:

$$\nabla \mathbf{v}(\mathbf{u}) \equiv \nabla_{\mathbf{u}} \mathbf{v} \qquad (\nabla_{\mathbf{u}} \omega)(\mathbf{v}) = \mathbf{u}[\omega(\mathbf{v})] - \omega(\nabla_{\mathbf{u}} \mathbf{v}). \tag{2.12}$$

When the continuum is endowed with a metric, classical theory of physics uses a particular connection satisfying the basic relation $\nabla \mathbf{g} \equiv 0$; i.e., we demand the metric tensor to be covariantly constant. The affine connection is said to be metric compatible and is called a metric connection.

2.2.3 Lie–Jacobi bracket and exterior derivative

Definition 2.2.1 *(Lie–Jacobi bracket) The bracket of two vector fields $\mathbf{u}$ and $\mathbf{v}$ on B is a vector field denoted $[\mathbf{u}, \mathbf{v}]$ and defined by the following, where ψ is any scalar field on a continuum B:*

$$[\mathbf{u}, \mathbf{v}](\psi) \equiv \mathbf{u}\mathbf{v}(\psi) - \mathbf{v}\mathbf{u}(\psi). \tag{2.13}$$

At each material point M of B, the constants of structure (Cartan), associated to a basis $(\mathbf{u}_1, \mathbf{u}_2, \mathbf{u}_3)$ and denoted $\aleph^c_{0ab}$, are defined by [22], [23]:

$$[\mathbf{u}_a, \mathbf{u}_b] \equiv \aleph^c_{0ab} \mathbf{u}_c. \tag{2.14}$$

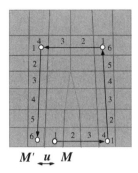

$M' \; \underleftrightarrow{u} \; M$

Figure 2.5. An affine connection may be visualized easily by considering a lattice structure. For the example of a crystal connection, we can proceed as follows. Consider a circuit surrounding a single dislocation (see internal triangle). The instructions "Go 3 steps, turn left, go 5 steps, turn left, go 3 steps, turn left, and finally, go 5 steps" generate a path that has a gap of exactly one lattice spacing, corresponding to one dislocation surrounded by this circuit. A large circuit will have an error of closure proportional to the number of dislocations that it contains (proportional to its area).

The basis $(\mathbf{u}_1, \mathbf{u}_2, \mathbf{u}_3)$ is associated to a system of coordinates if and only if for any two vectors of the basis $[\mathbf{u}_a, \mathbf{u}_b] = 0$, e.g., [23], [29]. The constants of structure are also called anholonomy of the basis $(\mathbf{u}_1, \mathbf{u}_2, \mathbf{u}_3)$, e.g., [212]. They are not the components of any tensor since their values are zero in any coordinate basis but do not vanish for an anholonomic basis.

The exterior derivative of a p-form ω may then be defined (intrinsically) by the relationship, for any vector fields $(\mathbf{u}_1, \ldots, \mathbf{u}_{p+1})$ on the continuum (vector with hat is omitted), e.g., [120], [123] (for more details, see Appendix C):

$$d\omega(\mathbf{u}_1, \ldots, \mathbf{u}_{p+1}) = \sum_{i=1}^{p+1} (-1)^{i+1} \mathbf{u}_i \left[\omega(\mathbf{u}_1, \ldots, \hat{\mathbf{u}}_i, \ldots, \mathbf{u}_{p+1}]\right.$$

$$+ \sum_{i<j} (-1)^{i+j} \omega([\mathbf{u}_i, \mathbf{u}_j], \mathbf{u}_1, \ldots, \hat{\mathbf{u}}_i, \ldots, \hat{\mathbf{u}}_j, \ldots \mathbf{u}_{p+1}). \tag{2.15}$$

Consider a 3-dimensional manifold. The p-forms in the 3-dimensional space are:

1. 0-form (scalar, $p = 0$), $\omega = \psi$

2. 1-form ($p = 1$), $\omega = \omega_1 \mathbf{u}^1 + \omega_2 \mathbf{u}^2 + \omega_3 \mathbf{u}^3$

3. 2-form ($p = 2$), $\omega = \omega_{12} \mathbf{u}^1 \wedge \mathbf{u}^2 + \omega_{23} \mathbf{u}^2 \wedge \mathbf{u}^3 + \omega_{31} \mathbf{u}^3 \wedge \mathbf{u}^1$

4. 3-form ($p = 3$), $\omega = \omega_{123} \mathbf{u}^1 \wedge \mathbf{u}^2 \wedge \mathbf{u}^3$.

In a coordinate base, associated to a system of parameters, the action of the exterior derivative on the previous p-forms is:

1. 0-form (scalar, $p = 0$)

$$d\omega = \mathbf{u}_1[\psi]\mathbf{u}^1 + \mathbf{u}_2[\psi]\mathbf{u}^2 + \mathbf{u}_3[\psi]\mathbf{u}^3$$

2. 1-form ($p = 1$)

$$
\begin{aligned}
d\omega &= (\mathbf{u}_1[\omega_2] - \mathbf{u}_2[\omega_1])\,\mathbf{u}^1 \wedge \mathbf{u}^2 + (\mathbf{u}_2[\omega_3] - \mathbf{u}_3[\omega_2])\,\mathbf{u}^2 \wedge \mathbf{u}^3 \\
&+ (\mathbf{u}_3[\omega_1] - \mathbf{u}_1[\omega_3])\,\mathbf{u}^3 \wedge \mathbf{u}^1
\end{aligned}
$$

3. 2-form ($p = 2$)

$$d\omega = (\mathbf{u}_1[\omega_{23}] + \mathbf{u}_2[\omega_{31}] + \mathbf{u}_3[\omega_{12}])\,\mathbf{u}^1 \wedge \mathbf{u}^2 \wedge \mathbf{u}^3$$

4. 3-form ($p = 3$)

$$d\omega = 0.$$

Remark. It is quite easy to identify the action of the exterior derivative as the gradient of a scalar, the rotational of a "vector field," the divergence of a "vector field." We shall use the exterior derivative to extend the definition of these differential operators on p-form fields occurring on an affinely connected manifold, not necessarily Euclidean.

2.3   Discontinuity of fields on continuum

2.3.1   Torsion and discontinuity of a scalar field

Consider on a continuum B an affine connection ∇ and a 1-form field ω and vector fields $\mathbf{u}_1$, $\mathbf{u}_2$, and $\mathbf{w}$.

Definition 2.3.1 (*Torsion*) *Associated to the affine connection ∇ defined on the continuum B, the field of torsion tensor $\aleph$, a 2-covariant 1-contravariant, is defined by the relation:*

$$\aleph(\mathbf{u}, \mathbf{v}, \omega) \equiv \omega(\nabla_{\mathbf{u}}\mathbf{v} - \nabla_{\mathbf{v}}\mathbf{u} - [\mathbf{u}, \mathbf{v}]) \qquad \forall \mathbf{u}, \mathbf{v} \in T_M B \qquad \forall \omega \in T_M B^*. \quad (2.16)$$

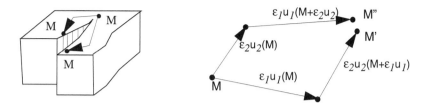

Figure 2.6. Field discontinuity in continuum (microcrack, Cartan path model). The path can be considered as an infinitesimal closed curve.

In any basis $(\mathbf{u}_1, \mathbf{u}_2, \mathbf{u}_3)$, the components of the torsion tensor are written, e.g., [29]:

$$\aleph = \aleph^c_{ab}\mathbf{u}^a \otimes \mathbf{u}^b \otimes \mathbf{u}_c \qquad \aleph^c_{ab} = (\Gamma^c_{ab} - \Gamma^c_{ba}) - \aleph^c_{0ab}. \qquad (2.17)$$

When the transformation is weakly continuous, internal slip surfaces (tangential discontinuity) may occur. For quantifying this singularity, we adopt a classical scheme introduced by Cartan [22], [23] by following a scalar field along two paths on B from a same point M and joining an extreme point along two directions given by the integral curves of two vector fields $\mathbf{u}_a$ and $\mathbf{u}_b$. The basic idea is to calculate the variation of θ for the two paths:

$$\theta' = \theta(M') - \theta(M) \qquad \theta'' = \theta(M'') - \theta(M). \qquad (2.18)$$

By using the connection affine properties and the directional derivative along a vector, we are able to calculate the variation:

$$\begin{aligned}
\frac{\theta' - \theta''}{\epsilon_1\epsilon_2} &= \nabla_{\mathbf{u}_2}(\nabla_{\mathbf{u}_1}\theta) - \nabla_{\mathbf{u}_1}(\nabla_{\mathbf{u}_2}\theta) + \nabla_{\nabla_{\mathbf{u}_1}\mathbf{u}_2}\theta - \nabla_{\nabla_{\mathbf{u}_2}\mathbf{u}_1}\theta \\
&+ [\epsilon_1\nabla_{\nabla_{\mathbf{u}_1}\mathbf{u}_2}(\nabla_{\mathbf{u}_1}\theta) - \epsilon_2\nabla_{\nabla_{\mathbf{u}_2}\mathbf{u}_1}(\nabla_{\mathbf{u}_2}\theta)].
\end{aligned} \qquad (2.19)$$

Hence, calculating the limit at the point M as ϵ_1 and ϵ_2 tends to 0, we obtain, given the definition of Lie–Jacobi bracket, [177]:

$$\lim_{\epsilon_1,\epsilon_2 \to 0} \frac{\theta' - \theta''}{\epsilon_1\epsilon_2} = \{\nabla_{\mathbf{u}_1}\mathbf{u}_2 - \nabla_{\mathbf{u}_2}\mathbf{u}_1 - [\mathbf{u}_1, \mathbf{u}_2]\}(\theta) = \aleph(\mathbf{u}_1, \mathbf{u}_2)[\theta]. \qquad (2.20)$$

Theorem 2.3.1 *(Jump of scalar field) Let θ be a scalar field on a continuum B. If the variation of a scalar θ from one point to another point depends on the path then the torsion field in the manifold B is not null and characterizes the jump of θ according to (2.20).*

The continuum theory presented here is in fact based on the celebrated Cartan circuit, wherein the body does not have lattice structure, and on Frank's Burgers circuit in the case of lattice defects.

2.3.2   Curvature and discontinuity of vector field

Definition 2.3.2 *(Curvature) Let ω be any 1-form field on the continuum B and $\mathbf{u}_1$, $\mathbf{u}_2$, and $\mathbf{w}$ any vector fields on B. The curvature tensor, denoted $\mathfrak{R}$, associated to the affine connection ∇ is defined by*

$$\mathfrak{R}(\mathbf{u}_1, \mathbf{u}_2, \mathbf{w}, \omega) \equiv \{\nabla_{\mathbf{u}_1}\nabla_{\mathbf{u}_2}\mathbf{w} - \nabla_{\mathbf{u}_2}\nabla_{\mathbf{u}_1}\mathbf{w} - \nabla_{[\mathbf{u}_1,\mathbf{u}_2]}\mathbf{w}\}(\omega). \tag{2.21}$$

Let $\mathbf{w}$ be a vector field on B. At a second order, we use the connection to transport all the calculus to M. Let $\mathbf{w}'$ be the value of the vector field in M':

$$\begin{aligned}
\mathbf{w}' &= \mathbf{w} + \epsilon_1 \nabla_{\mathbf{u}_1}\mathbf{w} + \epsilon_2 \nabla_{\mathbf{u}_2}\mathbf{w} + \epsilon_1\epsilon_2 \nabla_{\mathbf{u}_1}\nabla_{\mathbf{u}_2}\mathbf{w} + \epsilon_1\epsilon_2[\nabla_{\nabla_{\mathbf{u}_2}\mathbf{u}_1}\mathbf{w} \\
&+ \epsilon_2 \nabla_{\nabla_{\mathbf{u}_2}\mathbf{u}_1}(\nabla_{\mathbf{u}_2}\mathbf{w})].
\end{aligned} \tag{2.22}$$

Accordingly, the value at the point M'' is given by

$$\begin{aligned}
\mathbf{w}'' &= \mathbf{w} + \epsilon_1 \nabla_{\mathbf{u}_1}\mathbf{w} + \epsilon_2 \nabla_{\mathbf{u}_2}\mathbf{w} + \epsilon_1\epsilon_2 \nabla_{\mathbf{u}_2}\nabla_{\mathbf{u}_1}\mathbf{w} + \epsilon_1\epsilon_2[\nabla_{\nabla_{\mathbf{u}_1}\mathbf{u}_2}\mathbf{w} \\
&+ \epsilon_1 \nabla_{\nabla_{\mathbf{u}_1}\mathbf{u}_2}(\nabla_{\mathbf{u}_1}\mathbf{w})].
\end{aligned} \tag{2.23}$$

Therefore, the limit at the point M as ϵ_1 and ϵ_2 tend to 0, after application of a form ω, is [177]:

$$\lim_{\epsilon_1,\epsilon_2 \to 0} \frac{\omega(\mathbf{w}') - \omega(\mathbf{w}'')}{\epsilon_1\epsilon_2} = \mathfrak{R}(\mathbf{u}_1, \mathbf{u}_2, \mathbf{w}, \omega) - \omega(\nabla_{\aleph(\mathbf{u}_1,\mathbf{u}_2)}\mathbf{w}). \tag{2.24}$$

Theorem 2.3.2 *(Jump of vector field) Let $\mathbf{w}$ be a vector field on the continuum B. If the variation of $\mathbf{w}$ from one point to another point depends on the path then the torsion field $\aleph$ and the curvature field $\mathfrak{R}$ on the manifold B are not null and characterize the jump of $\mathbf{w}$ according to (2.24).*

The field singularity (here volume distribution of scalar and vector discontinuity) is entirely characterized by tensor fields (2.20) and (2.24) on B considered as constitutive primal variables. What can be stressed is that the curvature alone does not fully characterize the discontinuity of the vector field (disclinations). The jump of a vector quantity across an internal surface involves both torsion and curvature according to (2.24). A continuum is affinely equivalent to the referential body (Euclidean ambient space) if and only if torsion and curvature fields are identically null at every point of B, e.g., [159]. The use of differential geometry concepts allows us to replace the classical Euclidean (elastic) continuum in the presence of a defect distribution by an affinely connected manifold (not necessarily Euclidean) with smooth properties. Connections, besides vector fields, form other fundamental mathematical variables suitable to represent physical objects.

2.3.3 Compatibility equations: alternative approach

As for infinitesimal small strains of a continuum, the displacement vector has three components whereas there are six displacement-strain relations. The system is over-constrained and compatibility equations are needed. It is essential to notice that the continuum with a singularity distribution is nothing more than a continuum that violates the compatibility equations. In this subsection, an alternative proof of compatibility equations is sketched in two steps, e.g., [132]:

1. (**Torsion**) Let us consider the deformation of a continuum B. Starting with the placement $\mathbf{x} = \varphi(\mathbf{X})$, the triplet (X^1, X^2, X^3) constitutes, at any time, a system of curvilinear coordinates of the continuum. Therefore, $d\mathbf{x} = d\varphi(d\mathbf{X})$, or more familiarly $d\mathbf{x} = \mathbf{F}(d\mathbf{X}) = dX^i \mathbf{f}_i$ ($\mathbf{f}_i = \mathbf{F}(\mathbf{f}_{0i})$), is an exact differential with

$$\oint_C d\mathbf{x} = \oint_C dX^i \mathbf{f}_i = \int_S \left(\frac{\partial \mathbf{f}_i}{\partial X^j} - \frac{\partial \mathbf{f}_j}{\partial X^i} \right) dX^i dX^j = 0$$

in which, $C = \partial S$, we have applied the Stokes theorem. Since the curve C is arbitrary, the integrand also vanishes. If the triplet $(\mathbf{f}_1, \mathbf{f}_2, \mathbf{f}_3)$ is a basis of the tangent space, then there exists a set of real numbers Γ^k_{ij} (called coefficients of the connection) such that

$$\frac{\partial \mathbf{f}_i}{\partial X^j} = \Gamma^k_{ij} \mathbf{f}_k \Longrightarrow \left(\frac{\partial \mathbf{f}_i}{\partial X^j} - \frac{\partial \mathbf{f}_j}{\partial X^i} \right) = (\Gamma^k_{ij} - \Gamma^k_{ji}) \mathbf{f}_k.$$

We deduce that the tensor field, called the torsion tensor, $\aleph = (\Gamma^k_{ij} - \Gamma^k_{ji})\mathbf{f}^i \otimes \mathbf{f}^j \otimes \mathbf{f}_k$ vanishes on the continuum B. Explicitly, it is straightforward to calculate the coefficients of the connection as

$$\Gamma^k_{ij} = \frac{1}{2} g^{kn} \left(\frac{\partial g_{in}}{\partial x^j} + \frac{\partial g_{nj}}{\partial x^i} - \frac{\partial g_{ij}}{\partial x^n} \right) \qquad g_{ij} \equiv \mathbf{g}(\mathbf{f}_i, \mathbf{f}_j).$$

2. (**Curvature**) Let us consider a vector field $\mathbf{u} = u^i \mathbf{f}_i$ on B in which the components u^i are assumed of class C^2. The first and second derivatives are given by

$$\frac{\partial \mathbf{u}}{\partial X^m} = \left(\frac{\partial u^l}{\partial X^m} + u^k \Gamma^l_{km} \right) \mathbf{f}_l$$

$$\frac{\partial^2 \mathbf{u}}{\partial X^n \partial X^m} = \left[\frac{\partial}{\partial X^n} \left(\frac{\partial u^l}{\partial X^m} \right) + \frac{\partial u^k}{\partial X^m} \Gamma^l_{kn} + u^k \frac{\partial \Gamma^l_{kn}}{\partial X^m} \right] \mathbf{f}_l$$

$$+ \left(\frac{\partial u^l}{\partial X^m} + u^k \Gamma^l_{km} \right) \frac{\partial \mathbf{f}_l}{\partial X^m}.$$

By considering similar terms with inverted indices (m, n) and by making the difference of the two second-order derivatives, we obtain the following relation:

$$\frac{\partial^2 \mathbf{u}}{\partial X^n \partial X^m} - \frac{\partial^2 \mathbf{u}}{\partial X^m \partial X^n} = \left[\frac{\partial}{\partial X^n} \left(\frac{\partial u^l}{\partial X^m} \right) - \frac{\partial}{\partial X^m} \left(\frac{\partial u^l}{\partial X^n} \right) \right] \mathbf{f}_l$$

$$+ u^k \left(\frac{\partial \Gamma^l_{km}}{\partial X^n} - \frac{\partial \Gamma^l_{kn}}{\partial X^m} + \Gamma^r_{km} \Gamma^l_{rn} - \Gamma^r_{kn} \Gamma^l_{rm} \right) \mathbf{f}_l.$$

Permutation of the derivative is possible if the second term (second line) is identically equal to zero:

$$\Re = \left(\frac{\partial \Gamma^l_{km}}{\partial X^n} - \frac{\partial \Gamma^l_{kn}}{\partial X^m} + \Gamma^r_{km} \Gamma^l_{rn} - \Gamma^r_{kn} \Gamma^l_{rm} \right) \mathbf{f}^n \otimes \mathbf{f}^m \otimes \mathbf{f}^k \otimes \mathbf{f}_l \equiv 0.$$

In turn, this shows that the strain is compatible if the torsion and curvature tensors vanish on the continuum B.

2.3.4 Continuum with singularity

Definition 2.3.3 (*Continuum with singularity [163]*) *A continuum with singularity B is a medium in which the distribution of material points remains continuous but in which the existence of a continuous distribution of scalar and vector field discontinuity is permitted.*

It follows that the geometrical structure of a continuum with singularity (hereafter, a continuum with singularity may be also called a weakly continuous medium) is completely defined by, for any vector basis $(\mathbf{u}_1, \mathbf{u}_2, \mathbf{u}_3)$:

1. a metric tensor and volume form (usual variables of classical continuum models):

$$\mathbf{g} = g_{ab} \mathbf{u}^a \otimes \mathbf{u}^b \qquad \omega_0 = \det(\mathbf{u}_1, \mathbf{u}_2, \mathbf{u}_3) \mathbf{u}^1 \wedge \mathbf{u}^2 \wedge \mathbf{u}^3 \qquad (2.25)$$

2. an affine connection characterized by the torsion and curvature tensors (additional variables for a continuum with singularity):

$$\Gamma^c_{ab} = \mathbf{u}^c (\nabla_{\mathbf{u}_a} \mathbf{u}_b)$$
$$\aleph = [(\Gamma^c_{ab} - \Gamma^c_{ba}) - \aleph^c_{0ab}] \mathbf{u}^a \otimes \mathbf{u}^b \otimes \mathbf{u}_c$$
$$\Re = [-\mathbf{u}_b (\Gamma^c_{ad}) + \mathbf{u}_a (\Gamma^c_{bd}) + \Gamma^e_{da} \Gamma^c_{eb} - \Gamma^e_{db} \Gamma^c_{ea} - \aleph^e_{0ab} \Gamma^c_{ed}]$$
$$\mathbf{u}^a \otimes \mathbf{u}^b \otimes \mathbf{u}^d \otimes \mathbf{u}_c. \qquad (2.26)$$

A continuum B endowed with an affine but not necessarily a metric connection is called non-Riemaniann. A continuum B equipped with a connection ∇ is said to possess affine connection (it is an affinely connected continuum). Continua with(-out) singularity may be classified as follows according to the value or expression of the metric and affine connections:

1. $\mathbf{g} = c^{st}$, $\aleph = 0$, $\Re = 0$ Euclidean continuum ("flat" continuum with the possibility of defining a least one coordinate system where the connection coefficients identically vanish),

2. $\nabla \mathbf{g} = 0$, $\aleph = 0$, $\Re = 0$ or $\Re \neq 0$ Riemann continuum (continuum with a symmetric connection that is compatible with a metric, called a metric connection),

3. $\nabla \mathbf{g} = 0$, $\aleph \neq 0$, $\Re = 0$ or $\Re \neq 0$ Cartan continuum (continuum with a metric connection and a non-null torsion tensor) allowing vector field discontinuity,

4. $\exists \mathbf{v} \neq 0 \in T_M B$, $\nabla \mathbf{g} = \mathbf{v} \otimes \mathbf{g}$, $\aleph = 0$, $\Re = 0$ or $\Re \neq 0$ Weyl continuum (continuum with a nonmetric connection and a null torsion tensor),

5. $\exists \mathbf{v} \neq 0 \in T_M B$, $\nabla \mathbf{g} = \mathbf{v} \otimes \mathbf{g}$, $\aleph \neq 0$, $\Re = 0$ or $\Re \neq 0$ semimetric continuum (continuum with a nonmetric connection and a nonnull torsion tensor),

6. $\nabla \mathbf{g} \neq 0$, ∇, $\aleph = 0$ or $\aleph \neq 0$, $\Re = 0$ or $\Re \neq 0$ metric continuum (continuum endowed with a metric not necessarily compatible with the connection),

7. ∇, $\aleph = 0$ or $\aleph \neq 0$, $\Re = 0$ or $\Re \neq 0$ affinely connected continuum where vector and cotangent vector fields may evolve independently during deformation. Generally, torsion and curvature tensors are not null and their components in any basis are given by relations (2.26). There is no need to introduce a metric tensor $\mathbf{g}$ in such a continuum.

Remark. Physically, in the framework of dislocation theory, an infinitesimal closed path at any instant t_0 may be open at any other instant t. The amount of the jump may be characterized by the two types of discontinuity, e.g., [13], [137], [148]. Discontinuity of any scalar field causes the torsion tensor to vanish. The Burgers vector scheme as fictitious closing is sufficient to illustrate the distortion engendered by a dislocation edge (translation). It is however not enough to take into account the relative rotation of surfaces in contact, constituting the lips of a screw dislocation (rotation) and more generally mixed dislocation [61]. This is the reason we have to consider additionally the discontinuity of vector fields, e.g., [163]. For numerical simulation, it is reasonable to work with tensors obtained by contracting the torsion $\aleph$ and the curvature $\Re$. Systematic research on such tensors was conducted by Cartan, e.g., [22]. The most usual are, e.g., [116]:

1. Contraction of the torsion gives the 1-form:

$$\tilde{\aleph} \equiv [(\Gamma_{ab}^b - \Gamma_{ba}^b) - \aleph_{0ab}^b]\mathbf{u}^a. \tag{2.27}$$

2. Contraction of the curvature gives two 2-covariant symmetric (Ricci tensor) and skew-symmetric tensors, e.g., [123]:

$$\begin{aligned}
\Re_S &\equiv (\Re_{acb}^d - \Re_{abc}^d)\,\mathbf{u}^a \otimes \mathbf{u}^b &\tag{2.28}\\
\Re_A &\equiv \Re_{cab}^c\,\mathbf{u}^a \wedge \mathbf{u}^b. &\tag{2.29}
\end{aligned}$$

In the framework of Einstein–Cartan theory of general relativity, torsion was already viewed as translational holonomy per unit area, e.g., [155], a translational holonomy being defined as a nonclosure of a spacetime loop into a flat spacetime. It should be stressed that describing a continuum with singularity by simply measuring lengths with a metric tensor $\mathbf{g}$ is not enough. In usual continuum theory devoted to nondefected (noncracked solids, ideal fluids without vortices), it is common to choose a Euclidean affine connection to calculate derivatives of tensor fields. As soon as singularities are present in the bulk, this situation changes. Historically, Kondo (1958) first pointed out the uncertainty in the measurements of lengths in dislocated crystals. The elementary uncertainty is about the order of the modulus of the Burgers vector $\|\mathbf{b}\|$.

2.4   Deformation of continuum

In the following section, we consider a continuum with singularity B, the geometrical structure of which is defined by the variables (2.25) and (2.26) and the motion of which is defined by the placement $\varphi : B_0 \longrightarrow B$ with the velocity field $\mathbf{v}$ relative to a Galilean referential body. In classical continuum kinematics, it is always assumed that both the placement and its inverse are continuous functions of their arguments. In this work, that is no longer the case.

2.4.1   Linear tangent motions

Let us consider a strain gradient continuum B, the motion of which is defined by the placement $\varphi : B_0 \longrightarrow B$ with a velocity field $\mathbf{v}$.

Definition 2.4.1 *(Linear tangent motion) Let us consider a parametrized arc in the initial configuration B_0 of a continuum B, $\gamma : R \longrightarrow B_0$, $\xi \in R$ and let $\mathbf{u}_0$ be a vector at M_0 tangent to this arc, with $\mathbf{OM}_0 = \gamma(\xi_0)$. The linear tangent motion to φ*

is a differential map $d\varphi$ from the tangent space $T_{M_0}B_0$ to $T_M B$, which transforms a tangent vector $\mathbf{u}_0$ in M_0 according to, e.g., [159]:

$$\mathbf{u} = d\varphi(\mathbf{u}_0) \qquad d\varphi(\mathbf{u}_0) \equiv \left(\frac{d\varphi}{d\xi} \circ \gamma\right)(\xi_0). \tag{2.30}$$

Example: Elastic finite strain. Let B be a continuum undergoing large strain, $\varphi : B_0 \longrightarrow B$, in which B_0 is the initial configuration. The elementary definition of the linear tangent motion is given as

$$\begin{aligned}
\mathbf{OM} &= \varphi(\mathbf{X}) = \varphi(X^1, X^2, X^3) \\
d\mathbf{OM} &= \varphi(\mathbf{X} + d\mathbf{X}) - \varphi(\mathbf{X}) \\
&= \frac{\partial\varphi}{\partial\mathbf{X}}(d\mathbf{X})
\end{aligned}$$

in which we have implicitly used the Euclidean connection (this assumption is rarely mentioned in most textbooks on finite elasticity). The quantity $\mathbf{F} \equiv \frac{\partial\varphi}{\partial\mathbf{X}}$ is called the deformation gradient and is not a tensor. However, it may be decomposed on a tensorial basis as follows. Let us first assume that B evolves with respect to a Euclidean ambient space with a tangent basis (not necessarily Cartesian) $(\mathbf{u}_1, \mathbf{u}_2, \mathbf{u}_3)$. The deformation gradient is projected onto the tangent basis:

$$\mathbf{F} \equiv \frac{\partial\varphi}{\partial\mathbf{X}} = \frac{\partial\varphi^i}{\partial X^j}\,\mathbf{u}_i \otimes \mathbf{u}^j.$$

For instance, if the deformation of the continuum B has an axial symmetry and is analyzed with respect to cylindrical coordinates $(X^1 = r, X^2 = \theta, X^3 = z)$, the deformation gradient is written as follows:

$$\begin{aligned}
\mathbf{F} &= \frac{\partial\varphi_r}{\partial r}\mathbf{e}_r \otimes \mathbf{e}_r + \frac{1}{r}\frac{\partial\varphi_r}{\partial\theta}\mathbf{e}_r \otimes \mathbf{e}_\theta + \frac{\partial\varphi_r}{\partial z}\mathbf{e}_r \otimes \mathbf{e}_z \\
&+ \frac{\partial\varphi_\theta}{\partial r}\mathbf{e}_\theta \otimes \mathbf{e}_r + \frac{1}{r}\frac{\partial\varphi_\theta}{\partial\theta}\mathbf{e}_\theta \otimes \mathbf{e}_\theta + \frac{\partial\varphi_\theta}{\partial z}\mathbf{e}_\theta \otimes \mathbf{e}_z \\
&+ \frac{\partial\varphi_z}{\partial r}\mathbf{e}_z \otimes \mathbf{e}_r + \frac{1}{r}\frac{\partial\varphi_z}{\partial\theta}\mathbf{e}_z \otimes \mathbf{e}_\theta + \frac{\partial\varphi_z}{\partial z}\mathbf{e}_z \otimes \mathbf{e}_z
\end{aligned}$$

in which we have used the orthonormal basis $(\mathbf{e}_r, \mathbf{e}_\theta, \mathbf{e}_z)$ (notice that $\mathbf{u}_r = r\mathbf{e}_r$). Notice that the components of deformation gradient F^i_j do not involve any covariant derivatives. The reason is that $\varphi(\mathbf{X})$ is not a vector but rather a map from B_0 to the space Σ. It is worthwhile to notice that the projection of an "infinitesimal" superimposed motion $\delta\varphi$ on the motion φ onto the tangent basis of cylindrical coordinates system gives:

$$
\begin{aligned}
\delta \mathbf{F} = {}& \frac{\partial \delta \varphi_r}{\partial r} \mathbf{e}_r \otimes \mathbf{e}_r + \frac{1}{r} \left(\frac{\partial \delta \varphi_r}{\partial \theta} - \delta \varphi_\theta \right) \mathbf{e}_r \otimes \mathbf{e}_\theta + \frac{\partial \delta \varphi_r}{\partial z} \mathbf{e}_r \otimes \mathbf{e}_z \\
+{}& \frac{\partial \delta \varphi_\theta}{\partial r} \mathbf{e}_\theta \otimes \mathbf{e}_r + \frac{1}{r} \left(\frac{\partial \delta \varphi_\theta}{\partial \theta} + \delta \varphi_r \right) \mathbf{e}_\theta \otimes \mathbf{e}_\theta + \frac{\partial \delta \varphi_\theta}{\partial z} \mathbf{e}_\theta \otimes \mathbf{e}_z \\
+{}& \frac{\partial \delta \varphi_z}{\partial r} \mathbf{e}_z \otimes \mathbf{e}_r + \frac{1}{r} \frac{\partial \delta \varphi_z}{\partial \theta} \mathbf{e}_z \otimes \mathbf{e}_\theta + \frac{\partial \delta \varphi_z}{\partial z} \mathbf{e}_z \otimes \mathbf{e}_z.
\end{aligned}
$$

Definition 2.4.2 *(Embedded basis) Let* $(\mathbf{u}_{10}, \mathbf{u}_{20}, \mathbf{u}_{30})$ *be a vector basis in the initial configuration* B_0. *During the deformation, this basis deforms and is denoted* $(\mathbf{u}_1, \mathbf{u}_2, \mathbf{u}_3)$ *with* $\mathbf{u}_a = d\varphi(\mathbf{u}_{a0})$. *The vector basis* $(\mathbf{u}_1, \mathbf{u}_2, \mathbf{u}_3)$ *is said to be embedded in the motion of* B, *or simply embedded in* B.

Sometimes, it is called the material basis. The dual linear tangent motion $d\varphi^*$ is defined by, e.g., [159]:

$$
[d\varphi^* \omega](\mathbf{u}_0) \equiv \omega[d\varphi(\mathbf{u}_0)] \qquad \forall \mathbf{u}_0 \in T_{M_0} B_0 \qquad \forall \omega \in T_M B^* \tag{2.31}
$$

in which we can note that the existence of the inverse map is not necessary. The dual linear tangent motion $d\varphi^*$ transforms an element of the dual tangent space of the actual configuration into the dual tangent space of the initial configuration.

The dual of ω, a p-covariant tensor on B, is the tensor $d\varphi^* \omega$, a p-covariant on B_0, defined by:

$$
(d\varphi^* \omega)(\mathbf{u}_{10}, \ldots, \mathbf{u}_{p0}) \equiv \omega[d\varphi(\mathbf{u}_{10}), \ldots, d\varphi(\mathbf{u}_{p0})] \qquad \forall \mathbf{u}_{10}, \ldots, \mathbf{u}_{p0} \in T_{M_0} B_0. \tag{2.32}
$$

Starting from an initial configuration, we can observe that the Jacobian of a weakly continuous transformation satisfies the following condition, where the right-hand inequality is not strict:

$$
0 < J(d\varphi) \le \infty. \tag{2.33}
$$

The dual of $\mathbf{A}$, any q-contravariant tensor on the initial configuration B_0, to the actual configuration B of the continuum is the tensor $d\varphi \mathbf{A}$, which is q-contravariant on B, defined by:

$$
(d\varphi \mathbf{A})(\omega^1, \ldots, \omega^q) \equiv \mathbf{A}[d\varphi^*(\omega^1), \ldots, d\varphi^*(\omega^q)] \qquad \forall \omega^1, \ldots, \omega^q \in T_M B^*. \tag{2.34}
$$

Again, in all those previous definitions, it is not necessary to introduce the Jacobian inverse. Extended definitions of embedded 1-forms and embedded volume forms ω_0 are analogous to that of an embedded vector. Starting from the properties of the map and the properties of linear tangent motions (2.30), we can establish the classification below.

Continuous	φ	-	$d\varphi$	-
Immersion	φ	-	$d\varphi$ (1-to-1)	-
Submersion	φ	-	$d\varphi$ (onto)	-
Homeomorphism	φ	φ^{-1}	$d\varphi$	$d\varphi^{-1}$
Embedding	φ	φ^{-1}	$d\varphi$ (1-to-1)	$d\varphi^{-1}$ (onto)
Diffeomorphism	φ	φ^{-1}	$d\varphi$ (bijective)	$d\varphi^{-1}$ (bijective)

Table 2.1. Classes of maps based on the existence and properties of the transformation φ and its differential $d\varphi$.

By addressing the various classes of transformations, it is then possible to address the fundamental problem of modeling the irreversibility in continuum mechanics. Given the two configurations, initial B_0 and actual B, of the continuum, with a map between them, it is worthwhile to assess if it is possible to uniquely determine the tensorial rule over the actual configuration in terms of tensorial rule given over the initial state.

In rigid body dynamics, Euclidean geometry has dominated the rules of transformations, a rigid body being constructed from translations and rotations. Two basic invariants are associated to the metric tensor: size and shape. In classical elasticity, even undergoing nonlinear large strains, the diffeomorphism is the usual map and it can be merely considered as a coordinate transformation. Therefore, the use of the initial configuration has been successful for many centuries for it is always possible to make a correspondence between particles and their positions at any moment. Indeed, for the weakest homeomorphism (see Table 2.1), the linear tangent motions, either primal or dual, could each be used independently from the other for solving the problem. Both the initial configuration and actual configuration could be used, e.g., [160]. For diffeomorphic mappings, the problem resolution by means of contravariant and covariant tensor fields becomes indifferent. Recently Kiehn developed the notion of predictive and retroductive processes in [90] from a topological point of view. Namely, with an extended version of the previous table, he was able to point out that for nonhomeomorphic maps it is not possible to determine in a predictive manner some tensor fields. However, they could be determined in a retroductive manner, the retroductive determinism is resumed by the following basic question: Is it possible to determine (in a retroductive manner) the tensorial rule over the initial configuration, given the tensorial rule over the actual configuration?

Example. As an example of nontopological motion, consider two rigid plates of infinite length superimposed one over the other and sliding with a relative velocity

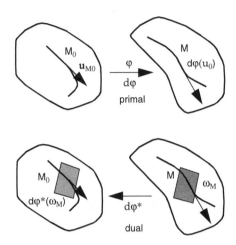

Figure 2.7. Linear tangent motion and its dual. Notice the arrow's direction, which is not the same for the linear map and for its dual.

w. The transformation is defined by the placement $\mathbf{y} = \varphi(\mathbf{x}, t)$, where t could be considered as a real parameter such that $0 \le t \le \infty$:

$$x^2 \ge h_+ \begin{cases} y^1 = x^1 + (v_0 + w)t \\ y^2 = x^2 \end{cases} \qquad x^2 \le h_- \begin{cases} y^1 = x^1 + v_0 t \\ y^2 = x^2 \end{cases} . \qquad (2.35)$$

The Jacobian of this transformation for a point located at the internal slip surface $J(\varphi)$ can be calculated as follows:

$$J(\varphi) = 1 + \lim_{h \to 0} \left(\frac{w}{h} t \right) = \infty \qquad (2.36)$$

showing that the Jacobian of a "weakly" continuous transformation (a transformation that is continuous in a weak way) may tend to infinity at some particular points. By inverting (2.35), we observe that the Jacobian of the inverse transformation also tends to infinity. The purpose of the next section is to address the effects of the topological change due to nucleation and development of singularity within the continuum.

2.4.2 Deformation of continuum with singularity

The deformation of a continuum B from an initial configuration assumed ideal (without field singularity distribution, without residual stress) to its deformed state includes: (a) a transformation of $\mathbf{g}$ (geometrically holonomic part) and (b) a transformation of

∇ (geometrically nonholonomic part), both parts observed internally by B. In other words, transformation is observed with respect to a basis $(\mathbf{u}_1, \mathbf{u}_2, \mathbf{u}_3)$ embedded in B. The nonholonomic transformation induces in fact a change of the local topology of the continuum by creating "holes" and "handles" which are singularity fields.

Let $(\mathbf{u}_{10}, \mathbf{u}_{20}, \mathbf{u}_{30})$ be a vector base such that $[\mathbf{u}_{a0}, \mathbf{u}_{b0}] \equiv 0$ for $a, b = 1, 2, 3$. During the body transformation, the initial base becomes $(\mathbf{u}_1, \mathbf{u}_2, \mathbf{u}_3)$ with $\mathbf{u}_a = d\varphi(\mathbf{u}_{a0})$ at each instant t. The following theorem then holds.

Theorem 2.4.1 *Let φ be a transformation $\varphi : B_0 \longrightarrow B$ and $d\varphi$ its differential map $d\varphi : T_{M_0} B_0 \longrightarrow T_M B$. Let $(\mathbf{u}_{10}, \mathbf{u}_{20}, \mathbf{u}_{30})$ be an initial vector base of $T_{M_0} B_0$ and $\mathbf{u}_a = d\varphi(\mathbf{u}_{a0})$. If the transformation φ is a C^∞ diffeomorphism then the image of a Lie–Jacobi bracket is the Lie–Jacobi of the image:*

$$d\varphi[\mathbf{u}_{a0}, \mathbf{u}_{b0}] = [d\varphi(\mathbf{u}_{a0}), d\varphi(\mathbf{u}_{b0})]. \tag{2.37}$$

Proof. See, e.g., [29]. In other words, $d\varphi$ is an isomorphism.

Remark. Starting with an initial coordinate (field of) base $[\mathbf{u}_{a0}, \mathbf{u}_{b0}] \equiv 0$, we can conclude, in the case where $d\varphi$ is a C^∞ diffeomorphism, that:

$$[d\varphi(\mathbf{u}_{a0}), d\varphi(\mathbf{u}_{b0})] = d\varphi[\mathbf{u}_{a0}, \mathbf{u}_{b0}] = 0. \tag{2.38}$$

Therefore, there are two classes of deformations.

1. The first class contains deformations (a) only for which the vector basis is embedded in the material so that $[\mathbf{u}_{a0}, \mathbf{u}_{b0}] = 0$. In this case, compatibility of strains is given by $\aleph = 0$ and $\Re = 0$, e.g., [130]. This case corresponds to the well-known material description of large strain elasticity.

2. The second class contains deformations (a) combined with (b) for which the affine equivalence of B with the Euclidean ambient space may be destroyed.

First of all, the constants of structure associated to the embedded basis $(\mathbf{u}_1, \mathbf{u}_2, \mathbf{u}_3)$ do not necessarily vanish at each state of deformation:

$$[\mathbf{u}_a, \mathbf{u}_b] = \aleph_{0ab}^c \mathbf{u}_c \qquad J_u = \omega_0(\mathbf{u}_1, \mathbf{u}_2, \mathbf{u}_3). \tag{2.39}$$

The constants of the structure $\aleph_{0ab}^c$ characterize the "irreversible" part of deformation (distortion) compared to a strongly continuous transformation for which these coefficients remain null if they were initially chosen null. For the reversible part of the transformation, for any vector $\mathbf{u}$ embedded in B, the directional derivative of $\|\mathbf{u}\|^2 = \mathbf{g}(\mathbf{u}, \mathbf{u})$ is expressed by means of the affine connection:

$$\nabla_w \|\mathbf{u}\|^2 = \frac{d}{d\varepsilon}(\|\mathbf{u}\|_{M+\varepsilon w}^2 - \|\mathbf{u}\|_M^2)\,|_{\varepsilon=0} = \mathbf{g}[(\nabla_{\mathbf{u}}\mathbf{w} + \nabla_{\mathbf{u}}\mathbf{w}^T), \mathbf{u}]. \tag{2.40}$$

An analogous calculation with the volume form ω_0, for any basis $(\mathbf{u}_1, \mathbf{u}_2, \mathbf{u}_3)$ embedded in B, gives

$$\frac{d}{d\varepsilon}[\omega_0(\mathbf{u}_1, \mathbf{u}_2, \mathbf{u}_3)_{M+\varepsilon\mathbf{w}} - \omega_0(\mathbf{u}_1, \mathbf{u}_2, \mathbf{u}_3)_M]\,|_{\varepsilon=0} = \text{tr}\,(\nabla\mathbf{w})\omega_0(\mathbf{u}_1, \mathbf{u}_2, \mathbf{u}_3). \quad (2.41)$$

For the irreversible part of the transformation, let $\tilde{\nabla}$ be another connection[2] for the continuum B. The variation of the affine connection denoted $\delta\nabla$, a 1-contravariant, 2-covariant tensor on B, is defined by [116]:

$$\delta\nabla(\mathbf{u}, \mathbf{v}, \omega) \equiv \omega(\tilde{\nabla}_{\mathbf{u}}\mathbf{v} - \nabla_{\mathbf{u}}\mathbf{v}) \qquad \forall\mathbf{u}, \mathbf{v} \in T_M B \qquad \forall\omega \in T_M B^*, \quad (2.42)$$

the projection of which onto $(\mathbf{u}_1, \mathbf{u}_2, \mathbf{u}_3)$ gives

$$\delta\nabla = \delta\Gamma^c_{ab}\mathbf{u}^a \otimes \mathbf{u}^b \otimes \mathbf{u}_c \qquad \delta\Gamma^c_{ab} = \tilde{\Gamma}^c_{ab} - \Gamma^c_{ab}.$$

The resulting variation of the torsion is defined by:

$$\delta\aleph(\mathbf{u}, \mathbf{v}, \omega) \equiv \frac{d}{d\varepsilon}[\aleph_{\nabla+\varepsilon\delta\nabla}(\mathbf{u}, \mathbf{v}, \omega) - \aleph_{\nabla}(\mathbf{u}, \mathbf{v}, \omega)]_{\varepsilon=0}. \quad (2.43)$$

In any basis $(\mathbf{u}_1, \mathbf{u}_2, \mathbf{u}_3)$ embedded or not in B, the variation of the torsion tensor is:

$$\delta\aleph^c_{ab} = \delta\Gamma^c_{ab} - \delta\Gamma^c_{ba}. \quad (2.44)$$

The variation of the curvature is defined by the directional derivative of the curvature tensor $\Re$ along the connection variation $\delta\nabla$. By analogy, let $(\mathbf{u}_1, \mathbf{u}_2, \mathbf{v})$ be any vector fields and ω any form on B. We can write

$$\delta\Re(\mathbf{u}_1, \mathbf{u}_2, \mathbf{v}, \omega) \equiv \frac{d}{d\varepsilon}[\Re_{\nabla+\varepsilon\delta\nabla}(\mathbf{u}_1, \mathbf{u}_2, \mathbf{v}, \omega) - \Re_{\nabla}(\mathbf{u}_1, \mathbf{u}_2, \mathbf{v}, \omega)]_{\varepsilon=0}. \quad (2.45)$$

This may not be expressed solely by means of the connection ∇. The components of the variation of curvature tensor in any basis $(\mathbf{u}_1, \mathbf{u}_2, \mathbf{u}_3)$ embedded in B are

$$\begin{aligned}
\delta\Re^c_{abd} &= [\mathbf{u}_a(\delta\Gamma^c_{bd}) + \Gamma^c_{ae}\delta\Gamma^e_{bd} + \Gamma^e_{bd}\delta\Gamma^c_{ae}] \\
&- [\mathbf{u}_b(\delta\Gamma^c_{ad}) + \Gamma^c_{be}\delta\Gamma^e_{ad} + \Gamma^e_{ad}\delta\Gamma^c_{be}].
\end{aligned} \quad (2.46)$$

---

[2]The continuum B, endowed with a metric tensor $\mathbf{g}$ that is symmetric positive definite (and a volume form ω_0 supposed positive), induced by the Euclidean referential body, is a Riemannian manifold (and oriented). A fundamental theorem of differential geometry may then be applied at each deformation state: Let B be a Riemannian continuum (class C^2), then there exists one and only one affine connection ∇ on B (usually called Riemannian connection) such that $\aleph = 0$ and $\Re = 0$. The proof is classical in differential geometry. Starting from a null torsion and from the covariant derivative of the metric tensor, the explicit expression of this Riemannian connection is directly obtained [29]:

$$\bar{\Gamma}^c_{ab} = \frac{1}{2}g^{cd}[\mathbf{u}_b(g_{ad}) + \mathbf{u}_a(g_{bd}) - \mathbf{u}_d(g_{ab})] - \frac{1}{2}(\aleph^c_{0ba} + g^{cd}g_{ae}\aleph^e_{0bd} + g^{cd}g_{eb}\aleph^e_{0ad}).$$

On the one hand, such a connection is not necessarily the transform by the motion φ of the initial connection, and on the other hand, the curvature tensor associated to this connection is not necessarily null.

Thanks to the expression of covariant derivative, we obtain, e.g., [116]:

$$\delta\mathfrak{R}^c_{abd} = \nabla_a(\delta\Gamma^c_{bd}) - \nabla_b(\delta\Gamma^c_{ad}) - \aleph^e_{ab}\delta\Gamma^c_{ed}. \tag{2.47}$$

Then, for calculating the the affine connection evolution from a given state of deformation, we vary ∇ by considering another connection $\nabla + \varepsilon\delta\nabla$. Variations of torsion and curvature tensors decomposed in the embedded basis $(\mathbf{u}_1, \mathbf{u}_2, \mathbf{u}_3)$ characterize the irreversibility.

Theorem 2.4.2 *(Singularity variation) Consider a nontopological transformation of a continuum B. The first relation (2.44) shows that the increase of scalar discontinuity density is additive. The change of torsion tensor induced by two deformations can be added. Conversely, the increase of the density of vector discontinuity (2.46) is not additive since it depends on the actual values of affine connection coefficients.*

These results could be related, although with great care, to the additive (Green–Nagdhi theory [68]) and multiplicative (Lee theory [112]) decomposition of plastic strain in solids and more recently to the decomposition of transformations in [108]. In the presence of nonholonomic geometrical transformations (variation of the affine connection), the affine equivalence of the continuum and the referential body is destroyed. It is no longer possible to find a homeomorphism to recover the initial configuration from the actual configuration. Le and Stumpf [108] have established for elastic-plastic continua the following result: The statement that the present configuration B is non-Euclidean is equivalent to the statement that plastic strain (defined as nongeometrically nonholonomic deformation) occurs in the body. The comparison of the three methods based respectively on the internal variables, the intermediate incompatible transformations (elastic and plastic), and the connections geometry developed here is certainly an extensive task but it should be clarified in the future.

Calculating the singularity field density in B is performed by updating the connection ∇ for each change of discontinuity density. Knowledge of the variations $\delta\aleph$ and $\delta\mathfrak{R}$ provides in principle the variation of the connection (2.42). This allows us to work at the evolution of the affine connection during the continuum deformation. However, the relation (2.26) shows that the torsion tensor $\aleph$ remains non-null even if we adopt an affine connection ∇ with vanishing coefficients, for instance those of the referential body (or symmetric $\overline{\Gamma}^c_{ab} = \overline{\Gamma}^c_{ba}$). Indeed, constants of the structure $\aleph^c_{0ab}$ associated to the basis $(\mathbf{u}_1, \mathbf{u}_2, \mathbf{u}_3)$ may not vanish.

2.4.3 Bianchi identities in a continuum

By analogy with the general relativistic spacetime geometry, some relations exist between the torsion tensor and the curvature tensor, e.g., [116]. These tensors are no longer arbitrary and must satisfy compatibility conditions called Bianchi identities, e.g., [200] the expressions of which with respect to any basis $(\mathbf{u}_1, \mathbf{u}_2, \mathbf{u}_3)$ embedded in B, e.g., [120] are given in the following paragraph.

The first identities of Bianchi relate the covariant derivative of the torsion tensor to the curvature as follows, $\forall \omega \in T_M B^*$:

$$\sum_{(abc)} (\nabla_{\mathbf{u}_a} \aleph)(\mathbf{u}_b, \mathbf{u}_c, \omega) + \aleph(\aleph(\mathbf{u}_a, \mathbf{u}_b), \mathbf{u}_c, \omega) = \sum_{(abc)} \Re(\mathbf{u}_a, \mathbf{u}_b, \mathbf{u}_c, \omega). \quad (2.48)$$

The second identities of Bianchi, conversely, relate the covariant derivative of the curvature tensor to the torsion tensor, $\forall \omega \in T_M B^*$, $\forall \mathbf{v} \in T_M B$:

$$\sum_{(abc)} (\nabla_{\mathbf{u}_c} \Re)(\mathbf{u}_a, \mathbf{u}_b, \mathbf{v}, \omega) + \sum_{(abc)} \Re(\aleph(\mathbf{u}_a, \mathbf{u}_b), \mathbf{u}_c, \mathbf{v}, \omega) = 0. \quad (2.49)$$

The derivation of these two sets of identities is reported in Appendix D. Another way of obtaining these identities could be to start from the classical Cartan structural equations and use exterior differentiation, e.g., [29], [120]. In this sense, Edelen and Lagoudas [47] called the Cartan structural equations and their exterior derivatives, respectively, kinematics equations of singularity and continuity equations of singularity. The usefulness of the Bianchi identities should not be overestimated. Still, in the framework of general relativity—wherein the torsion tensor vanishes—relations (2.49) may be used as a starting point to derive the acclaimed conservation equations of the Einstein tensor, e.g., [116]. The physical content of equations (2.48) and (2.49) is that any defect state with discontinuity of scalar and vector fields that satisfies these relations is admissible.

2.5   Kinematics of continuum

2.5.1   Motion and velocity

Definition 2.5.1 *(Continuum motion) Given a referential body, a motion of a continuum B, with continuous distribution of singularity, is a 1-parameter family of configurations indexed by the time t. For each time t, we have the location of the particle M:*

$$\varphi : (M, t) \longrightarrow \mathbf{OM}(t) = \varphi(M, t). \quad (2.50)$$

The set $B_0 \equiv \{\mathbf{OM}(t_0) = \varphi(M, t_0)\}$ defines the initial configuration of B. In addition, the motion of B includes the time evolution of the connection $\nabla(M, t)$, characterized by the evolution of the torsion tensor $\aleph(M, t)$, and that of the curvature tensor $\Re(M, t)$.

The kinematics of a continuum with a continuous distribution of singularity includes the history of the connection $\nabla(t)$. In the existing literature, homeomorphisms can be associated to various reference configurations. They should not be confused with referential, which is a referential rigid body associated with the ambient space. The mapping φ associates the reference configuration to the actual (deformed) configuration at a time t. As previously shown, the displacement of the particle M at time t is defined by the vector

$$\mathbf{u}(M, t) \equiv \varphi(M, t) - \varphi(M, t_0). \tag{2.51}$$

Let $(\mathbf{e}_1, \mathbf{e}_2, \mathbf{e}_3)$ be any vector base field embedded in the referential body (ambient space) Σ. A vector field $\mathbf{u}(M, t)$ is fixed with respect to the referential body (rigid body) if and only if

$$\frac{d}{dt}[\mathbf{u}(M, t) \cdot \mathbf{e}_i] = 0 \qquad i = 1, 2, 3. \tag{2.52}$$

In the case where the referential body is not endowed with a metric tensor (however, we assume that the space is endowed with a metric tensor in this work), the more general definition holds for time-independent vector field:

$$\frac{d}{dt}\left[\mathbf{u}(M, t) \cdot (\mathbf{e}^i)\right] = 0 \qquad \mathbf{e}^j(\mathbf{e}_i) = \delta_i^j \qquad i = 1, 2, 3. \tag{2.53}$$

Definition 2.5.2 *(Velocity field) Consider a continuum B that deforms with respect to a referential body $\sum$ with the displacement field $\mathbf{u}(M, t)$. Let $(\mathbf{e}_1, \mathbf{e}_2, \mathbf{e}_3)$ be any vector base field embedded in $\sum$. The velocity field of B with respect to $\sum$ is defined as the vector field*

$$\mathbf{v}(M, t) \equiv \frac{d}{dt}[\mathbf{u}(M, t)(\mathbf{e}^i)]\mathbf{e}_i. \tag{2.54}$$

The motion $\varphi(M, t)$ can be also used instead of the displacement field to define the velocity. Notice that this definition requires neither the notion of "material velocity" nor of "spatial velocity." The definition is merely an extension of the very classical definition of particle velocity in rigid body kinematics. The key point of the definition lies in the properties of embedding and the relativity of the motion with respect to a referential body is essential. Before embarking in the definition of objective time derivative, let us recall two classical time derivatives in mechanics.

1. **Rigid body motion, Poisson's theorem.** Let us consider rigid motion of a continuum S with a velocity field $\mathbf{v}(M, t)$. If a vector is embedded in the solid S, then $\exists A, B \in S \mid \mathbf{u} = \mathbf{AB}$. The time derivative gives

$$\frac{d\mathbf{u}}{dt} = \frac{d\mathbf{OB}}{dt} - \frac{d\mathbf{OA}}{dt} = \mathbf{v}_B - \mathbf{v}_A.$$

Since S is a rigid body, we can write

$$\|\mathbf{AB}\|^2 = \mathbf{AB} \cdot \mathbf{AB} = \mathbf{AB}_0 \cdot \mathbf{AB}_0$$
$$\frac{d\|\mathbf{AB}\|^2}{dt} = 2\,\mathbf{AB} \cdot (\mathbf{v}_B - \mathbf{v}_A) = 0.$$

Then $\exists\, \Omega \mid v_B - v_A = \Omega(\mathbf{AB})$. Ω is called the spin rate tensor of the rigid body. It is easily shown that the spin rate tensor Ω is a uniform field on the rigid body. We obtain Poisson's theorem

$$\frac{d\mathbf{u}}{dt} = \Omega(\mathbf{u}) \tag{2.55}$$

in which $\Omega = \Omega(t)$ is a skew-symmetric second-order tensor. Since the velocity field is of the form $\mathbf{v} = \mathbf{v}_G + \Omega(\mathbf{GM})$, it is easy to check that $\Omega \equiv \frac{1}{2}(\nabla\mathbf{v} - \nabla\mathbf{v}^T)$. If we seek a time derivative of a vector field $\mathbf{u}$ such that it vanishes for any field embedded within the solid S a natural candidate would be

$$\frac{d^S}{dt}\mathbf{u} \equiv \frac{d}{dt}\mathbf{u} - \Omega(\mathbf{u}).$$

2. **Elastic finite strain.** Let us consider a continuum B undergoing finite deformation. The transformation of a vector $\mathbf{u}_0 \in T_{M_0}B_0$ (of the tangent space of the continuum) gives the vector $\mathbf{u} \in T_M B$:

$$\begin{aligned}
\mathbf{u} &= d\varphi(\mathbf{u}_0) = \frac{\partial\varphi}{\partial\mathbf{X}}(\mathbf{u}_0) \\
\frac{d\mathbf{u}}{dt} &= \frac{\partial}{\partial t}\frac{\partial\varphi}{\partial\mathbf{X}}(\mathbf{u}_0) = \frac{\partial}{\partial\mathbf{X}}\frac{\partial\varphi}{\partial t}(\mathbf{u}_0) = \frac{\partial\mathbf{v}}{\partial\mathbf{X}}(\mathbf{u}_0) = \frac{\partial\mathbf{v}}{\partial\mathbf{x}}(\mathbf{u}).
\end{aligned}$$

We deduce the classical relation of the continuum finite deformation for any embedded vector $\mathbf{u}$ (derivatives with respect to space $\mathbf{X}$ and time t were inverted to provide sufficient regularity):

$$\frac{d\mathbf{u}}{dt} = \nabla\mathbf{v}(\mathbf{u}) = \Omega(\mathbf{u}) + \mathbf{D}(\mathbf{u}) \tag{2.56}$$

in which we have defined the spin rate tensor $\Omega \equiv \frac{1}{2}(\nabla\mathbf{v} - \nabla\mathbf{v}^T)$ and the strain rate tensor $\mathbf{D} \equiv \frac{1}{2}(\nabla\mathbf{v} + \nabla\mathbf{v}^T)$. In elastic finite strain theory it is usual, although not necessary, to define the Green–Lagrange strain measure:

$$\varepsilon(\mathbf{u}_{0a}, \mathbf{u}_{0b}) \equiv \frac{1}{2}\left[\mathbf{g}(\mathbf{u}_a, \mathbf{u}_b) - \mathbf{g}(\mathbf{u}_{0a}, \mathbf{u}_{0b})\right]. \tag{2.57}$$

This 2-covariant tensor field on the initial configuration of the continuum B_0 allows us to calculate the change of length of any tangent vector and the change of angle between two vectors during the deformation. It reduces to the classical Cauchy strain when infinitesimal transformations occur. Its components projected onto the initial tangent basis are given by

$$\varepsilon_{ab} = \frac{1}{2}\mathbf{g}\left(\mathbf{u}_{0a}, \frac{\partial\varphi}{\partial\mathbf{X}}^T\frac{\partial\varphi}{\partial\mathbf{X}}(\mathbf{u}_{0b}) - \mathbf{u}_{0b}\right).$$

2.5.2 *Time derivative with respect to a continuum*

The concept of the objective time derivative is not simple for continua with singularity since the basic philosophy for building the frame indifference was usually based solely on the existence of a metric tensor (rigid body motion). A continuum with singularity, as for dislocated crystals, has of course the property of metricity. However, additionally, the affine connection may include information concerning not only the metricity (such as Christoffel symbols) but also the distortion of the initial material basis ($\aleph^c_{0ab} \neq 0$).

For kinematics, we define a time derivative operator with respect to B such that any tensor (vector $\mathbf{u}$, form $\aleph$, volume form ω_0) embedded in B must have a vanishing derivative. By the way, a total derivative is a derivative with respect to the referential body Σ.

Definition 2.5.3 *(B-derivative) Consider a continuum B with a velocity field* $\mathbf{v}$. *The time derivatives with respect to the continuum B are defined as the following derivatives respectively for a vector* $\mathbf{u}$, *a form* ω, *and a volume form* ω_0:

$$\frac{d^B}{dt}\mathbf{u} \equiv \frac{d}{dt}\mathbf{u} - \nabla\mathbf{v}(\mathbf{u}) \tag{2.58}$$

$$\frac{d^B}{dt}\omega \equiv \frac{d}{dt}\omega + \nabla\mathbf{v}^T(\omega) \tag{2.59}$$

$$\frac{d^B}{dt}\omega_0 \equiv \frac{d}{dt}\omega_0 + \mathrm{tr}\,(\nabla\mathbf{v})\omega_0. \tag{2.60}$$

These three definitions are compatible. A vector, 1-form, and volume form are said to be instantaneously embedded in the motion of B during the lapse of time $[t, t + dt]$ if they satisfy respectively the relations:

$$\frac{d^B}{dt}\mathbf{u} = 0 \qquad \frac{d^B}{dt}\omega = 0 \qquad \frac{d^B}{dt}\omega_0 = 0. \tag{2.61}$$

Definition 2.5.4 *(B-derivative) Let* $\mathbf{A}$ *be a mixed tensor field of the type* (p, q) *defined on B. The time derivative of* $\mathbf{A}$ *with respect to the continuum B is a tensor of the same type as* $\mathbf{A}$, *satisfying for any p-plet of vectors* $(\mathbf{u}_1, \ldots, \mathbf{u}_p)$ *and any q-plet of 1-forms* $(\omega_1, \ldots, \omega_q)$, *embedded in B, the condition:*

$$\left(\frac{d^B}{dt}\mathbf{A}\right)(\mathbf{u}_1, \ldots, \mathbf{u}_p, \omega^1, \ldots, \omega^q) \equiv \frac{d}{dt}[\mathbf{A}(\mathbf{u}_1, \ldots, \mathbf{u}_p, \omega^1, \ldots, \omega^q)]. \tag{2.62}$$

As an illustration, we can obtain the derivative of a 2-covariant tensor (resp. 2-contravariant) with respect to B, the forms of which are like the classical derivatives of Rivlin (resp. Olroyd), e.g., [125]:

$$\frac{d^B}{dt}\mathbf{A} = \frac{d}{dt}\mathbf{A} + \mathbf{A}\nabla\mathbf{v} + \nabla\mathbf{v}^T\mathbf{A} \qquad \frac{d^B}{dt}\mathbf{A} = \frac{d}{dt}\mathbf{A} - \nabla\mathbf{v}\mathbf{A} - \mathbf{A}\nabla\mathbf{v}^T. \qquad (2.63)$$

Problems related to the objectivity of derivative operators may have effects on the invariant formulation of irreversible transformations, e.g., [39], [108], [131], [150]. Although formulae (2.63), and their analogues for tensors of different order, have exactly the same forms as convective derivatives (Jaumann, Olroyd, Rivlin, Truesdell, ...), e.g., [125], [130], the essential difference comes from the nonsymmetric part (lower indices) of the coefficients of the affine connection (nonnull torsion and curvature tensors). The nonsymmetric part may be associated to local slippage of matter.[3]

Example. Elastic finite strain. The above definitions conform to the classical definition in elasticity. Let B be an elastic continuum undergoing large deformation with respect to a Euclidean ambient space endowed with a Cartesian frame $(O, \mathbf{e}_1, \mathbf{e}_2, \mathbf{e}_3)$:

$$\begin{aligned} \mathbf{OM}(t) &= \varphi(\mathbf{X}, t) & \mathbf{u}_i &= d\varphi(\mathbf{u}_{i0}) \\ J_e &= \omega_0(\mathbf{e}_1, \mathbf{e}_2, \mathbf{e}_3) = 1 \end{aligned}$$

in which we set $\mathbf{u}_{i0} \equiv \mathbf{e}_i$ for the sake of simplicity. The B-derivative of the volume form (uniform over all space) and the metric tensor (also uniform over all space) on the "material basis" $(\mathbf{u}_1, \mathbf{u}_2, \mathbf{u}_3)$ gives

$$\begin{aligned} \frac{d^B}{dt}\omega_0(\mathbf{u}_1, \mathbf{u}_2, \mathbf{u}_3) &= \frac{d}{dt}[\omega_0(\mathbf{u}_1, \mathbf{u}_2, \mathbf{u}_3)] = \mathrm{tr}\,(\nabla\mathbf{v})\,\omega_0(\mathbf{u}_1, \mathbf{u}_2, \mathbf{u}_3) = \mathrm{div}\,\mathbf{v}J_u \\ \frac{d^B\mathbf{g}}{dt}(\mathbf{u}_i, \mathbf{u}_j) &= \frac{d}{dt}[\mathbf{g}(\mathbf{u}_i, \mathbf{u}_j)] = \mathbf{g}\left(\frac{d\mathbf{u}_i}{dt}, \mathbf{u}_j\right) + \mathbf{g}\left(\mathbf{u}_i, \frac{d\mathbf{u}_j}{dt}\right). \end{aligned}$$

---

[3]Consider a continuum B and a vector field $\mathbf{u}$ on B. Previous studies, e.g., [149] defined the equations of motion for such a vector field as:

$$\frac{d\mathbf{u}}{dt} = \nabla\mathbf{v}(\mathbf{u}) + \mathbf{A}(\mathbf{u})$$

where $\mathbf{A}(\mathbf{u})$ represents the slippage of the vector $\mathbf{u}$ with respect to the continuum (the second-order tensor $\mathbf{A}$ is called the slip tensor). Comparison with relation (2.58) shows that the derivative with respect to the continuum also characterizes the slippage of the vector $\mathbf{u}$ with respect to the matter. The "slippage" theory has been used to study the constitutive laws of polymeric fluids consisting in flexible macromolecules, rodlike molecules suspended and interacting with the fluid, e.g., [105].

In this last equation, we can proceed as follows to obtain:

$$
\frac{d u_i}{dt} = \frac{d}{dt}\frac{\partial \varphi}{\partial \mathbf{X}}(\mathbf{u}_{0i}) = \frac{\partial}{\partial \mathbf{X}}\frac{d\varphi}{dt}(\mathbf{u}_{0i}) = \frac{\partial}{\partial \mathbf{x}}\left(\frac{d\varphi}{dt}\right)\frac{\partial \varphi}{\partial \mathbf{X}}(\mathbf{u}_{0i})
$$

$$
= \frac{\partial \mathbf{v}}{\partial \mathbf{x}}(\mathbf{u}_i)
$$

$$
\frac{d^B \mathbf{g}}{dt}(\mathbf{u}_i, \mathbf{u}_j) = \mathbf{g}\left(\frac{\partial \mathbf{v}}{\partial \mathbf{x}}(\mathbf{u}_i), \mathbf{u}_j\right) + \mathbf{g}\left(\mathbf{u}_i, \frac{\partial \mathbf{v}}{\partial \mathbf{x}}(\mathbf{u}_j)\right).
$$

Projection of this relation onto a system of Cartesian coordinates (x_1, x_2, x_3) may be treated as follows. In the case where $\mathbf{u}_i = \mathbf{e}_i$ at the instant t, we find:

$$
\frac{d^B}{dt}\omega_0(\mathbf{e}_1, \mathbf{e}_2, \mathbf{e}_3) = \frac{\partial v_i}{\partial x_i}
$$

$$
\frac{1}{2}\frac{d^B \mathbf{g}}{dt}(\mathbf{e}_i, \mathbf{e}_j) = \frac{1}{2}\left(\frac{\partial v_i}{\partial x_j} + \frac{\partial v_j}{\partial x_i}\right).
$$

Let us consider any 1-form field $\mathbf{u}^a$ embedded within a continuum B and element of a dual basis. The definition of the dual basis holds $\mathbf{u}^a(\mathbf{u}_b) = \delta^a_b$. We thus have:

$$
\frac{d}{dt}[\mathbf{u}^a(\mathbf{u}_b)] = \frac{d\mathbf{u}^a}{dt}(\mathbf{u}_b) + \mathbf{u}^a\left(\frac{d\mathbf{u}_b}{dt}\right).
$$

Since $\mathbf{u}_b$ is also an embedded vector, we can easily deduce:

$$
\frac{d\mathbf{u}_b}{dt} = \nabla\mathbf{v}(\mathbf{u}_b)
$$

$$
\frac{d\mathbf{u}^a}{dt} = -\nabla\mathbf{v}^T(\mathbf{u}^a).
$$

This second relation allows us to check the compatibility of the above three definitions on the embedded vector, form, and volume form.

Definition 2.5.5 (*Velocity of continuum*) *Consider a continuum B undergoing holomic and nonholomic deformations, characterized by the displacement field* $\mathbf{u}(M, t)$ *and the connection* $\nabla(M, t)$. *For the sake of covariance and frame indifference, kinematics is described by the metric* $\mathbf{g}(M, t)$, *the torsion* $\aleph(M, t)$, *and the curvature* $\Re(M, t)$. *Kinematics is defined by the objective time derivatives:*

$$
\zeta_\mathbf{g} \equiv \frac{1}{2}\frac{d^B}{dt}\mathbf{g} \tag{2.64}
$$

$$
\zeta_\aleph \equiv \frac{d^B}{dt}\aleph \tag{2.65}
$$

$$
\zeta_\Re \equiv \frac{d^B}{dt}\Re. \tag{2.66}
$$

For completeness, we can add the derivative of the volume form:

$$\zeta_{\omega_0} \equiv \frac{d^B}{dt}\omega_0$$

which can be deduced by contraction of $\zeta_{\mathbf{g}}$ in most cases.

2.5.3 Theorems of transposition of derivatives

Let B be a continuum with singularity, the motion of which with respect to a reference is written as $\varphi = \varphi_{t-t_0}$. Details of the proof of the following theorems on transpositions may be found in, e.g., [163].

Theorem 2.5.1 *Let ω be a p-form on the actual configuration B. The dual of the derivative of ω with respect to B is equal to the total derivative of the dual of ω:*

$$d\varphi^*\left(\frac{d^B}{dt}\omega\right) = \frac{d}{dt}\left(d\varphi^*\omega\right). \tag{2.67}$$

Proof. From the definition, the dual of the p-form ω by the motion φ to the initial configuration B_0 gives, for any p-plet of vectors $(\mathbf{u}_{10}, \ldots, \mathbf{u}_{p0})$ in B_0:

$$d\varphi^*\omega(\mathbf{u}_{10}, \ldots, \mathbf{u}_{p0}) = \omega[d\varphi(\mathbf{u}_{10}), \ldots, d\varphi(\mathbf{u}_{p0})].$$

The derivative of the p-form ω on B with respect to B is then written as:

$$\frac{d^B}{dt}\omega(d\varphi\mathbf{u}_{10}, \ldots, d\varphi\mathbf{u}_{p0}) = \frac{d}{dt}[\omega(d\varphi\mathbf{u}_{10}, \ldots, d\varphi\mathbf{u}_{p0})]$$

$$= \frac{d}{dt}[d\varphi^*\omega(\mathbf{u}_{10}, \ldots, \mathbf{u}_{p0})].$$

For any p-plet $(\mathbf{u}_{10}, \ldots, \mathbf{u}_{p0})$ in B_0 not depending on t, we then obtain by using the dual definition:

$$d\varphi^*\left(\frac{d^B}{dt}\omega\right)(\mathbf{u}_{10}, \ldots, \mathbf{u}_{p0}) = \frac{d}{dt}(d\varphi^*\omega)(\mathbf{u}_{10}, \ldots, \mathbf{u}_{p0}).$$

Remark. The relation (2.67) allows us to calculate the time derivative with respect to the continuum of a p-form, as the metric tensor, and the volume form by means of the total derivative and via the transposition.

Theorem 2.5.2 *Let $\mathbf{A}$ be a q-contravariant tensor field on the initial configuration B_0. If the image of $\mathbf{A}$ in the actual configuration B, denoted $d\varphi\mathbf{A}$, is applied on 1-forms embedded in B, then the derivative of $d\varphi\mathbf{A}$ with respect to B is equal to the image of the total derivative of $\mathbf{A}$:*

$$\frac{d^B}{dt}(d\varphi\mathbf{A}) = d\varphi\left(\frac{d}{dt}\mathbf{A}\right). \tag{2.68}$$

Proof. Consider the q-plet of 1-forms $(\omega^1, \ldots, \omega^q)$ in the actual configuration B. Then let us apply the tensor $d\varphi\mathbf{A}$ on these forms:

$$d\varphi\mathbf{A}(\omega^1, \ldots, \omega^q) = \mathbf{A}(d\varphi^*\omega^1, \ldots, d\varphi^*\omega^q)$$

Again, by applying on the q-plet of forms the derivative with respect to the continuum B of the q-contravariant tensor $\mathbf{A}$, we obtain

$$\frac{d^B}{dt}(d\varphi\mathbf{A})(\omega^1, \ldots, \omega^q) = \frac{d}{dt}[d\varphi\mathbf{A}(\omega^1, \ldots, \omega^q)]$$
$$= \frac{d}{dt}[\mathbf{A}(d\varphi^*\omega^1, \ldots, d\varphi^*\omega^q)]$$

First, we can write from (2.67):

$$\frac{d}{dt}(d\varphi^*\omega^k) = d\varphi^*\left(\frac{d^B}{dt}\omega^k\right) = d\varphi^*\left[\frac{d}{dt}\omega^k + \nabla\mathbf{v}^T(\omega^k)\right].$$

Then, when distributing the time derivative, the right-hand side term becomes

$$\frac{d^B}{dt}(d\varphi\mathbf{A})(\omega^1, \ldots, \omega^q) = d\varphi\left(\frac{d}{dt}\mathbf{A}\right)(\omega^1, \ldots, \omega^q) + d\varphi\mathbf{A}\left(\frac{d^B}{dt}\omega^1, \ldots, \omega^q\right)$$
$$+ \cdots + d\varphi\mathbf{A}\left(\omega^1, \ldots, \frac{d^B}{dt}\omega^q\right).$$

If the 1-forms ω^k are embedded forms in the continuum B, then their derivatives with respect to B vanish. Therefore, we obtain

$$\frac{d^B}{dt}(d\varphi\mathbf{A})(\omega^1, \ldots, \omega^q) = d\varphi\left(\frac{d}{dt}\mathbf{A}\right)(\omega^1, \ldots, \omega^q).$$

Remark. The theorem (2.68) has been widely used for elastic solids undergoing large deformations, e.g., [130]. Indeed, it was implicitly used in classical theories of nonlinear elasticity.

2.5.4 Spatial and material accelerations

Consider a body B the velocity field of which is denoted $\mathbf{v}$. The total variation of a scalar field $\psi(M, t)$ on B between two instants t and $t + dt$ is (total variation means change due to both time and position):

$$d\psi(M, t) = \lim_{\varepsilon \to 0} \frac{1}{\varepsilon}[\psi(M + \varepsilon\mathbf{v}dt, t + \varepsilon dt) - \psi(M, t)]. \tag{2.69}$$

A straightforward calculationh shows that the total derivative of ψ is, ∇ denoting the affine connection on the continuum:

$$\frac{d}{dt}\psi(M, t) = \frac{\partial}{\partial t}\psi(M, t) + \nabla_{\mathbf{v}(M,t)}\psi(M, t). \tag{2.70}$$

Let ω be a 1-form uniform and constant field on the continuum B, meaning that $\frac{d\omega}{dt} = 0$. Substituting the scalar field $\psi(M, t) = \omega[\mathbf{u}(M, t)]$ in (2.70), we obtain

$$\frac{d}{dt}\mathbf{u}(M, t) = \frac{\partial}{\partial t}\mathbf{u}(M, t) + \nabla_{\mathbf{v}(M,t)}\mathbf{u}(M, t). \tag{2.71}$$

Relations (2.70) and (2.71) are the classical formulae of Euler, extended to continuum with singularity, by using non (sym)-metric connection. Let $\psi = \omega(\mathbf{u})$ now be a scalar field defined on B where ω and $\mathbf{u}$ are respectively form and vector fields. The total derivative of ψ yields

$$\frac{d}{dt}[\omega(\mathbf{u})] = \left(\frac{\partial}{\partial t}\omega + \nabla_{\mathbf{v}}\omega\right)\mathbf{u} + \omega\left(\frac{\partial}{\partial t}\mathbf{u} + \nabla_{\mathbf{v}}\mathbf{u}\right) \tag{2.72}$$

provided we have the following relationships:

$$\frac{d}{dt}\omega = \frac{\partial}{\partial t}\omega + \nabla_{\mathbf{v}}\omega \qquad \frac{d}{dt}\mathbf{u} = \frac{\partial}{\partial t}\mathbf{u} + \nabla_{\mathbf{v}}\mathbf{u}.$$

1. **Spatial acceleration.** Suppose that the form ω is embedded in the referential body (spatial). Then we have the same relation as (2.71):

$$\frac{d}{dt}\omega = 0 \quad \Longrightarrow \quad \frac{d}{dt}[\omega(\mathbf{u})] = \omega\left(\frac{\partial}{\partial t}\mathbf{u} + \nabla_{\mathbf{v}}\mathbf{u}\right). \tag{2.73}$$

When $\mathbf{u} = \mathbf{v}$ we obtain the spatial acceleration in the theory of fluid mechanics, where we recognize the D'Alembert–Euler formula for acceleration:

$$\frac{d}{dt}[\omega(\mathbf{v})] = \omega\left(\frac{\partial}{\partial t}\mathbf{v} + \nabla_{\mathbf{v}}\mathbf{v}\right). \tag{2.74}$$

2. **Material acceleration.** If the form ω is embedded in B (material), we obtain

$$\frac{d^B}{dt}\omega = 0 \quad \Longrightarrow \quad \frac{d}{dt}\omega = -\nabla_{\omega}\mathbf{v}^T = -\nabla\mathbf{v}^T(\omega).$$

The total derivative (2.72) then gives

$$\frac{d}{dt}[\omega(\mathbf{u})] = \omega\left(\frac{d^B}{dt}\mathbf{u}\right) = \omega\left(\frac{\partial}{\partial t}\mathbf{u} + \nabla_{\mathbf{v}}\mathbf{u} - \nabla_{\mathbf{u}}\mathbf{v}\right)$$

or

$$\frac{d}{dt}[\omega(\mathbf{u})] = \omega\left(\frac{\partial}{\partial t}\mathbf{u} + \aleph(\mathbf{v}, \mathbf{u}) + [\mathbf{v}, \mathbf{u}]\right). \tag{2.75}$$

When $\mathbf{u} = \mathbf{v}$, the material acceleration in solid mechanics takes the form

$$\frac{d}{dt}[\omega(\mathbf{v})] = \omega\left(\frac{d^B}{dt}\mathbf{u}\right) = \omega\left(\frac{\partial}{\partial t}\mathbf{v}\right). \tag{2.76}$$

Various referential bodies could be used to measure spacetime quantities as accelerations in different ways, e.g., [130]. The metric tensor, volume form, and affine connection associated with the referential bodies are the basic tools for working at accelerations. In the present study, the 1-form in (2.74) and (2.76) is the mathematical variable to do that. Regarding the relation (2.75), the use of a defected referential body—not affinely equivalent to a Euclidean referential—induces much more difficulties from the kinematics point of view. Analysis of continuum motion with respect to a referential body involves indeed the comparison of spacetime structure by means of affine connection [40].

2.5.5 Remarks on the objective time derivatives

For many years, and more particularly over the last twenty years, the subject of objective time derivatives has been a controversial one in continuum thermomechanics. The interest, for many authors, in looking for appropriate objective rates of tensors originated in the work of Dienes [41] when he analyzed the plastic deformation of the simple shear with kinematic hardening using the Zaremba–Jaumann rate and found an oscillatory stress response.

In the domain of elastic-plastic large deformations, various objective time derivatives of tensor variables have been introduced and used more or less successfully. Most of the previous studies started with the matrix representation of the linear tangent motion $d\varphi$. They called it the deformation gradient $\mathbf{F}$ such that $0 < \det(\mathbf{F}) < \infty$. The fundamental theorem of polar decomposition, attributed to Cauchy, allows us to write the matrix decomposition $\mathbf{F} = \mathbf{R}\mathbf{U}$ where $\mathbf{R}$ is an orthogonal matrix $\mathbf{R}^T\mathbf{R} = \mathbf{I}$ and $\mathbf{U}$ a positive definite symmetric matrix, e.g., [130]. By considering an initial vector base $(\mathbf{u}_{10}, \mathbf{u}_{20}, \mathbf{u}_{30})$ locally, we can derive the matrix $\mathbf{U}^2$ whose components are given by the relation $U_{ab}^2 \equiv \mathbf{g}(\mathbf{u}_a, \mathbf{u}_b)$. The Green–Nagdhi rate is defined by the following relation [68]:

$$\frac{d^{GN}}{dt}\mathbf{A} \equiv \frac{d}{dt}\mathbf{A} - \Omega\mathbf{A} + \mathbf{A}\Omega \qquad \Omega = \frac{d}{dt}\mathbf{R}\mathbf{R}^T \tag{2.77}$$

Considering the actual tangent motion (gradient of the velocity field), another spin tensor can be defined from the Stokes–Euler–Cauchy decomposition, e.g., [191]:

$$\omega = \frac{1}{2}(\nabla\mathbf{v} - \nabla\mathbf{v}^T).$$

The Zaremba–Jaumann rate, sometimes called corotational rate, is defined by the relation, e.g., [191]:

$$\frac{d^{ZJ}}{dt}\mathbf{A} \equiv \frac{d}{dt}\mathbf{A} - \omega\mathbf{A} + \mathbf{A}\omega \tag{2.78}$$

where previous authors indeed have restricted the covariant derivative to the metric derivative (meaning a use of metric connection). Schieck and Stumpf recently [169] proposed a corotational rate by considering first the multiplicative decomposition [112] of the deformation gradient matrix:

$$\mathbf{F} = \mathbf{F}^e\mathbf{F}^p$$

into an elastic part $\mathbf{F}^e$ and a plastic part $\mathbf{F}^p$ such that $0 < \det\mathbf{F}^e < \infty$ and $0 < \det\mathbf{F}^p < \infty$. Neither $\mathbf{F}^e$ nor $\mathbf{F}^p$ is a true linear tangent motion. Applying the polar decomposition theorem, Schieck and Stumpf proposed:

$$\mathbf{F}^e = \mathbf{R}^e\mathbf{U}^e \qquad \mathbf{F}^p = \mathbf{R}^p\mathbf{U}^p.$$

Provided the previous decompositions, a back-rotated elastic stretched tensor has been defined as $\mathbf{U}^e \equiv \mathbf{R}^{pT}\mathbf{U}^e\mathbf{R}^p$ and the composition of elastic and plastic matrices holds $\mathbf{Q} \equiv \mathbf{R}^e\mathbf{R}^p$. Then, a new spin tensor has been defined as the Schieck–Stumpf rate [169]:

$$\frac{d^S}{dt}\mathbf{A} \equiv \frac{d}{dt}\mathbf{A} - \tilde{\Omega}\mathbf{A} + \mathbf{A}\tilde{\Omega} \qquad \tilde{\Omega} \equiv \frac{d\mathbf{Q}}{dt}\mathbf{Q}^T. \tag{2.79}$$

It has been shown that this rate reduces to the Green–Nagdhi rate for hypoelastic and rigid plastic solids. Further, the Zaremba–Jaumann rate remains valid only for moderate deformation (moderate elastic, moderate plastic strains). The Schieck–Stumpf rate has been shown to be valid for the whole range of finite elastic-plastic deformations. After the present brief review, it should be pointed out that the main drawback of the Zaremba–Jaumann rate is that it does not account for the existence of singularity within the continuum. Indeed, the affine connection involved in the covariant derivative should account for the nucleation and the evolution of the singularity density (scalar and vector). Therefore, the use of affine connection with nonvanishing torsion and curvature is more appropriate when applied to time derivative as the Zaremba–Jaumann rate [156]. Splitting the connection according to (2.42) into a symmetric part and nonsymmetric part, the objective time derivative (2.63) 1 holds and extends the classical Zaremba–Jaumann rate (2.78) as follows:

$$\frac{d^B}{dt}\mathbf{A} \equiv \frac{d}{dt}\mathbf{A} - \omega\mathbf{A} + \mathbf{A}\omega + \zeta_g\mathbf{A} + \mathbf{A}\zeta_g \qquad \zeta_g \equiv \frac{1}{2}\frac{d^B}{dt}\mathbf{g}. \tag{2.80}$$

This derivative resembles the classical convected rate. However, the inclusion of the irreversible part in the derivative process, quantified by the nonsymmetric part of the connection ∇, will be developed in detail in the chapter devoted to vortex theory and is quite similar to the creation of such vortices in the fluid volume during flow.

3

Conservation Laws

The basic roots and meaning of conservation laws are reminded in Appendix B in relation with the concept of the invariance group. The scope of this book is rather limited to classical thermomechanics laws. The basic problem in deriving the conservation laws focuses on calculating the evolution of any p-form field during a continuum deformation. For the sake of simplicity, consider a continuum B of dimension three, with an affine connection ∇, a metric $\mathbf{g}$, and a volume form ω_0, in motion with respect to a Galilean reference. The motion is defined by the application φ_{t-t_0} the velocity field of which is transcribed as $\mathbf{v}(M, t)$.

3.1 Introduction

3.1.1 Thermodynamic process (local variables)

By adopting Coleman and Noll's definition [33], we assume that the thermodynamic process in a continuum, with respect to a given referential body, is described by eight functions depending on the position of a particle M and time t [193]:

1. the vector position $\mathbf{OM}(t)$ of M with respect to the referential body, the metric tensor $\mathbf{g}(M, t)$, the volume form $\omega_0(M, t)$, the affine connection $\nabla(M, t)$ characterized by the torsion $\aleph(M, t)$, and the curvature $\Re(M, t)$ as a gauge for irreversible deformation,

2. the absolute temperature $\theta(M, t)$,

3. the stress tensor $\sigma(M, t)$,

4. the entropy $s(M, t)$,

5. the internal energy $e(M, t)$ or, alternatively,
   the Helmholtz free energy $\phi(M, t)$,

6. the heat flux vector $\mathbf{q}(M, t)$,

7. the body force $\rho\mathbf{b}(M, t)$,

8. the volume heat source $\rho r(M, t)$.

These eight tensor fields (thermomechanic fields) must satisfy the conservation laws for any part of B, e.g., [194], [201]. The constitutive variables 1 and 2 are the primal variables and from 3 to 6 the dual variables.

Body forces and volume heat source are of secondary interest in continuum thermomechanics. In the present section we deal with the formulation of contact mechanical action and contact heat action at the continuum boundary. From the axiomatic point of view, the existence of a stress tensor σ and a heat flux vector $\mathbf{J}_q$ is not postulated a priori.

The present work rather leans on the assumption of the existence of a boundary mechanical action that is modeled by a 2-form field ω_C on the boundary ∂B and a boundary heat action also modeled by a 2-form field ω_H on ∂B. In the present monograph, the formulation of the cut principle of Euler and Cauchy starts by considering, within the shape of any body B at any given time, a smooth and closed surface. The action of the part outside that surface is equivalent to that of a field of 2-forms (one mechanical, one thermal) defined on the surface. The equivalent form of Cauchy's theorem will be developed in the following section within the framework of an affinely connected manifold. To this end, some basic mathematical tools should be introduced.

3.1.2 Intrinsic divergence operator

Consider a continuum B, with an affine connection ∇, in motion with respect to a Galilean referential body endowed with a metric tensor $\mathbf{g}$ and a volume form ω_0.

Definition 3.1.1 (g-isomorphism) *For any vector field* $\mathbf{v}$ *on B, we define the 1-form field* $\mathbf{v}^*$ *on B, g-isomorphic of* $\mathbf{v}$, *by*

$$\forall \mathbf{v} \in T_M B \qquad \mathbf{v}^*(\mathbf{u}) \equiv \mathbf{g}(\mathbf{u}, \mathbf{v}). \qquad (3.1)$$

The components of this 1-form in any vector base are related to those of the vector field as:

$$v_i^* = g_{ij} v^j.$$

Remark. The three elements v^i are the contravariant components of the vector **v**, and the three elements v_i the covariant components of **v**. In some sense, the **g**-isomorphism is a process for the lowering or rising of indices. We remark that the position of indices is indifferent for an orthonormal vector base.

Definition 3.1.2 *(Interior product) Let ω be a p-form on B, **v** a vector field, and $(\mathbf{u}_1, \ldots, \mathbf{u}_{p-1})$ any $(p-1)$-plet of vector fields on a continuum B. The interior product of **v** by the form ω is the $(p-1)$-form field on B defined by:*

$$\forall (\mathbf{u}_1, \ldots, \mathbf{u}_{p-1}) \in T_M B \qquad i_\mathbf{v} \omega(\mathbf{u}_1, \ldots, \mathbf{u}_{p-1}) \equiv \omega(\mathbf{v}, \mathbf{u}_1, \ldots, \mathbf{u}_{p-1}). \qquad (3.2)$$

Let **v** be a vector field and ω be a p-form on B. Their interior product $(p-1)$-form field is given by:

1. $i_\mathbf{v} \psi = 0$, ψ is a scalar field (0-form)

2. $i_\mathbf{v} \omega = \omega(\mathbf{v})$, ω is a 1-form.

Definition 3.1.3 *(ω_0-isomorphism) Consider a continuum B endowed with a volume form ω_0. Let us define an isomorphism h_0 by the volume form from the tangent space $T_M B$ to the set of $(m-1)$-forms on B by*

$$\forall \mathbf{v} \in T_M B \qquad h_0(\mathbf{v}) = i_\mathbf{v} \omega_0. \qquad (3.3)$$

Using the h_0 definition and the interior product of **v** with the volume form ω_0, one can write for the tridimensional continuum and the correspondence between the vector field and the 2-form field:

$$\forall (\mathbf{u}_1, \mathbf{u}_2) \in T_M B \qquad h_0(\mathbf{v})(\mathbf{u}_1, \mathbf{u}_2) = \omega_0(\mathbf{v}, \mathbf{u}_1, \mathbf{u}_2). \qquad (3.4)$$

Example. Let $(\mathbf{u}_1, \mathbf{u}_2, \mathbf{u}_3)$ be a vector base (possibly nonorthogonal) on the continuum B. Application of the vector product allows us to write

$$\mathbf{u}_1 \times \mathbf{u}_2 = J_u \mathbf{u}^3 \qquad \mathbf{u}_2 \times \mathbf{u}_3 = J_u \mathbf{u}^1 \qquad \mathbf{u}_3 \times \mathbf{u}_2 = J_u \mathbf{u}^2,$$

where $J_u = \omega_0(\mathbf{u}_1, \mathbf{u}_2, \mathbf{u}_3)$ and $(\mathbf{u}^1, \mathbf{u}^2, \mathbf{u}^3)$ the dual vector base.

To define the divergence operator, it is necessary to use the exterior derivative of a p-form ω by the relationship, for any vector fields $(\mathbf{u}_1, \ldots, \mathbf{u}_{p+1})$ on the continuum B (vector with hat is omitted), e.g., [120], [123]:

$$d\omega(\mathbf{u}_1, \ldots, \mathbf{u}_{p+1}) = \sum_{i=1}^{p+1} (-1)^{i+1} \mathbf{u}_i [\omega(\mathbf{u}_1, \ldots, \hat{\mathbf{u}}_i, \ldots, \mathbf{u}_{p+1}]$$
$$+ \sum_{i<j} (-1)^{i+j} \omega([\mathbf{u}_i, \mathbf{u}_j], \mathbf{u}_1, \ldots, \hat{\mathbf{u}}_i, \ldots, \hat{\mathbf{u}}_j, \ldots \mathbf{u}_{p+1}). (3.5)$$

Example: 2-form. Let ω be a 2-form field on the continuum B. The exterior derivative of ω is calculated as follows, when projected onto the tangent basis $(\mathbf{u}_1, \mathbf{u}_2, \mathbf{u}_3)$:

$$
\begin{aligned}
\omega &= \omega_{23}\, \mathbf{u}^2 \wedge \mathbf{u}^3 + \omega_{31}\, \mathbf{u}^3 \wedge \mathbf{u}^1 + \omega_{12}\, \mathbf{u}^1 \wedge \mathbf{u}^2 \\
d\omega(\mathbf{u}_1, \mathbf{u}_2, \mathbf{u}_3) &= \mathbf{u}_1[\omega(\mathbf{u}_2, \mathbf{u}_3)] - \mathbf{u}_2[\omega(\mathbf{u}_1, \mathbf{u}_3)] + \mathbf{u}_3[\omega(\mathbf{u}_1, \mathbf{u}_2)] \\
&\quad - \omega([\mathbf{u}_1, \mathbf{u}_2], \mathbf{u}_3) + \omega([\mathbf{u}_1, \mathbf{u}_3], \mathbf{u}_2) - \omega([\mathbf{u}_2, \mathbf{u}_3], \mathbf{u}_1)
\end{aligned}
$$

in which we have:

$$
\begin{aligned}
[\mathbf{u}_1, \mathbf{u}_2] &= \aleph^1_{012}\mathbf{u}_1 + \aleph^2_{012}\mathbf{u}_2 + \aleph^3_{012}\mathbf{u}_3 \\
[\mathbf{u}_2, \mathbf{u}_3] &= \aleph^1_{023}\mathbf{u}_1 + \aleph^2_{023}\mathbf{u}_2 + \aleph^3_{023}\mathbf{u}_3 \\
[\mathbf{u}_3, \mathbf{u}_1] &= \aleph^1_{031}\mathbf{u}_1 + \aleph^2_{031}\mathbf{u}_2 + \aleph^3_{031}\mathbf{u}_3.
\end{aligned}
$$

We easily obtain in component form the exterior derivative on the tangent basis:

$$
\begin{aligned}
d\omega(\mathbf{u}_1, \mathbf{u}_2, \mathbf{u}_3) &= \mathbf{u}_1\,(\omega_{23}) + \mathbf{u}_2\,(\omega_{31}) + \mathbf{u}_3\,(\omega_{12}) \\
&\quad + \omega_{23}\left(\aleph^2_{021} + \aleph^3_{031}\right) \\
&\quad + \omega_{31}\left(\aleph^3_{031} + \aleph^1_{012}\right) \\
&\quad + \omega_{12}\left(\aleph^1_{013} + \aleph^2_{023}\right).
\end{aligned}
$$

In the framework of general relativity theory, Cartan has defined the following 1-form:

$$
\tilde{\aleph}_0 \equiv \left(\aleph^2_{021} + \aleph^3_{031}\right)\mathbf{u}^1 + \left(\aleph^3_{031} + \aleph^1_{012}\right)\mathbf{u}^2 + \left(\aleph^1_{013} + \aleph^2_{023}\right)\mathbf{u}^3.
$$

For the sake of simplicity, let us define $\Omega^i \equiv \varepsilon^{ijk}\,\omega_{jk}$. The exterior derivative projected onto the tangent basis has the form of:

$$
\begin{aligned}
d\omega(\mathbf{u}_1, \mathbf{u}_2, \mathbf{u}_3) &= \mathbf{u}_i\left(\Omega^i\right) + \Omega^i\,\tilde{\aleph}_{0i} \\
&= \mathbf{u}_1\left(\Omega^1\right) + \mathbf{u}_2\left(\Omega^2\right) + \mathbf{u}_3\left(\Omega^3\right) \\
&\quad + \Omega^1\,\tilde{\aleph}_{01} + \Omega^2\,\tilde{\aleph}_{02} + \Omega^3\,\tilde{\aleph}_{03}.
\end{aligned}
$$

Definition 3.1.4 (*Divergence of a vector*) *For any vector field* $\mathbf{v}$ *on a continuum* B, *its isomorphic image by the volume form from* $T_M B$ *to the set of* $(m-1)$-*forms is written as* $h_0(\mathbf{v})$. *The exterior derivative of* $h_0(\mathbf{v})$ *is an m-form. Hence, there exists a scalar field on* B, *called the divergence of* $\mathbf{v}$ *on* B *and defined by the relation:*

$$
\forall \mathbf{v} \in T_M B \qquad dh_0(\mathbf{v}) \equiv (\mathrm{div}\ \mathbf{v})\omega_0 \tag{3.6}
$$

Example: Divergence of a vector. Let $\mathbf{v}$ be a vector field on the strain gradient continuum B. Its isomorph $h_0(\mathbf{v})$ by the volume form ω_0 is a second-order form the components of which on the tangent base are

$$
\begin{aligned}
h_0(\mathbf{v})(\mathbf{u}_2, \mathbf{u}_3) &= \omega_{23} = \omega_0(\mathbf{v}, \mathbf{u}_2, \mathbf{u}_3) = J_u\, v^1 \\
h_0(\mathbf{v})(\mathbf{u}_3, \mathbf{u}_1) &= \omega_{31} = \omega_0(\mathbf{v}, \mathbf{u}_3, \mathbf{u}_1) = J_u\, v^2 \\
h_0(\mathbf{v})(\mathbf{u}_1, \mathbf{u}_2) &= \omega_{12} = \omega_0(\mathbf{v}, \mathbf{u}_1, \mathbf{u}_2) = J_u\, v^3 .
\end{aligned}
$$

The divergence of a vector field on a strain gradient continuum is thus given by the following relation:

$$
\operatorname{div} \mathbf{v} = \frac{1}{J_u}\, \mathbf{u}_i \left(J_u\, v^i \right) + v^i\, \tilde{\aleph}_{0i} . \tag{3.7}
$$

Definition 3.1.5 *(Divergence of tensor) Let σ be any 2-contravariant tensor on B. The divergence of σ, denoted $\operatorname{div} \sigma$, is a contravariant vector implicitly defined by the relation*

$$
\forall \omega \in T_M B^* \qquad (\operatorname{div} \sigma)(\omega) \equiv \operatorname{div}\, [\sigma^T(\omega)] . \tag{3.8}
$$

The tensor σ^T means a transpose of the tensor σ.

Example: Divergence of tensor. Let σ be a 2-contravariant tensor field on a strain gradient continuum B, and ω a uniform 1-form field on B. Their decomposition on a tangent basis gives

$$
\begin{aligned}
\sigma &= \sigma^{ij}\, \mathbf{u}_i \otimes \mathbf{u}_j \\
\omega &= \omega_k\, \mathbf{u}^k \\
\sigma^T(\omega) &= \sigma^{ij}\, \omega_i\, \mathbf{u}_j .
\end{aligned}
$$

By applying the tensor divergence definition, we can write

$$
(\operatorname{div} \sigma)\,(\omega) \equiv \operatorname{div}\, [\sigma^T(\omega)] = \frac{1}{J_u}\, \mathbf{u}_i \left(J_u \sigma^{ji} \omega_j \right) + \tilde{\aleph}_{0i} \left(\sigma^{ji} \omega_j \right) .
$$

By writing $(\operatorname{div} \sigma)\,(\omega) = (\operatorname{div} \sigma)^j\, \omega_j$, we directly find that the components of the tensor divergence on a tangent basis are given by:

$$
(\operatorname{div} \sigma)^j = \frac{1}{J_u}\, \mathbf{u}_i \left(J_u \sigma^{ji} \right) + \tilde{\aleph}_{0i} \left(\sigma^{ji} \right) .
$$

The first term of the rhs of the previous equation is the classical "macroscopic" divergence, usually applied for the classical continuum whereas the second term is due to the gradient effects (internal forces due to defects).

3.2   Boundary actions and Cauchy's theorem

3.2.1   Vector-valued p-form [29]

Scalar integration allows us to calculate the integral physical variables such as the entropy, the internal energy, the Helmholtz free energy, and the volume heat source. For vector constitutive variables of the thermodynamic process, we have to use the concept of vector-valued differential forms. For this purpose, we consider first the Euclidean vector space Σ underlying the referential body.

Definition 3.2.1 *(Vector-valued p-form) A vector-valued p-form field is a p-form field ω at the point M of the continuum B (of finite dimension) with values in the vector space Σ and is thus a totally antisymmetric p-linear map from $T_M B$ into Σ:*

$$\omega : (\mathbf{u}_1, \ldots, \mathbf{u}_p) \longrightarrow \omega(\mathbf{u}_1, \ldots, \mathbf{u}_p).$$

Consider now a base of Σ denoted $(\mathbf{e}_i)$. We can write

$$\omega(\mathbf{u}_1, \ldots, \mathbf{u}_p) = \omega^i(\mathbf{u}_1, \ldots, \mathbf{u}_p)\mathbf{e}_i$$

where $\omega^i(\mathbf{u}_1, \ldots, \mathbf{u}_p)$ are real numbers. In some sense, we have to introduce the projection of the form on a 1-form according to ω^i. Thanks to the properties of ω, we can define a set of p-form fields ω^i which are related by

$$\omega = \sum_{i=1}^{n} \mathbf{e}_i \otimes \omega^i. \tag{3.9}$$

For completeness (to apply the Stokes theorem), the exterior derivative of a vector-valued p-form ω is by definition a $(p + 1)$-form with values in Σ:

$$\omega = \sum_{i=1}^{n} \mathbf{e}_i \otimes \omega^i \qquad d\omega = \sum_{i=1}^{n} \mathbf{e}_i \otimes d\omega^i. \tag{3.10}$$

The vector variables of the thermodynamic process can be rewritten as follows:

$$\rho b = \rho b^i \mathbf{e}_i \otimes \omega_0 \qquad \rho v = \rho v^i \mathbf{e}_i \otimes \omega_0. \tag{3.11}$$

3.2.2   Cauchy's fundamental theorem

Most continuum thermomechanics theories assume the existence of a contact force intensity, e.g., [195] (a vector) which is therefore assumed not only to be a function of the particle M and of the instant t but additionally to depend on the oriented surface across which it acts (Cauchy's postulate). Instead, in the present work we assume the existence of boundary actions (mechanical, thermal).

Definition 3.2.2 *(Boundary action) Consider the continuum B with boundary ∂B. The density of the boundary action is assumed to be modeled by a covariant 2-form field ω_C (and ω_H) on ∂B.*

This allows us to write a geometric version of Cauchy's theorem.

Theorem 3.2.1 *(Cauchy's theorem) Consider a continuum B with boundary ∂B. We assume the existence of a field of 2-forms supplying the flux term ω_C on ∂B. Then, there exists a vector field on this boundary called the flux vector $\mathbf{J}_C$. The supplying flux may be written as a scalar product of the flux vector and the normal unit vector at the boundary.*

Proof. By using the ω_0-isomorphism, we respectively have for any two independent tangential vectors $(\mathbf{u}_1, \mathbf{u}_2)$ of the boundary ∂B:[1]

$$
\begin{aligned}
h_0(\mathbf{J}_C)(\mathbf{u}_1, \mathbf{u}_2) &= \omega_C(\mathbf{u}_1, \mathbf{u}_2) \\
\mathbf{J}_C &= J_C^i \mathbf{e}_i \\
h_0(\mathbf{J}_C)(\mathbf{u}_1, \mathbf{u}_2) &= \omega_0(J_C^i \mathbf{e}_i, \mathbf{u}_1, \mathbf{u}_2) = J_C^i u_1^j u_2^k \omega_0(\mathbf{e}_i, \mathbf{e}_j, \mathbf{e}_k).
\end{aligned}
$$

The latter relation shows that

$$
\omega_C(\mathbf{u}_1, \mathbf{u}_2) = h_0(\mathbf{J}_e)(\mathbf{u}_1, \mathbf{u}_2) = \mathbf{J}_e \cdot \mathbf{n} da \qquad \mathbf{n} da \equiv u_1^j u_2^k \omega_0(\mathbf{e}_i, \mathbf{e}_j, \mathbf{e}_k) \mathbf{e}^i \quad (3.12)
$$

demonstrating Cauchy's fundamental theorem.

For thermics, the boundary heat action may then be written as follows:

$$
\int_{\partial B} \omega_H = \int_{\partial B} h_0(\mathbf{J}_q).
$$

For mechanics, the basic idea is to assume first a vector-valued 2-form field as the boundary mechanical flux ω_C. Then the boundary mechanical action may be also decomposed as:

$$
\int_{\partial B} \omega_C = \sum_{i=1}^{n} \int_{\partial B} \mathbf{e}_i \otimes \omega_C^i = \sum_{i=1}^{n} \int_{\partial B} \mathbf{e}_i \otimes h_0(\mathbf{p}_n^i).
$$

3.2.3 Existence of (second-order) stress tensor

The main consequence of the assumption of the existence of a 2-form boundary action is the existence of a stress tensor field.

---

[1]The usual version of Cauchy's theorem starts with a scalar-valued flux. Let there be three quantities $\rho e, r_e$, and J_e respectively of class C^1, C^0, and C^1 verifying the conservation laws. Assume that the flux J_e is a function of the unit normal vector $\mathbf{n}$ at ∂B: $J_e = J_e(\mathbf{n})$. Then a unique vector field $\mathbf{J}_e$ exists such that (this vector is called the "flux density vector") $J_e(\mathbf{n}) = \mathbf{J}_e.\mathbf{n}$.

Theorem 3.2.2 *(Existence of stress tensor) Let B be a continuum with its boundary*
∂B. If we assume on ∂B the occurrence of a mechanical boundary (contact) action
modeled by a 2-form field ω_C, then the boundary mechanical action (density) can be
modeled by a second-order tensor field σ.

Proof. First, for each (scalar-valued) 2-form ω_C^i, the existence of a stress vector $\mathbf{p}_n^i$
holds. The boundary mechanical action may be written as the scalar product with the
unit normal vector $\mathbf{n}$:

$$\omega_C = \sum_{i=1}^{n} \mathbf{e}_i \otimes h_0(\mathbf{p}_n^i) \qquad h_0(\mathbf{p}_n^i)(\mathbf{u}_1, \mathbf{u}_2) = \mathbf{p}_n^i \cdot \mathbf{n}. \tag{3.13}$$

Second, by definition, the application of the vector-valued 2-form (mechanical action)
on any boundary element defined by two tangential vectors produces the following:

$$\omega_C(\mathbf{u}_1, \mathbf{u}_2) = \sum_{i=1}^{n}(\mathbf{p}_n^i \cdot \mathbf{n})\mathbf{e}_i = (\sum_{i=1}^{n} \mathbf{e}_i \otimes \mathbf{p}_n^i)(\mathbf{n}) = \sigma(\mathbf{n}). \tag{3.14}$$

The right-hand side is a linear function with respect to $\mathbf{n}$. Hence, it defines a second-
order tensor field, called the stress tensor. From (3.14), we have:

$$\sigma(\mathbf{e}^j) = \mathbf{p}_n^{ij}\mathbf{e}_i = \mathbf{p}_n^j \qquad i = 1, 2, 3. \tag{3.15}$$

Remarks.

1. Provided the Stokes theorem and the definition of the divergence operator (3.8),
   it is clear that the resultant boundary actions $\int_{\partial B} \omega_C$ and $\int_{\partial B} \omega_H$ do not necessarily define the stress vector $\mathbf{p}_n^i$ and the heat flux $\mathbf{J}_q$ uniquely. Indeed, we may
   add either to the stress tensor σ or to the heat flux vector $\mathbf{J}_q$ any tensor fields
   $\tilde{\sigma}$ and $\tilde{\mathbf{J}}_q$ such that div $\tilde{\sigma} = 0$ and div $\tilde{\mathbf{J}}_q = 0$. In other words, the problem of
   unicity should not be overlooked.

   Gurtin and Williams have studied the problem in depth and proposed a theory
   that admitted continuity assumptions such as the existence of a positive number
   ε. This number ε is such that for any separated parts B_1 and B_2 of the whole
   body B, the intensity of the boundary actions of B_1 to B_2 is lower or equal
   to the area of contact surface between them multiplied by ε. Thus, they have
   rigorously deduced as theorem [75] that, provided global actions (mechanical
   C and thermal H) from B_1 to B_2, there are essentially bounded densities. We
   have called them ω_C and ω_H in the present context, such that:

$$C = \int_S \omega_C \qquad H = \int_S \omega_H \qquad S = \partial B_1 \cap \partial B_2.$$

2. **Stress tensor.** From an engineering point of view, the definition of stress needs an area element of arbitrary size and orientation. On the rear side of this element there is material and on its front side there is an empty space (principle of cut). We choose in this course an outer normal pointing into the empty space. By considering the six faces of a unit cube, the engineering approach of elasticity theory allows us to define stress as force per unit area and the stress tensor components as the components of stress onto the axes directed by the cube edges. In such a case, the stress components reduce to "physical components." For an arbitrary shape of the volume under consideration, we may use the contravariant components σ^{ij} with $\sigma = \sigma^{ij} \mathbf{u}_i \otimes \mathbf{u}_j$, in which $(\mathbf{u}_1, \mathbf{u}_2, \mathbf{u}_3)$ is the tangent basis. The scalar σ^{ij} is not a force per unit area but is connected with such a quantity by a conversion factor $J_u |\mathbf{u}_j|$, different for each component.

3.3   Conservation laws

We consider a continuum with a continuous distribution of singularity whose motion is denoted φ.

3.3.1   Integral invariance of Poincaré

Definition 3.3.1 *(Integral invariance of Poincaré) A p-form field ω on a continuum B is an integral invariant by the motion φ if for any p-chain of B we have:*

$$\int_c \omega = \int_{\varphi(c)} \omega. \tag{3.16}$$

Let $\mathbf{v}$ be the velocity field on B engendered by the local group φ. The p-form field is an integral invariant by $\mathbf{v}$ if it is an integral invariant for every element of the group $\varphi = \varphi_{t-t_0}$ for any t. We recall the following lemma.

Lemma 3.3.1 *Let U be any open set of R^m and let ω be a p-form field on U ($p \leq m$). Then, we can write, c being any p-cube of U,*

$$\int_c \omega = 0 \quad \forall c \quad \Longrightarrow \quad \omega = 0. \tag{3.17}$$

Theorem 3.3.1 *A p-form field ω on B is an integral invariant by φ if and only if, e.g., [153]:*

$$d\varphi^* \omega = \omega. \tag{3.18}$$

Theorem 3.3.2 *(Localization of the integral invariance) Consider the continuum B of velocity field $\mathbf{v}$ and a p-form field ω. Then ω is an integral invariant of the velocity* *field* **v** *if and only if:*

$$\frac{d^B}{dt}\omega = 0. \tag{3.19}$$

Proof. It can be proved in two steps.

1. If the p-form field is an integral invariant by **v**, one has

$$d\varphi^*_{t-t_0}\omega = \omega. \tag{3.20}$$

We deduce, by remarking that φ_0 is an identity,

$$\lim_{t \to t_0} \frac{d\varphi^*_{t-t_0}\omega - \omega}{t - t_0} = 0. \tag{3.21}$$

This means that for any p-plet $(\mathbf{u}_1, \ldots, \mathbf{u}_p)$ of embedded vectors in B, we have

$$\lim_{t \to t_0} \frac{\omega(d\varphi_{t-t_0}\mathbf{u}_1, \ldots, d\varphi_{t-t_0}\mathbf{u}_p) - \omega(\mathbf{u}_1, \ldots, \mathbf{u}_p)}{t - t_0} = 0. \tag{3.22}$$

2. If the derivative of ω with respect to the continuum vanishes, then for any p-plet of vectors $(\mathbf{u}_1, \ldots, \mathbf{u}_p)$ embedded in B, the following relation holds:

$$\frac{d^B}{dt}\omega(\mathbf{u}_1, \ldots, \mathbf{u}_p) = \frac{d}{dt}[\omega(\mathbf{u}_1, \ldots, \mathbf{u}_p)] \tag{3.23}$$

That implies, provided the theorem on the derivative (2.67),

$$d\varphi^*_{t-t_0}\left(\frac{d^B}{dt}\omega\right) = \frac{d}{dt}(d\varphi^*_{t-t_0}\omega) = 0 \qquad \forall t \geq t_0. \tag{3.24}$$

Finally, we have the following relation, since for $t = t_0$ the motion φ_0 is an identity:

$$d\varphi^*_{t-t_0}\omega = \omega \qquad \forall t \geq t_0. \tag{3.25}$$

3.3.2 Global formulation of conservation laws

First, let us recall the integral invariance of a physical quantity. Let **v** be the velocity field on B engendered by the local group φ. The p-form field is an integral invariant by **v** if it is an integral invariant for every element of the group $\varphi = \varphi_{t-t_0}$ for any t.

Theorem 3.3.3 *(Localization of the integral invariance [163]) Consider a continuum B of velocity field* **v** *and a p-form field ω. Then ω is an integral invariant of the velocity field* **v** *if and only if*

$$\frac{d^B}{dt}\omega = 0. \tag{3.26}$$

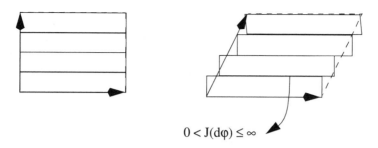

$$0 < J(d\varphi) \leq \infty$$

Figure 3.1. The integral invariance must take into account the occurrence of internal slip surfaces in a continuum with singularity. The existence of internal surface discontinuity does not permit us to neglect the torsion tensor (or accordingly the Lie–Jacobi brackets [163]).

For the sake of simplicity, consider now a 3-form field $\rho e \omega_0$ defined on B, modeling any physical variable and assumed to be an integral invariant according to Poincaré's definition during the motion φ, e.g., [153]. Such an invariance allows us to write:

$$\int_B \rho e \omega_0 = \int_{\varphi(B)} \rho e \omega_0 \qquad \Longleftrightarrow \qquad \frac{d^B}{dt}(\rho e \omega_0) = 0.$$

Second, consider the causes of the changes of this physical quantity. Conservation laws for continuum thermomechanics are usually written in an integral form requiring the integration of scalar and vector fields on a manifold. For the sake of clarity, integration of a so-called scalar function, say ρe, is in fact a simplified manner for expressing the integration of the 3-form $\rho e \omega_0$ on the body B, e.g., [29]. The general form of conservation laws of quantities such as density, linear momentum, angular momentum, energy, and entropy (inequality) may then be written as:

$$\int_B \frac{d^B}{dt}(\rho e \omega_0) = \int_B r_e \omega_0 + \int_{\partial B} h_0(\mathbf{J}_e) \tag{3.27}$$

where $r_e \omega_0$ is the volume source (3-form) and $h_0(\mathbf{J}_e)$ the flux at the boundary (scalar- or vector-valued 2-form). By applying Stokes' theorem, we obtain the generic conservation equation:

$$\int_B \frac{d^B}{dt}(\rho e \omega_0) = \int_B r_e \omega_0 + \int_B dh_0(\mathbf{J}_e) = \int_B r_e \omega_0 + \int_B \text{div} (\mathbf{J}_e) \omega_0. \tag{3.28}$$

We can interpret the balance equations (3.28) as asserting that the rate of increase of the quantity ρe in the body B (and more generally in any subbody) may be expressed as the sum of two effects: inflow through the boundary ∂B and growth inside the bulk B. We assume that ρe is continuously differentiable in B and continuous on ∂B, but the motion itself may present internal discontinuity (scalar or vector). In these conditions, conservation laws of continuum physics can be derived by means of (3.28). Starting from conservation equation (3.28), we have respectively:

1. **For mass.** By replacing $e = 1$, $r_e = 0$, $\mathbf{J}_e = 0$, mass conservation holds:

$$\int_B \frac{d^B}{dt}(\rho\omega_0) = 0. \tag{3.29}$$

Axiom 3.3.1 *(Mass conservation) The mass of any part B of a continuum does not change as B moves with the continuum.*

2. **For linear momentum.** By using the definition of a vector-valued 1-form, we can consider the linear momentum (per unit mass) by introducing the variables $e = \rho v^i$, $r_e = \rho b^i$, and $\mathbf{J}_e = \mathbf{p}_n^i$. The conservation law for linear momentum then holds:

$$\int_B \frac{d^B}{dt}(\rho v^i \omega_0) = \int_B \rho b^i \omega_0 + \int_B h_0(\mathbf{p}_n^i). \tag{3.30}$$

Axiom 3.3.2 *(Conservation of linear momentum) Let B be any part of a moving continuum. The rate of change of B's linear momentum equals the resultant of the body and contact forces on B as B moves with the continuum.*

3. **For angular momentum.** By considering now the angular momentum about any fixed point A in the ambient space, we introduce the variables:

$$e = \omega_0(\mathbf{AM}, \mathbf{v}, \mathbf{e}_i) = l_i \qquad r_e = \omega_0(\mathbf{AM}, \rho\mathbf{b}, \mathbf{e}_i) \qquad \mathbf{J}_e = \mathbf{AM} \wedge \mathbf{p}_{ni}.$$

Angular momentum's conservation law holds:

$$\int_B \frac{d^B}{dt}(\rho l_i \omega_0) = \int_B \rho(\mathbf{AM} \times \mathbf{b})_i \omega_0 + \int_B \operatorname{div}(\mathbf{AM} \times \mathbf{p}_{ni})\omega_0. \tag{3.31}$$

Axiom 3.3.3 *(Conservation of angular momentum) Let B be any part of a moving continuum. The change rate of B's angular momentum equals the moment of the body and contact forces on B as B moves with the continuum.*

4. **For internal energy.** The global conservation law takes the form of (where the power density of the boundary action is naturally defined as $\omega_C(\mathbf{v})$):[2]

$$\int_B \frac{d^B}{dt}\left[\rho\left(e+\frac{\mathbf{v}^2}{2}\right)\omega_0\right] = \int_B \rho\mathbf{b}\cdot\mathbf{v}\,\omega_0 + \int_B \text{div }(\mathbf{v}\cdot\sigma)\omega_0$$
$$+ \int_B r\,\omega_0 + \int_B \text{div }\mathbf{J}_q\,\omega_0. \qquad (3.32)$$

Axiom 3.3.4 *(Energy conservation) Let B be any part of a moving continuum. The rate of change of B's total energy—kinetic and internal energies—is equal to the rate at which work is being done on B, plus the heat produced within B, plus the rate at which heat is conducted into the volume, as B moves with the continuum.*

5. **For entropy.** The global dissipation inequality takes the following form:

$$\int_B \frac{d^B}{dt}(\rho s\,\omega_0) \geq \int_B \frac{r}{\theta}\,\omega_0 - \int_B \text{div }\left(\frac{\mathbf{J}_q}{\theta}\right)\omega_0. \qquad (3.33)$$

Axiom 3.3.5 *(Dissipation principle) For all members of the class of thermomechanical processes that are admissible for an internal constitutive assumption, the rate of entropy production must have nonnegative values.*

Remark. For linear and angular momentum equations, a priori use of a base associated to the referential body does not allow them to have the status of covariant tensorial equations, e.g., [130]. It is quite interesting to notice that the metric tensor is associated to the linear momentum while the volume form is associated to the angular momentum.

3.3.3 Local formulation of conservation laws

The integral form of conservation laws has only limited usefulness in problem solutions because they can only be used to solve those types of problems that exhibit

---

[2]Cauchy's theorem (3.12) implies that there exists a vector field $\mathbf{J}_C$ (contact force intensity) such that $\omega_C = h_0(\mathbf{J}_C)$ and consequently $dh_0(\mathbf{J}_C) = \text{div }\mathbf{J}_C\,\omega_0$. Moreover, it can be shown that $\mathbf{J}_C = \mathbf{v}\cdot\sigma$. The exterior derivative of the mechanical action density is written:

$$d\omega_C = dh_0(\mathbf{p}_n^i)\otimes\mathbf{e}_i = \text{div }(\mathbf{p}_n^i)\omega_0\otimes\mathbf{e}_i.$$

complete symmetry, such as spherical or cylindrical symmetry. To be useful in more general cases the conservations laws must be recast (localized) into a form that describes the relation between thermomechanic fields at any point M and any time t. Now the balance equations (3.27) and (3.28) are asserted to hold for all shapes of all parts B_i of a given continuum B. Then (3.28) holds for all parts $B_i \subseteq B$ the shape of which at the time t are sufficiently small open disks about the position of M in the interior of B. Then, if the integral of a continuous function Φ over every small disk about M vanishes, the function itself vanishes in M. Conversely, if $\Phi \equiv 0$ at all points M of B, so does its integral over every region contained in its domain, e.g., [191]. Then we can deduce from (3.28):

$$\frac{d^B}{dt}(\rho e \, \omega_0) = r_e \, \omega_0 + \text{div } \mathbf{J}_e \, \omega_0. \tag{3.34}$$

Hence for any embedded basis $(\mathbf{u}_1, \mathbf{u}_2, \mathbf{u}_3)$ in B, the vanishing of the time derivative with respect to the continuum yields:

$$\frac{d^B}{dt}(\rho e \, \omega_0)(\mathbf{u}_1, \mathbf{u}_2, \mathbf{u}_3) = \frac{d}{dt}[\rho e \, \omega_0(\mathbf{u}_1, \mathbf{u}_2, \mathbf{u}_3)].$$

By introducing the derivative of scalar field and that of volume form ω_0, we obtain:

$$\left[\frac{\partial}{\partial t}(\rho e) + \nabla_\mathbf{v}(\rho e) + \rho e \, \text{tr}(\nabla \mathbf{v})\right]\omega_0(\mathbf{u}_1, \mathbf{u}_2, \mathbf{u}_3) = 0$$

$$\implies \quad \frac{\partial}{\partial t}(\rho e) + \text{div } (\rho e \mathbf{v}) = 0. \tag{3.35}$$

Remark. The time derivative with respect to the continuum B is more appropriate and should be used for either the global laws or the localization instead of the total derivative with respect to the referential body [163].

The search for a truly invariant formulation in continuum mechanics has been suggested by previous authors, e.g., [133] where the best formulation was clearly proposed as the one that is related to the material itself (canonical projection of fields equations). The results presented here conform to this basic idea. We deduce the local conservation laws in their intrinsic form.

1. **Mass.** The conservation of mass (the volume source Υ_ρ vanishes if the mass is conservative):

$$\frac{d^B}{dt}(\rho e \, \omega_0) = \Upsilon_\rho \implies \frac{\partial \rho}{\partial t} + \text{div } (\rho \mathbf{v}) = \Upsilon_\rho. \tag{3.36}$$

In deriving the local mass conservation (3.36), it was assumed that there are no external sources or sinks of mass inside the continuum. For fluids, an example of a mass sink would be a thin tube through which the fluid is sucked with a defined mass flux $\Upsilon_\rho \neq 0$. Nevertheless, the situation when mass is created in the whole bulk of the continuum (fluid or solid) is rather unusual.

2. **Momentum.** The conservation of linear momentum and that of angular momentum (in the absence of volume couple stress) (Cauchy's equations of motion):

$$\frac{d^B}{dt}(\rho v^i \, \omega_0) = e^i (\operatorname{div} \sigma + \rho \mathbf{b}) \, \omega_0$$

$$\rho \left(\frac{\partial \mathbf{v}}{\partial t} + \nabla_{\mathbf{v}} \mathbf{v} \right) + \mathbf{v} \left[\frac{\partial \rho}{\partial t} + \operatorname{div}(\rho \mathbf{v}) \right] = \operatorname{div} \sigma + \rho \mathbf{b}$$

$$\sigma^* = \sigma. \tag{3.37}$$

3. **Energy.** The conservation of total energy holds:

$$\frac{d^B}{dt}(\rho e \omega_0) + \frac{d^B}{dt} \left(\rho \frac{\mathbf{v}^2}{2} \omega_0 \right) = -\operatorname{div} \mathbf{J}_q \, \omega_0 + [(\operatorname{div} \sigma + \rho \mathbf{b}) \cdot \mathbf{v}] \omega_0$$

$$+ \ (\sigma \, \nabla \mathbf{v}) \, \omega_0 + r \, \omega_0$$

$$\rho \left(\frac{\partial e}{\partial t} + \nabla_{\mathbf{v}} e \right) + \left(e + \frac{\mathbf{v}^2}{2} \right) \left[\frac{\partial \rho}{\partial t} + \operatorname{div}(\rho \mathbf{v}) \right]$$

$$= -\operatorname{div} \mathbf{J}_q + \sigma : \nabla \mathbf{v} + r \tag{3.38}$$

4. **Entropy.** The entropy inequality may be first written as follows:

$$\frac{d^B}{dt}(\rho s \, \omega_0) + \operatorname{div} \left(\frac{\mathbf{J}_q}{\theta} \right) \omega_0 - \frac{r}{\theta} \, \omega_0 \geq 0. \tag{3.39}$$

Substituting the Helmholtz free energy $\varphi \equiv e - \theta s$, we obtain the inequality of Clausius–Duhem:

$$-\rho s \left(\frac{d^B \theta}{dt} \right) \omega_0 - \frac{d^B}{dt}(\rho \varphi \, \omega_0) + \sigma : \nabla \mathbf{v} \, \omega_0 - \frac{\mathbf{J}_q}{\theta} \cdot \nabla \theta \, \omega_0 \geq 0. \tag{3.40}$$

Finally, we have a more classical formulation of this inequality:

$$-\rho s \left(\frac{\partial \theta}{\partial t} + \nabla_{\mathbf{v}} \theta \right) - \rho \left(\frac{\partial \varphi}{\partial t} + \nabla_v \varphi \right) - \varphi \left[\frac{\partial \rho}{\partial t} + \operatorname{div}(\rho \mathbf{v}) \right]$$

$$+\sigma : \nabla \mathbf{v} - \frac{\mathbf{J}_q}{\theta} \cdot \nabla \theta \geq 0. \tag{3.41}$$

Theorem 3.3.4 (*Intrinsic conservation laws*) *Conservation laws (3.36)–(3.41) have exactly the same form as for the classical spatial description, except that the gradient and the divergence operators must be defined in the intrinsic sense (3.6) and (3.8) to be consistent with the distribution of singularity.*

Remark. The local conservation laws developed in this chapter will be used to derive the conservation equations governing either fluid-like or solid-like materials.

4

Continuum with Singularity

Constitutive relations are mathematical relations between dual variables and primal variables. The essence of constitutive relations is that they define idealized materials. The histories of dual variables at any point M and any time t are determined by the histories of the primal variables in all points of the body up to the time t. When only the nth order time derivatives of primal variables are involved instead of their entire histories, materials are said to be of the rate type n, e.g., [194]. The goal of this chapter is to propose the basic framework for continuum with singularity distribution. The essential points of this chapter were developed in a previous paper [163].

4.1 Introduction

4.1.1 Principles for constitutive laws

As with classical continua, constitutive laws of continuum with distribution of singularity should account for behaviors such as elasticity, anelasticity, and viscosity. Physically, any internal process within materials has in principle a natural time. This time is defined as the measure of the time needed for the internal process to move to a new equilibrium after a change of the macroscopic external loading. The different mechanical behaviors of materials can then be classified as a function of their natural time scale, e.g., [158]:

1. **Elasticity.** The medium is said to be elastic if the time scale of the observer, defined as the interval between two macroscopic typical observations, is infinitely smaller than any natural time of all internal processes in the medium. An elastic medium remembers entirely its previous state.

2. **Anelasticity.** The medium has a finite time memory (sometimes called anelastic) if the observer perceives the medium to return to the equilibrium state with a time delay. The time scale of the observer is of the same order of any natural time of all internal processes.

3. **Viscosity.** The medium has a long-term memory (sometimes called viscous) if it never returns to its original reference configuration meaning that the time scale of the observer is infinitely greater than any time scale of all internal processes. A viscous fluid does not remember its initial state.

In this book, we limit ourselves to elastic and viscous effects and then a priori neglect any influence of the finite time memory. However, introduction of two geometrical variables such as torsion and curvature may indirectly account for "some" history of deformation.

Following the idea of Coleman, a comprehensive theory of continuum with singularity may be based on the hypothesis that the dual variables, such as the stress tensor, heat flux vector, internal energy, and entropy, are determined by the histories of deformation and temperature. Constitutive functionals are allowed to depend on the $\nabla\theta$, $\aleph$, and $\Re$ as parameters (principle of determinism):

$$\Im = \tilde{\Im}_{s=0}^{s=\infty} \left[\omega_0(t-s), \mathbf{g}(t-s), \theta(t-s), \aleph, \Re, \nabla\theta \right].$$

They extend the usual thermo-viscoelasticity by including the histories of deformation and temperature [158]. They also replace the viscoelastic laws of the rate type by accounting for the effects of memory. The concept of a simple material was proposed by Noll in the fifties. The main principles used to formulate constitutive equations have been laid down by Noll in 1958, e.g., [146], [194], [201]:

1. **Principle of determinism:** Knowledge of the motion and temperature histories of all particles of the body is enough to determine the values of the dual variables of particle M at instant t.

2. **Principle of material frame indifference:** A referential body with a clock defines an observer. If two observers consider the same motion and temperature histories in a given body, they find the same state for all dual variables. Basic roots underlying the principle of material frame indifference (objectivity) in relation with various invariance groups are recalled and discussed in Appendix B.

3. **Principle of local action:** The motion and temperature outside an arbitrarily small neighborhood of a particle may be disregarded in determining the dual variables of that particle. For continua with singularity, local action restricted the arguments to the metric tensor, volume form, torsion, and curvature tensors.

4. **Principle of fading memory:** The motion and temperature histories outside an arbitrarily small interval of time around the present instant t may be disregarded in determining the dual variables at that instant (for any particle M). This was later proposed by Truesdell in 1949 when studying the theory of rarefied gases.

4.1.2 Uniformity and homogeneity

The further development of continuum thermomechanics in this book falls within the fundamental axioms laid down in, e.g., [146]: determinism, local action, and frame indifference. The theory of constitutive laws developed in this chapter is broad enough to include various materials ranging from solid to fluid for simple materials. However, the theory presented here is slightly different from that of Noll. Noll's theory was mainly based on the concept of material isomorphism [148]: If each point M of a body B is materially isomorphic to every other, all material points have exactly the same physical properties, and B is called materially uniform [201]. For any pair of material points M and M', two different configurations may be necessary to exhibit isomorphism. In the particular case where a single configuration is enough for the isomorphism of all material points of a materially uniform body, then the body is called homogeneous. Indeed, in a nonhomogeneous body that is not materially uniform, infinitely many different configurations associated to each material point may be required to check that all points have the same physical properties. In other words, we have to break the body into an infinite number of small pieces (microcosms). In the framework of Noll's theory on uniform body [148], the fact that every material point is (materially) isomorphic to every other asserts that an affine connection is necessarily defined over the continuum B before any comparison of properties. The simple material of Noll considers the particular class of connection with vanishing curvature.

The starting point of the present work is slightly different since we first consider the discontinuity of scalar and vector fields [163]. Dislocations in translations may be associated to scalar field discontinuity while dislocations in rotations to the vector field discontinuity. From equation (2.24), we can notice that only a restricted class of vector field discontinuity is captured by the nonvanishing curvature tensor. Both the curvature and torsion tensors are necessary to quantify the complete vector field discontinuity. The class of material we consider hereafter could then include Noll's simple materials [163]. However, for practical purposes, we further restrict the material models to those that satisfy automatically the Clausius–Duhem inequality, called generalized standard materials.

4.2   Continuum with singularity of the rate type

The generic model developed in the present monograph basically assumes that all dual constitutive variables at a point M at given time t are entirely determined by the primal variables ω_0, $\mathbf{g}$, $\aleph$, $\Re$, θ, $\nabla\theta$ and their objective rates ζ_g, $\zeta_\aleph$, $\zeta_\Re$, ζ_θ, $\zeta_{\nabla\theta}$ at that point M at that time t. Entire primal variable histories are not needed.

4.2.1   Definition

Definition 4.2.1 *(Continuum with singularity) A continuum with singularity of the rate type $n = 1$ is a continuum of the rate type for which the constitutive laws are defined by the tensor functions:*

$$\Im = \tilde{\Im}(\omega_0, \mathbf{g}, \aleph, \Re, \theta, \nabla\theta, \zeta_g, \zeta_\aleph, \zeta_\Re, \zeta_\theta, \zeta_{\nabla\theta}) \tag{4.1}$$

in which the arguments are the primal tensor variables and their first order time derivatives with respect to the continuum :

$$\zeta_{\omega_0} \equiv \frac{d^B}{dt}\omega_0 \qquad \zeta_g \equiv \frac{1}{2}\frac{d^B}{dt}\mathbf{g} \qquad \zeta_\aleph \equiv \frac{d^B}{dt}\aleph \qquad \zeta_\Re \equiv \frac{d^B}{dt}\Re \tag{4.2}$$

and the rates of thermic variables:

$$\zeta_\theta \equiv \frac{d^B}{dt}\theta \qquad \zeta_{\nabla\Theta} \equiv \frac{d^B}{dt}\nabla\Theta. \tag{4.3}$$

Remark. To some extent, the introduction of the volume form ω_0 as an argument of constitutive functions $\Im$ could be associated to the theorem of Cauchy–Weyl on the representation of isotropic tensor function (see Appendix E). By reducing the orthogonal group to the group of rotations (positive determinant), the volume form logically appears as an explicit argument of the tensor constitutive functions.

4.2.2   Free energy and entropy inequality

In order to apply the classical method of Coleman and Noll [31], we first consider the constitutive functions of the type (4.1) not depending on the second-order derivatives of the primal variables:

$$\frac{d^B}{dt}\zeta_g \qquad \frac{d^B}{dt}\zeta_\aleph \qquad \frac{d^B}{dt}\zeta_\Re \qquad \frac{d^B}{dt}\zeta_\theta \qquad \frac{d^B}{dt}\zeta_{\nabla\theta}. \tag{4.4}$$

Then one can arbitrarily choose the second-order derivatives of the primal variables (4.4) at time t^+ (brusque variation) and at the same time satisfy the entropy inequality:[1]

$$\left(\sigma - \rho \frac{\partial \phi}{\partial \omega_0} : \omega_0 \mathbf{i} - 2\rho \frac{\partial \phi}{\partial \mathbf{g}}\right) : \zeta_g - \rho \frac{\partial \phi}{\partial \aleph} : \zeta_\aleph - \rho \frac{\partial \phi}{\partial \Re} : \zeta_\Re$$

$$-\rho \frac{\partial \phi}{\partial \zeta_g} : \frac{d^B}{dt} \zeta_g - \rho \frac{\partial \phi}{\partial \zeta_\aleph} : \frac{d^B}{dt} \zeta_\aleph - \rho \frac{\partial \phi}{\partial \zeta_\Re} : \frac{d^B}{dt} \zeta_\Re$$

$$-\left(s + \frac{\partial \phi}{\partial \theta}\right) \zeta_\theta - \rho \frac{\partial \phi}{\partial \nabla \theta} : \zeta_{\nabla\theta} - \rho \frac{\partial \phi}{\partial \zeta_\theta} : \frac{d^B}{dt} \zeta_\theta - \rho \frac{\partial \phi}{\partial \zeta_{\nabla\theta}} : \frac{d^B}{dt} \zeta_{\nabla\theta}$$

$$-\frac{\mathbf{J}_q}{\theta} \cdot \nabla \theta \geq 0. \tag{4.5}$$

The coefficients of the second-order derivatives (4.4) must then vanish. Otherwise the inequality (4.5) no longer holds. It implies:

$$\frac{\partial \phi}{\partial \zeta_g} = 0 \qquad \frac{\partial \phi}{\partial \zeta_\aleph} = 0 \qquad \frac{\partial \phi}{\partial \zeta_\Re} = 0 \qquad \frac{\partial \phi}{\partial \zeta_\theta} = 0 \qquad \frac{\partial \phi}{\partial \nabla \theta} = 0. \tag{4.6}$$

In short, if the free energy ϕ and the entropy s do not depend on ζ_θ, then one can take arbitrarily ζ_θ and its gradient $\zeta_{\nabla\theta}$ at time t^+ without consequences on the entropy inequality. The coefficient of ζ_θ and that of $\zeta_{\nabla\theta}$ in (4.5) must also vanish:

$$\frac{\partial \phi}{\partial \theta} = -s \qquad \frac{\partial \phi}{\partial \zeta_{\nabla\theta}} = o. \tag{4.7}$$

We deduce that:

1. The free energy ϕ of a material defined by the constitutive functions (4.1) should be written as follows:

$$\phi = \tilde{\phi}(\omega_0, \mathbf{g}, \aleph, \Re, \theta). \tag{4.8}$$

2. The entropy inequality may be written as follows, where the last two terms characterize the irreversibility due to field singularity in the continuum:

$$\mathbf{J}_q \cdot \zeta_q + \mathbf{J}_g : \zeta_g + \mathbf{J}_\aleph : \zeta_\aleph + \mathbf{J}_\Re : \zeta_\Re \geq 0 \tag{4.9}$$

---

[1] From relation (4.1), the expansion of the objective time derivative of the free energy with respect to the continuum reads:

$$\frac{d^B \varphi}{dt} = \frac{\partial \varphi}{\partial \omega_0} : \omega_0 \operatorname{tr} \zeta_g + 2 \frac{\partial \varphi}{\partial \mathbf{g}} : \zeta_g + \cdots .$$

where we have defined the following variables, primal and dual:

$$\zeta_q \equiv \theta \nabla \left(\frac{1}{\theta} \right)$$

$$\mathbf{J}_q \equiv \sigma - \rho \frac{\partial \phi}{\partial \omega_0} : \omega_0 \mathbf{i} - 2\rho \frac{\partial \phi}{\partial \mathbf{g}}$$

$$\mathbf{J}_\aleph \equiv -\rho \frac{\partial \phi}{\partial \aleph}$$

$$\mathbf{J}_\Re \equiv -\rho \frac{\partial \phi}{\partial \Re}. \tag{4.10}$$

Theorem 4.2.1 *(Singularity and irreversible dissipation) Let B be a continuum with a continuous distribution of singularity (discontinuity of scalar and vector fields). Its free energy is written $\phi = \tilde{\phi}(\omega_0, \mathbf{g}, \aleph, \Re, \theta)$. The entropy inequality (4.9) includes the thermal and viscous dissipations and dissipation due to singularity (due to scalar and vector discontinuity).*

Remarks. Some comments should be made:

1. **Eshelbian stresses.** The dissipation due to the singularity field is related to the irreversibility induced by the change in the affine structure of the continuum. This dissipation is due to the change of topology. The two tensor fields $\mathbf{J}_\aleph$ and $\mathbf{J}_\Re$ may be interpreted as a physical response to the presence of singularity (scalar and vector discontinuity), e.g., [102]. These dual variables are analogous to the Eshelby stress tensor e.g. [132] in the theory of dislocation elasticity. Primal variables, $\zeta_\aleph$ and $\zeta_\Re$, represent objective (in the tensor sense) extension of the classical inhomogeneity velocity gradient, and as the evolution rate of the dislocation pattern, e.g., [51]. By the way, relation (4.10)2 extends the Doyle–Ericksen formula in finite hyperelasticity [42], [183].

2. **Uncoupled dissipations.** To uncouple the dissipations, one can define first the thermal dissipation $\mathbf{J}_q \cdot \zeta_q$. Then, by maintaining constant the metric and by getting the temperature gradient to vanish, we can rewrite the nonnegativity of the intrinsic dissipation as follows:

$$\mathbf{J}_g : \zeta_g + \mathbf{J}_\aleph : \zeta_\aleph + \mathbf{J}_\Re : \zeta_\Re \geq 0. \tag{4.11}$$

More restrictive assumptions are often proposed by uncoupling dissipations in thermal, viscous, e.g., [39], [66], and singularity parts, e.g., [197]:

$$\mathbf{J}_q \cdot \zeta_q \geq 0 \qquad \mathbf{J}_g : \zeta_g \geq 0 \qquad \mathbf{J}_\aleph : \zeta_\aleph + \mathbf{J}_\Re : \zeta_\Re \geq 0. \tag{4.12}$$

The two first inequalities are an extension of Curie's Principle (excluding, for an isotropic medium, the coupling effects between primal and dual variables the tensor orders of which differ by one), e.g., [103]. Curie's Principle could then permit us to couple the effects of thermal torsion and viscosity curvature. Nevertheless, application of this principle to continuum with singularity distribution requires a deeper analysis in a nonlinear situation since coupling of scalar and vector discontinuity exists de facto with relations (2.20) and (2.24).

3. **Gradient of internal variables.** It is well known that the presence of an internal variable and its covariant derivative in constitutive functions engenders supplemental difficulty in the application of Coleman and Noll's method. In fact, localization and imposition of the entropy inequality induce severe restrictions on the evolution of an internal variable, e.g., [119] and references herein. In the present approach, we do not have these severe restrictions since the torsion tensor $\aleph(M, t)$ is neither the covariant derivative nor any the partial derivatives of the metric tensor $\mathbf{g}(M, t)$. The affine connection $\nabla(M, t)$ is fully an independent variable and the torsion and curvature tensors are not obtained by derivatives of the metric tensor. In the particular case where the continuum is Euclidean (such is the case in nonlinear elasticity without nonholonomic deformation), the problem of course arises.

4.3 Operators on continuum with singularity

4.3.1 Divergence operator

The basic steps for deriving the complete set of equations governing continuum thermomechanics may be summarized as follows:

1. First, we lay down generic principles, the conservation laws for the broadest class of bodies.

2. Second, we propose constitutive laws to define various (idealized) materials: dual variables in terms of primal variables.

Taken individually, conservation laws are indeterminate, as are the constitutive laws. Usually, by choosing one constitutive law (among the infinite number of possible ones), we can define a material theory giving a certain result and then can fit the theoretical data to data provided by some experiment. If theoretical results do not conform to the experimental ones (in the range of admissible accuracy), we question the constitutive law and rarely the formulation of conservation laws. Some authors such as Noll [148] and Kröner [102] have gone beyond this reasoning and have reformulated the classical equilibrium equations of Cauchy for defected continuum. Conforming to these works, we attempt, in the present section, to rederive the divergence operator (of vector and tensor fields) appropriate for the continuum model with distributed singularity. The development of the basic theory underlying the present section has been conducted in a previous work [163]. To describe large deformations of continuum with singularity, it is worthwhile to adopt an appropriate vector basis and affine connection. Practically, there exist at least three possibilities to project conservation and constitutive laws, e.g., [163]:

1. **Spatial basis, spatial connection.** The first description is based on the use of $(\mathbf{e}_1, \mathbf{e}_2, \mathbf{e}_3)$ directed by the referential body Σ and on the use of its Euclidean connection; such a method is essentially used to study fluid-like material for its convenience in velocity description (lack of $\mathbf{g}$ in equations).

2. **Embedded basis, nonsymmetric connection.** A second possibility is to choose a basis $(\mathbf{u}_1, \mathbf{u}_2, \mathbf{u}_3)$ embedded in B and to update step-by-step the affine connection ∇ by integrating the evolution laws of torsion and curvature combined with (2.44) and (2.46).

3. **Embedded basis, (sym)-metric connection.** A third method consists in choosing the same basis $(\mathbf{u}_1, \mathbf{u}_2, \mathbf{u}_3)$ embedded in B and to adopt the Riemanian connection $\overline{\nabla}$ compatible with the metric $\mathbf{g}$ the coefficients of which are expressed entirely by means of $g_{ab} = \mathbf{g}(\mathbf{u}_a, \mathbf{u}_b)$ and $[\mathbf{u}_a, \mathbf{u}_b] = \aleph^c_{0ab}\mathbf{u}_c$. With such a method, neither constants of structure nor curvature tensors are null, and torsion components are $\aleph^c_{ab} = -\aleph^c_{ba}$ since the connection $\overline{\nabla}$ is symmetric.

For methods 2 and 3, it is worthwhile to choose initially a vector basis $(\mathbf{u}_{10}, \mathbf{u}_{20}, \mathbf{u}_{30})$ the Lie–Jacobi bracket of which is null. This allows us to define an initial parametrization, i.e., a coordinate basis. This base is transformed into $(\mathbf{u}_1, \mathbf{u}_2, \mathbf{u}_3)$ with $\mathbf{u}_a \equiv d\varphi(\mathbf{u}_{a0})$ and then may become a noncoordinate basis, e.g., [144] at the deformed state. Actually, since B deforms in both holonomic ($\mathbf{g}$ and ω_0) and nonholonomic ($\aleph$ and $\Re$) manners, the constants of structure are not necessarily null, the deformation being weakly continuous ($0 < J(\varphi) \leq \infty$):[2]

$$[\mathbf{u}_a, \mathbf{u}_b] = \nabla_{\mathbf{u}_a}\mathbf{u}_b - \nabla_{\mathbf{u}_b}\mathbf{u}_a - \aleph(\mathbf{u}_a, \mathbf{u}_b) = \aleph^c_{0ab}\mathbf{u}_c$$

$$\mathbf{u}_a \times \mathbf{u}_b = \omega_0(\mathbf{u}_a, \mathbf{u}_b, \mathbf{u}_c)\mathbf{u}^c.$$

Starting with the intrinsic definition (3.6), we recall the general form of the divergence of a vector field on B (vector with hat is omitted):

$$\text{div}(\mathbf{v})\omega_0(\mathbf{u}_1, \ldots, \mathbf{u}_m) = \sum_{i=1}^{m}(-1)^{i+1}\mathbf{u}_i[\omega_0(\mathbf{v}, \mathbf{u}_1, \ldots, \hat{\mathbf{u}}_i, \ldots, \mathbf{u}_m)]$$

$$+ \sum_{i<j=1}^{m}(-1)^{i+j}\omega_0(\mathbf{v}, [\mathbf{u}_i, \mathbf{u}_j], \mathbf{u}_1, \ldots, \hat{\mathbf{u}}_i, \ldots, \hat{\mathbf{u}}_j, \ldots, \mathbf{u}_m). \qquad (4.13)$$

---

[2]For crystalline solids, e.g., [201], the triplet $(\mathbf{u}_1, \mathbf{u}_2, \mathbf{u}_3)$ is called the crystallographic basis and the constants of structure characterize the crystallographic nonhomogeneity of the solid.

Recall the following relations in the tridimensional case:

$$[\mathbf{u}_a, \mathbf{u}_b] \times \mathbf{u}_c = \aleph^d_{0ab}\omega_0(\mathbf{u}_d, \mathbf{u}_c, \mathbf{u}_e)\mathbf{u}^e \qquad J_u = \omega_0(\mathbf{u}_1, \mathbf{u}_2, \mathbf{u}_3). \qquad (4.14)$$

In this case, use of a basis not associated to a parametrization and the definition of a vector product $\omega_0(\mathbf{v}, \mathbf{u}_2, \mathbf{u}_3) \equiv (\mathbf{u}_2 \times \mathbf{u}_3)(\mathbf{v})$ gives the divergence of a vector field, projected in any basis (circular permutation for abc):

$$\text{div } \mathbf{v} = \frac{1}{J_u}[\nabla_{\mathbf{u}_a} J_u \mathbf{v}(\mathbf{u}^a) + \sum_{(abc)} \mathbf{v} \, \aleph^d_{0ab}\omega_0(\mathbf{u}_d, \mathbf{u}_c, \mathbf{u}_e)\mathbf{u}^e]. \qquad (4.15)$$

The divergence is split into two distinct contributions. Say the usual divergence is

$$\overline{\text{div}} \, \mathbf{v} \equiv \frac{1}{J_u}\nabla_{\mathbf{u}_a} J_u \mathbf{v}(\mathbf{u}^a),$$

and the influence of the singularity distribution is:

$$\tilde{\aleph}_0(\mathbf{v}) \equiv \frac{1}{J_u} \sum_{(abc)} \mathbf{v} \, \aleph^d_{0ab}\omega_0(\mathbf{u}_d, \mathbf{u}_c, \mathbf{u}_e)\mathbf{u}^e,$$

in which we have defined the 1-form, originally proposed by Cartan in general relativity theory [22], to be:

$$\tilde{\aleph}_0 \equiv \aleph^b_{0ab} \, \mathbf{u}^a.$$

Finally, the divergence of a vector field on a continuum with volumic distribution of singularity takes on the form:

$$\text{div } \mathbf{v} = \overline{\text{div}} \, \mathbf{v} + \tilde{\aleph}_0(\mathbf{v}).$$

For the divergence of a tensor, given the definition of h_0, the right-hand side yields:

$$\text{div }[\sigma^T(\omega)]\omega_0(\mathbf{u}_1, \mathbf{u}_2, \mathbf{u}_3) = dh_0[\sigma^T(\omega)](\mathbf{u}_1, \mathbf{u}_2, \mathbf{u}_3). \qquad (4.16)$$

By using the exterior derivative, we obtain an explicit formula for the divergence (symbol $\times$ being here a vector product operation):

$$\begin{aligned}
\text{div }(\sigma)\omega_0(\mathbf{u}_1, \mathbf{u}_2, \mathbf{u}_3) = \quad & \nabla_{\mathbf{u}_1}[\sigma(\mathbf{u}_2 \times \mathbf{u}_3)] + \nabla_{\mathbf{u}_2}[\sigma(\mathbf{u}_3 \times \mathbf{u}_1)] \\
+ \quad & \nabla_{\mathbf{u}_3}[\sigma(\mathbf{u}_1 \times \mathbf{u}_2)] + \sigma([\mathbf{u}_1, \mathbf{u}_2] \times \mathbf{u}_3) \\
+ \quad & \sigma([\mathbf{u}_2, \mathbf{u}_3] \times \mathbf{u}_1) + \sigma([\mathbf{u}_3, \mathbf{u}_1] \times \mathbf{u}_2). \qquad (4.17)
\end{aligned}$$

We deduce the general form of the divergence of a 2-contravariant tensor projected in any basis $(\mathbf{u}_1, \mathbf{u}_2, \mathbf{u}_3)$:

$$\text{div } \sigma = \frac{1}{J_u}[\nabla_{\mathbf{u}_a} J_u \sigma(\mathbf{u}^a) + \sum_{(abc)} \sigma \, \aleph^d_{0ab}\omega_0(\mathbf{u}_d, \mathbf{u}_c, \mathbf{u}_e)\mathbf{u}^e]. \qquad (4.18)$$

As for the vector divergence, calculation of the second term shows that the divergence of a second-order tensor is also split into two contributions:

$$\operatorname{div} \sigma = \overline{\operatorname{div}} \, \sigma + \sigma(\tilde{\aleph}_0)$$

where we have defined a "smooth" divergence and the contribution of the singularity distribution respectively as

$$\overline{\operatorname{div}} \, \sigma \;\equiv\; \frac{1}{J_u} \nabla_{\mathbf{u}_a} J_u \sigma(\mathbf{u}^a)$$

$$\sigma(\tilde{\aleph}_0) \;\equiv\; \frac{1}{J_u} \sum_{(abc)} \sigma \, \aleph^d_{0ab} \omega_0(\mathbf{u}_d, \mathbf{u}_c, \mathbf{u}_e) \mathbf{u}^e.$$

Theorem 4.3.1 *(Piola identity) Relations (4.15) and (4.18) provide the expression of vector and tensor divergence on a continuum with a singularity field. Relation (4.18) extends the Piola identity, e.g., [130] used in large strain elasticity to continuum undergoing irreversible large deformations.*

Example of equilibrium equation. Let B be a continuum in equilibrium not submitted to any external body force. The continuum B is assumed to have been deformed with holonomic and nonholonomic strain. In the following formulae, $(\mathbf{e}_1, \mathbf{e}_2, \mathbf{e}_3)$ is the Cartesian basis of the ambient space. The cylindrical coordinates (r, θ, z) are associated to a local tangent basis $(\mathbf{u}_r, \mathbf{u}_\theta, \mathbf{u}_z)$:

$$\begin{aligned}
\mathbf{u}_r &= \cos\theta \mathbf{e}_1 + \sin\theta \mathbf{e}_2 \\
\mathbf{u}_\theta &= -r\sin\theta \mathbf{e}_1 + r\cos\theta \mathbf{e}_2 \\
\mathbf{u}_z &= \mathbf{e}_3
\end{aligned}$$

and to a dual basis $(\mathbf{u}^r, \mathbf{u}^\theta, \mathbf{u}^z)$:

$$\begin{aligned}
\mathbf{u}^r &= \cos\theta \mathbf{e}_1 + \sin\theta \mathbf{e}_2 \\
\mathbf{u}^\theta &= -\frac{1}{r}\sin\theta \mathbf{e}_1 + \frac{1}{r}\cos\theta \mathbf{e}_2 \\
\mathbf{u}^z &= \mathbf{e}_3.
\end{aligned}$$

The metric tensor and the volume form are given by their components:

$$[g_{ij}] = \begin{pmatrix} g_{rr} = 1 & g_{r\theta} = 0 & g_{rz} = 0 \\ g_{\theta r} = 0 & g_{\theta\theta} = r^2 & g_{\theta z} = 0 \\ g_{zr} = 0 & g_{z\theta} = 0 & g_{zz} = 1 \end{pmatrix}$$

$$J_u = \omega_0(\mathbf{u}_r, \mathbf{u}_\theta, \mathbf{u}_z) = r.$$

Without going into details, projecting the "macroscopic" divergence onto the orthonormal basis $(\mathbf{e}_r, \mathbf{e}_\theta, \mathbf{e}_z)$ gives the standard divergence of a tensor in cylindrical coordinates:

$$
\begin{aligned}
\sigma \;=\;& \sigma_{rr}\,\mathbf{e}_r \otimes \mathbf{e}_r + \sigma_{r\theta}\,\mathbf{e}_r \otimes \mathbf{e}_\theta + \sigma_{rz}\,\mathbf{e}_r \otimes \mathbf{e}_z \\
+\;& \sigma_{\theta r}\,\mathbf{e}_\theta \otimes \mathbf{e}_r + \sigma_{\theta\theta}\,\mathbf{e}_\theta \otimes \mathbf{e}_\theta + \sigma_{\theta z}\,\mathbf{e}_\theta \otimes \mathbf{e}_z \\
+\;& \sigma_{zr}\,\mathbf{e}_z \otimes \mathbf{e}_r + \sigma_{z\theta}\,\mathbf{e}_z \otimes \mathbf{e}_\theta + \sigma_{zz}\,\mathbf{e}_z \otimes \mathbf{e}_z
\end{aligned}
$$

$$
\begin{aligned}
\overline{\mathrm{div}}\sigma \;=\;& \left(\frac{\partial \sigma_{rr}}{\partial r} + \frac{1}{r}\frac{\partial \sigma_{r\theta}}{\partial \theta} + \frac{\partial \sigma_{rz}}{\partial z} + \frac{\sigma_{rr} - \sigma_{\theta\theta}}{r} \right) \mathbf{e}_r \\
+\;& \left(\frac{\partial \sigma_{\theta r}}{\partial r} + \frac{1}{r}\frac{\partial \sigma_{\theta\theta}}{\partial \theta} + \frac{\partial \sigma_{\theta z}}{\partial z} + \frac{\sigma_{r\theta} + \sigma_{\theta r}}{r} \right) \mathbf{e}_\theta \\
+\;& \left(\frac{\partial \sigma_{zr}}{\partial r} + \frac{1}{r}\frac{\partial \sigma_{z\theta}}{\partial \theta} + \frac{\partial \sigma_{zz}}{\partial z} + \frac{\sigma_{zr}}{r} \right) \mathbf{e}_z.
\end{aligned}
$$

In addition to this classical ("macroscopic") divergence, we have to introduce the strain gradient effects due to nonholonomic strains, captured by the Cartan 1-form:

$$
\tilde{\aleph}_0 \;=\; \left(\aleph^\theta_{0\theta r} + \aleph^z_{0zr} \right) \mathbf{e}_r + \frac{1}{r}\left(\aleph^r_{0r\theta} + \aleph^z_{0z\theta} \right) \mathbf{e}_\theta + \left(\aleph^r_{0rz} + \aleph^\theta_{0\theta z} \right) \mathbf{e}_z
$$

$$
\begin{aligned}
\sigma\left(\tilde{\aleph}_0\right) \;=\;& \sigma_{rr}\left(\aleph^\theta_{0\theta r} + \aleph^z_{0zr} \right) \mathbf{e}_r + \sigma_{\theta r}\left(\aleph^\theta_{0\theta r} + \aleph^z_{0zr} \right) \mathbf{e}_\theta \\
+\;& \sigma_{zr}\left(\aleph^\theta_{0\theta r} + \aleph^z_{0zr} \right) \mathbf{e}_z + \frac{\sigma_{r\theta}}{r}\left(\aleph^r_{0r\theta} + \aleph^z_{0z\theta} \right) \mathbf{e}_r \\
+\;& \frac{\sigma_{\theta\theta}}{r}\left(\aleph^r_{0r\theta} + \aleph^z_{0z\theta} \right) \mathbf{e}_\theta + \frac{\sigma_{z\theta}}{r}\left(\aleph^r_{0r\theta} + \aleph^z_{0z\theta} \right) \mathbf{e}_z \\
+\;& \sigma_{rz}\left(\aleph^r_{0rz} + \aleph^\theta_{0\theta z} \right) \mathbf{e}_r + \sigma_{\theta z}\left(\aleph^r_{0rz} + \aleph^\theta_{0\theta z} \right) \mathbf{e}_\theta \\
+\;& \sigma_{zz}\left(\aleph^r_{0rz} + \aleph^\theta_{0\theta z} \right) \mathbf{e}_z.
\end{aligned}
$$

It is obvious that the use of the tensor full contravariant components permits us to avoid the division by the vector radius r. This strain gradient effect us vanishes when nonholonomic deformations occur.

4.3.2 Rotational operator

To characterize a flow field, it is usual in classical fluid mechanics to introduce the vorticity vector defined as $\Omega \equiv \mathrm{rot}\,(\mathbf{v})$ where $\mathbf{v}$ is the velocity vector field. The vorticity vector characterizes the local rotation of a fluid element around its center of mass and has been shown to be equal to twice the angular velocity of rotation of the small element. The vorticity vector allows us to define rotational flows ($\Omega \neq 0$) and irrotational flows ($\Omega = 0$). For continuum with continuous distribution of singularity, it is necessary to extend the classical definition of the rotational operator first and then to analyze the consequences of such a definition physically.

Definition 4.3.1 *(Rotational) For any vector field* **v** *on a continuum B, its isomorphic image by the volume form (from $T_M B$ to the set of 2-forms) and its isomorphic image by the metric tensor (a 1-form) are denoted respectively $h_0(\mathbf{v})$ and $\mathbf{v}^*$. The exterior derivative of $h_0(\mathbf{v})$ is a 3-form. Hence, there exists a vector field on B, called the rotational of* **v** *on B defined by the relation:*

$$h_0(\text{rot } \mathbf{v}) \equiv d\mathbf{v}^*. \tag{4.19}$$

Using the definition of the ω_0-isomorphism, we obtain:

$$h_0(\text{rot } \mathbf{v})(\mathbf{u}_a, \mathbf{u}_b) = \omega_0(\text{rot } \mathbf{v}, \mathbf{u}_a, \mathbf{u}_b) \qquad \forall \mathbf{u}_a, \mathbf{u}_b \in T_M B. \tag{4.20}$$

By applying the exterior derivative, we obtain

$$d\mathbf{v}^*(\mathbf{u}_a, \mathbf{u}_b) = \mathbf{u}_a[\mathbf{v}^*(\mathbf{u}_b)] - \mathbf{u}_b[\mathbf{v}^*(\mathbf{u}_a)] - \mathbf{v}^*([\mathbf{u}_a, \mathbf{u}_b]). \tag{4.21}$$

Finally, we deduce the intrinsic (implicit) formulation of the rotational of a vector as follows, for any vector fields $\forall \mathbf{u}_a, \mathbf{u}_b \in T_M B$:

$$\omega_0(\text{rot } \mathbf{v}, \mathbf{u}_a, \mathbf{u}_b) = \mathbf{u}_a[\mathbf{g}(\mathbf{v}, \mathbf{u}_b)] - \mathbf{u}_b[\mathbf{g}(\mathbf{v}, \mathbf{u}_a)] - \mathbf{g}(\mathbf{v}, [\mathbf{u}_a, \mathbf{u}_b]). \tag{4.22}$$

Again, as for divergence, an extra term that includes the constants of structure is observed. Calculation of the rotational components is straightforward using the formula (4.22). Provided the formula (2.12), the extra term of the rotational may be related to the torsion tensor:

$$\mathbf{u}_a[\mathbf{v}^*(\mathbf{u}_b)] = (\nabla_{\mathbf{u}_a} \mathbf{v}^*)(\mathbf{u}_b) + \mathbf{v}^*(\nabla_{\mathbf{u}_a} \mathbf{u}_b). \tag{4.23}$$

Permutation and subtraction of the analogous second term in (4.22) allows us to write (4.21) as

$$\begin{aligned} d\mathbf{v}^*(\mathbf{u}_a, \mathbf{u}_b) &= (\nabla_{\mathbf{u}_a} \mathbf{v}^*)(\mathbf{u}_b) - (\nabla_{\mathbf{u}_b} \mathbf{v}^*)(\mathbf{u}_a) \\ &+ \mathbf{v}^*(\nabla_{\mathbf{u}_a} \mathbf{u}_b - \nabla_{\mathbf{u}_b} \mathbf{u}_a) - \mathbf{v}^*([\mathbf{u}_a, \mathbf{u}_b]) \end{aligned} \tag{4.24}$$

and then

$$d\mathbf{v}^*(\mathbf{u}_a, \mathbf{u}_b) = (\nabla_{\mathbf{u}_a} \mathbf{v}^*)(\mathbf{u}_b) - (\nabla_{\mathbf{u}_b} \mathbf{v}^*)(\mathbf{u}_a) + \aleph(\mathbf{u}_a, \mathbf{u}_b, \mathbf{v}^*). \tag{4.25}$$

This allows us to propose the following theorem for practical calculation of the rotational vector field within a continuum with singularity.

Theorem 4.3.2 *(Rotational) Let* **v** *be the velocity field of a continuum B with singularity field. The rotational of a vector field may be calculated with the relation:*

$$\omega_0(\text{rot } \mathbf{v}, \mathbf{u}_a, \mathbf{u}_b) = (\nabla_{\mathbf{u}_a}\mathbf{v}^*)(\mathbf{u}_b) - (\nabla_{\mathbf{u}_b}\mathbf{v}^*)(\mathbf{u}_a) + \aleph(\mathbf{u}_a, \mathbf{u}_b, \mathbf{v}^*)$$
$$\mathbf{v}^*(\mathbf{u}) = \mathbf{g}(\mathbf{u}, \mathbf{v}) \qquad \forall \mathbf{u} \in T_M B \qquad \forall \mathbf{u}_a, \mathbf{u}_b \in T_M B. \tag{4.26}$$

Given a velocity field **v** on the continuum B, the vorticity vector field Ω is defined by:

$$\Omega \equiv \text{rot } \mathbf{v}. \tag{4.27}$$

Equation (4.26) shows that the scalar discontinuity affects the vorticity distribution. It is worthwhile to understand the effects of the irreversible deformation on the vorticity field. From a kinematics point of view, consider the relative velocity of two particles separated by a vector **u**:

$$\nabla_{\mathbf{u}}\mathbf{v}^* = \mathbf{v}^*(M + \mathbf{u}, t) - \mathbf{v}^*(M, t)$$

where, for convenience, we use the **g**-isomorph $\mathbf{v}^*$ of the velocity field for calculation. The classical decomposition of Stokes–Euler–Cauchy, e.g., [201]:

$$(\nabla_{\mathbf{u}_a}\mathbf{v}^*)(\mathbf{u}_b) = \frac{1}{2}[(\nabla_{\mathbf{u}_a}\mathbf{v}^*)(\mathbf{u}_b) + (\nabla_{\mathbf{u}_b}\mathbf{v}^*)(\mathbf{u}_a)]$$
$$+ \frac{1}{2}[(\nabla_{\mathbf{u}_a}\mathbf{v}^*)(\mathbf{u}_b) - (\nabla_{\mathbf{u}_b}\mathbf{v}^*)(\mathbf{u}_a)] \tag{4.28}$$

shows that the right-hand side contributions correspond to the rate of deformation and to a rigid body rotation. By combining the relations (4.26) and (4.28), we deduce the important relationship between the velocity gradient, the strain rate tensor, and the vorticity and the torsion tensors:

$$(\nabla_{\mathbf{u}_a}\mathbf{v}^*)(\mathbf{u}_b) \quad = \quad \frac{1}{2}[(\nabla_{\mathbf{u}_a}\mathbf{v}^*)(\mathbf{u}_b) + (\nabla_{\mathbf{u}_b}\mathbf{v}^*)(\mathbf{u}_a)]$$
$$+ \quad \frac{1}{2}[\omega_0(\text{rot } \mathbf{v}, \mathbf{u}_a, \mathbf{u}_b) - \aleph(\mathbf{u}_a, \mathbf{u}_b, \mathbf{v})]. \tag{4.29}$$

Theorem 4.3.3 *(Nucleation of vorticity) Vorticity may be produced by a local rigid body rotation around the particle M and additionally by the creation of a singularity as a scalar discontinuity field (nonvanishing torsion tensor).*

Remark. For both the Zaremba–Jaumann rate (2.78) and the convected rate (2.80), it should be stressed from (4.29) that the spin includes the vorticity tensor and the torsion tensor. From relation (4.26) and (4.29), we can extend the definition of a vortex. As a reminder, in classical fluid dynamics (in linear elasticity too), it is recognized that in a simply connected domain the velocity field consists of two parts: a pure gradient component (rot $\mathbf{v} = 0$) and a pure rotational component (div $\mathbf{v} = 0$). Usually vorticity is associated with the rigid body rotational motion of a fluid particle around its center of mass. From the relations (4.26) and (4.29) we can conclude that there is another source of vorticity characterized by the torsion tensor. This supplementary source induces in fact a change in topology (irreversible evolution) and both these sources could be assimilated to harmonic components of the velocity field. The domain becomes multi-connected.

4.3.3   Theorem on vorticity

Let B be a continuum with singularity in motion with respect to a referential body. Definition of a material vector, 1-form, and volume form can be entirely characterized by the time derivative with respect to a continuum (2.58); here "material" means "embedded." Consider any integral of a physical variable field over a material region (point, line, surface, volume). Its value generally changes during the motion for two reasons: (a) the field itself changes and (b) the domain of integration changes as a consequence of the deformation of B. In the following part, we focus on the change of a vector circulation along a curve and of a vector flux across a surface.

Theorem 4.3.4 *(Zorawski's criterion[3]) Let ω be any 2-form field on a continuum B, the velocity field of which is denoted $\mathbf{v}$. The flux of ω across any material facet defined by the couple of vectors embedded in B, $(\mathbf{u}_a, \mathbf{u}_b)$, vanishes if and only if:*

$$\frac{d^B}{dt}\omega \equiv 0. \tag{4.30}$$

Proof. Consider the flux $\Phi = \omega(\mathbf{u}_a, \mathbf{u}_b)$. Its time derivative, in the case where $(\mathbf{u}_a, \mathbf{u}_b)$ are embedded in the motion of B, is written

$$\frac{d}{dt}[\omega(\mathbf{u}_a, \mathbf{u}_b)] = \left(\frac{d^B}{dt}\omega\right)(\mathbf{u}_a, \mathbf{u}_b) = 0 \qquad \forall \mathbf{u}_a, \mathbf{u}_b \in T_M B$$

which implies

$$\frac{d^B}{dt}\omega \equiv 0.$$

Specific theorems applied to fluids and solids will be developed in the two next chapters. Hence if we apply Zorawski's criterion for the ω_0-isomorphic of the vorticity $\Omega = $ rot $\mathbf{v}$, we can recover a classic vorticity theorem (Helmholtz's third vorticity theorem) by stating that the flux of vorticity through each material surface remains

---

[3] In fact, this is a particular case of the localization theorem of integral invariance (3.26). Here, we have sketched an elementary proof.

constant in time if and only if:

$$\frac{d}{dt}[h_0(\Omega)(\mathbf{u}_a, \mathbf{u}_b)] = 0 \quad \Longrightarrow \quad \frac{d^B}{dt}h_0(\Omega) = 0. \tag{4.31}$$

By using the isomorphism, one obtains

$$\frac{d^B}{dt}\Omega = 0. \tag{4.32}$$

4.4 General equations of continuum

4.4.1 Conservation laws

Given the extended form of the divergence and rotational operators in the presence of singularity, we can now derive the local form of conservation laws. To this end, we project conservation laws on the basis $(\mathbf{u}_1, \mathbf{u}_2, \mathbf{u}_3)$ keeping in mind that torsion and curvature tensors do not necessarily vanish. By using the divergence operator (4.15), the mass conservation (3.36) reads

$$\frac{\partial \rho}{\partial t} + \overline{\mathrm{div}}\,(\rho\mathbf{v}) + \tilde{\aleph}_0(\rho\mathbf{v}) = 0 \tag{4.33}$$

where a mass source-like $-\tilde{\aleph}_0(\rho\mathbf{v})$ appears as in an open system (flux of singularity distribution). By using the divergence operator on the tensor field (4.18), the conservation of linear momentum (3.37) yields

$$\rho\left(\frac{\partial\mathbf{v}}{\partial t} + \nabla_{\mathbf{v}}\mathbf{v}\right) = \overline{\mathrm{div}}\,\sigma + \sigma(\tilde{\aleph}_0) + \rho\mathbf{b}. \tag{4.34}$$

There is also a creation of internal forces $\sigma(\tilde{\aleph}_0)$ due to nonhomogeneity (singularity distribution) of medium, e.g., [132], [148], [194]. It must be emphasized that it is not always possible to eliminate such a supplementary force field by merely using an "appropriate" affine connection and an "appropriate" basis, e.g., [3].[4] Because Noll first extended Cauchy's classical equations of motion to defected material in [148],

---

[4]Eshelby (1951), e.g., [54] first introduced the concept of force acting on a singularity or defect in continuum mechanics. A review of such material forces has been conducted in [133]. Although using a classical nondistorted Riemannian manifold, in [133] a unified concept for different singularity forces has been proposed, such as: material forces, e.g., [132], configurational forces, e.g., [74], [143], force on a singularity, e.g., [54], force on a elastic defect, e.g., [10], [213], force on a dislocation [152], inhomogeneity forces, e.g., [148], [194], g-invariant integral [26], and J-integral fracture [167]. The present work extended

Wang in [202] called relation (4.34) Noll's equations of motion. A four dimensional space-time structure has been investigated earlier by means of affine connection e.g. [40]. Physical interpretation of connection coefficients and namely their relation with the inertial forces have been highlighted as early as in [21] by Cartan and up to recently in the area of elastic large deformations [130].

Starting with the expression of the internal energy function with its list of arguments $e(\omega_0, \mathbf{g}, \aleph, \Re)$ and decomposing (3.38) on an arbitrary basis $(\mathbf{u}_1, \mathbf{u}_2, \mathbf{u}_3)$, we obtain

$$
\begin{aligned}
\rho C \left(\frac{\partial \theta}{\partial t} + \nabla_{\mathbf{v}} \theta \right) &= -\overline{\mathrm{div}}\, \mathbf{J}_q - \tilde{\aleph}_0(\mathbf{J}_q) + \rho \theta \frac{\partial}{\partial \theta} \left(\frac{\sigma}{\rho} \right) \\
&+ \left[\mathbf{J}_g - \rho \theta \frac{\partial}{\partial \theta} \left(\frac{\mathbf{J}_g}{\rho} \right) \right] : \zeta_g \\
&+ \left[\mathbf{J}_\aleph - \rho \theta \frac{\partial}{\partial \theta} \left(\frac{\mathbf{J}_\aleph}{\rho} \right) \right] : \zeta_\aleph \\
&+ \left[\mathbf{J}_\Re - \rho \theta \frac{\partial}{\partial \theta} \left(\frac{\mathbf{J}_\Re}{\rho} \right) \right] : \zeta_\Re + r \qquad (4.35)
\end{aligned}
$$

where we have defined the heat capacity C and in which we have introduced the entropy s as:

$$
C \equiv \frac{\partial}{\partial \theta} e(\omega_0, \mathbf{g}, \aleph, \Re, \theta) \qquad s = -\frac{\partial}{\partial \theta} \phi(\omega_0, \mathbf{g}, \aleph, \Re, \theta) \qquad (4.36)
$$

Hence, we extend the heat propagation equation for continuum in the presence of singularity distribution (scalar and vector discontinuity). The local variation of temperature at M varies due to: (a) the heat conduction with a convective contribution because of singularity field distribution; (b) the viscous friction and the heat generation provoked by singularity; (c) the entropy variation and the volume source of heat, e.g., [164]. As for mass and linear momentum conservation, we again have a volume source of heat. Indeed, equation (4.35) shows that heat sources may appear in regions where the density of singularity varies. Physically, this represents a distribution of heat sources concentrated in singularity regions.

---

the concept of material forces to affinely connected manifolds (continuum with singularity [163]). Indeed, the material forces in (4.34) capture distribution of singularity fields in a continuum. Cartan (1923) [21] propounded the requirement of a change in the affine structure of spacetime to incorporate gravity forces in the framework of non relativistic space-time. His proposition was independent from relativization of time since the phenomena in question did not necessarily involve large speeds of matter, e.g., [48]. According to Einstein–Cartan's theory of gravitation, spacetime corresponding to a distribution of matter should be represented by an affinely connected manifold with nonvanishing torsion and curvature, the torsion field being related to the density of spin (intrinsic angular momentum).

4.4.2 Normal dissipation, homogeneous potential

For completeness, to the conservation laws (4.33), (4.34), and (4.35) must be added the constitutive laws. The latter may be defined with a free energy potential ϕ and a dissipation potential ψ. Such a medium is called a generalized standard material, e.g., [66]. Both of them are expressed in a tangent basis $(\mathbf{u}_1, \mathbf{u}_2, \mathbf{u}_3)$, meaning that the practical formulation of constitutive equations depends on the referential configuration. To extend the theory of the thermo-elastic continua to continuum with singularity, we focus on a particular class of dissipating materials [65]. Hereafter, let us define the space of rates $E_\zeta = \{\zeta_g, \zeta_\aleph, \zeta_\Re, \zeta_q\}$ and the space of dual variables $E_J = \{\mathbf{J}_g, \mathbf{J}_\aleph, \mathbf{J}_\Re, \mathbf{J}_q\}$.

Among constitutive functions (4.1), we limit ourselves to the particular class of the type:

$$\Im = \tilde{\Im}(\omega_0, \mathbf{g}, \aleph, \Re, \theta, \zeta_\aleph, \zeta_\Re, \zeta_g, \zeta_q) \qquad \Im = \rho, \sigma, s, \phi, \mathbf{J}_g, \mathbf{J}_\aleph, \mathbf{J}_\Re, \mathbf{J}_q. \quad (4.37)$$

With the balance equations (4.33), (4.34), and (4.35) and constitutive equations (4.37), we recover the two particular models (a) a strongly continuous solid and (b)a fluid the functions of which are respectively of the form:

$$\Im = \tilde{\Im}(\omega_0, \mathbf{g}, \zeta_g, \theta, \zeta_g, \zeta_q) \qquad \Im = \tilde{\Im}(\omega_0, \zeta_g, \theta, \zeta_g, \zeta_q).$$

The equations in the spatial description, e.g., [39], [194] (resp. material, e.g., [36], [71]) are totally recovered by a simple choice of a vector base embedded in space (resp. in material). This method was already proposed in, e.g., [160] to show the equivalence between classical Lagrangian description and the concept of embedded basis in nonlinear elasticity.

We now focus on expressing the dual variables $\mathbf{J}_\alpha$ in terms of primal ones ζ_α. The Clausius–Duhem inequality (4.9) is a constraint the constitutive functions (4.1) must satisfy (thermodynamic admissibility). The hypothesis of normal dissipation restricts the class of constitutive laws although remaining a relatively general framework for continuum models satisfying the second principle of thermodynamics, e.g., [39], [64].

Axiom 4.4.1 *(Normal dissipation) Let $\zeta = (\zeta_g, \zeta_\aleph, \zeta_\Re, \zeta_q)$ be the rates that characterize the internal transformations of a continuum B. The normal dissipation hypothesis assumes the existence of an internal parametrization by the variables ζ and the existence of a dissipation potential function $\tilde{\psi}$, continuous positive (or null) and convex, such that:*

$$\psi = \tilde{\psi}(\zeta_g, \zeta_\aleph, \zeta_\Re, \zeta_q) \qquad \tilde{\psi}(0, 0, 0, 0) = 0$$

$$\mathbf{J}_g = \frac{\partial}{\partial \zeta_g}\psi \qquad \mathbf{J}_\aleph = \frac{\partial}{\partial \zeta_\aleph}\psi \qquad \mathbf{J}_\Re = \frac{\partial}{\partial \zeta_\Re}\psi \qquad \mathbf{J}_q = \frac{\partial}{\partial \zeta_q}\psi. \quad (4.38)$$

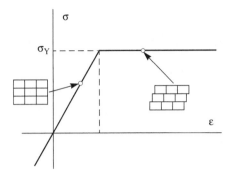

Figure 4.1. In a uniaxial test a material is called perfectly plastic (history dependent) if stress σ and strain ε are related according to the following rule: (a) For stress below the yield stress $\sigma < \sigma_Y$ they obey Hooke's law, (b) At the yield stress $\sigma = \sigma_Y$ the strain can increase indefinitely, without restriction of the strain rate $\frac{d\varepsilon}{de}$ except that this strain rate can never be negative, (c) the stress magnitude cannot be increased beyond the yield stress σ_Y. We can observe two main states: (a) no holonomic deformation occurs, (b) nonholonomic deformation appears within the bulk.

Remark. The potential of dissipation may admit the quantities ω_0, **g**, $\aleph$, $\Re$, and θ as parameters. Hereafter, we do not mention explicitly this dependence except when necessary.

The relations (4.38)2 can be considered as the extension of "phenomenological laws" of linear irreversible thermodynamics, e.g., [72], [103], in which the dissipation potential is a quadratic function. They express the material reaction to strain rate and dislocation and disclination (objective) rates. Convexity and relations (4.38) induce nonnegativity (4.9) of the dissipation, e.g., [39]. In the following the torsion and curvature tensors are assumed to be internal variables [34], not introduced empirically but rather by deduction from the geometrical structure of the continuum [163].

Definition 4.4.1 *(Conjugate potential) For a continuously differentiable dissipation potential, the conjugate potential is defined by the partial transform of Legendre:*

$$\tilde{\psi}^*(\zeta_g, \mathbf{J}_\aleph, \mathbf{J}_\Re, \zeta_q) \equiv \mathbf{J}_\aleph : \zeta_\aleph + \mathbf{J}_\Re : \zeta_\Re - \tilde{\psi}(\zeta_g, \zeta_\aleph, \zeta_\Re, \zeta_q). \qquad (4.39)$$

The derivative of (4.39) with respect to the dual variables provides the evolution laws of field singularity:

$$\zeta_\aleph = \frac{\partial}{\partial \mathbf{J}_\aleph} \tilde{\psi}^*(\zeta_g, \mathbf{J}_\aleph, \mathbf{J}_\Re, \zeta_q) \qquad \zeta_\Re = \frac{\partial}{\partial \mathbf{J}_\Re} \tilde{\psi}^*(\zeta_g, \mathbf{J}_\aleph, \mathbf{J}_\Re, \zeta_q). \qquad (4.40)$$

For a material characterized by the existence of a stress threshold, e.g., [64], constitutive functions depend on the history of external applied forces. On making a first approximation, one can observe macroscopically that the behavior of such a material may change abruptly when the intensity of applied forces exceeds a certain critical value. This abrupt variation requires a noncontinuously differentiable model [65]. In such a situation, the conjugate dissipation potential is defined by the partial Legendre–Fenchel transform, e.g., [140], an extension of (4.39):

$$\tilde{\psi}^*(\zeta_g, \mathbf{J}_\aleph, \mathbf{J}_\Re, \zeta_q) \equiv \mathrm{Sup}_{\zeta_\aleph, \zeta_\Re \in E_\zeta}[\mathbf{J}_\aleph : \zeta_\aleph + \mathbf{J}_\Re : \zeta_\Re - \tilde{\psi}(\zeta_g, \zeta_\aleph, \zeta_\Re, \zeta_q)] \quad (4.41)$$

where $\psi^* : E_J \longrightarrow R \cup \{+\infty\}$ and $\forall \mathbf{J}_\aleph, \mathbf{J}_\Re \in E_J$. Equation (4.41) is equivalent to Young's inequality:

$$\tilde{\psi}^*(\zeta_g, \mathbf{J}_\aleph, \mathbf{J}_\Re, \zeta_q) + \tilde{\psi}(\zeta_g, \zeta_\aleph, \zeta_\Re, \zeta_q) - \mathbf{J}_\aleph : \zeta_\aleph - \mathbf{J}_\Re : \zeta_\Re \geq 0$$
$$\forall \zeta_\aleph, \zeta_\Re, \mathbf{J}_\aleph, \mathbf{J}_\Re. \quad (4.42)$$

One can in principle resolve the particular values $(\zeta_\aleph^*, \zeta_\Re^*)$, maximizing the term of Legendre–Fenchel transform, which is between the brackets in (4.41), to give:

$$\tilde{\psi}^*(\zeta_g, \mathbf{J}_\aleph, \mathbf{J}_\Re, \zeta_q) \equiv \mathbf{J}_\aleph : \zeta_\aleph^* + \mathbf{J}_\Re : \zeta_\Re^* - \tilde{\psi}(\zeta_g, \zeta_\aleph^*, \zeta_\Re^*, \zeta_q).$$

By introducing this formula in the inequality (4.42), one then obtains $\forall \zeta_\aleph, \zeta_\Re \in E_\zeta$:

$$\tilde{\psi}(\zeta_g, \zeta_\aleph, \zeta_\Re, \zeta_q) - \tilde{\psi}(\zeta_g, \zeta_\aleph^*, \zeta_\Re^*, \zeta_q) - \mathbf{J}_\aleph : (\zeta_\aleph - \zeta_\aleph^*)$$
$$-\mathbf{J}_\Re : (\zeta_\Re - \zeta_\Re^*) \geq 0. \quad (4.43)$$

In this case, quantities $\mathbf{J}_\aleph$ and $\mathbf{J}_\Re$ are subgradients of ψ and they are called elements of the subdifferential (partial) of the potential. For convenience, we write, e.g., [140]:

$$\mathbf{J}_\aleph \in \partial \tilde{\psi}(\zeta_g, \zeta_\aleph, \zeta_\Re, \zeta_q) \qquad \mathbf{J}_\Re \in \partial \tilde{\psi}(\zeta_g, \zeta_\aleph, \zeta_\Re, \zeta_q). \quad (4.44)$$

Inversion of (4.44) with the Fenchel inequality, e.g., [140] directly provides the evolution laws of torsion and curvature tensors, extending the differentiable relations (4.40) to the noncontinuously differentiable case:

$$\zeta_\aleph \in \partial \tilde{\psi}^*(\zeta_g, \mathbf{J}_\aleph, \mathbf{J}_\Re, \zeta_q) \qquad \zeta_\Re \in \partial \tilde{\psi}^*(\zeta_g, \mathbf{J}_\aleph, \mathbf{J}_\Re, \zeta_q). \quad (4.45)$$

Remarks. Some remarks should be made:

1. These laws (4.45) connect the evolution of the patterns of the dislocations and disclinations to the variables similar to the Eshelby stress-like tensor. In fact, (4.45) extends similar laws developed in, e.g., [51] which were restricted to a dislocation pattern evolution.

2. Previous studies on dislocation mechanics, e.g., [89] already developed the thermodynamic equations governing the elastic-viscous-plastic solids by assuming, first, that deformation and entropy depends on stress and temperature

and, second, that the dislocation densities and the mobile fractions of each kind of dislocation are governed by certain evolutionary equations. The main advantage of the present study is that the evolution equations (4.45) are obtained by assuming the existence of a convex dissipative potential $\tilde{\psi}^*(\zeta_g, \mathbf{J_\aleph}, \mathbf{J_\Re}, \zeta_q)$ with sufficient regularity and convexity conditions to satisfy all the thermodynamic requirements for constitutive laws.

3. **Integrability conditions.** We must observe that the evolution laws are designed to calculate independently the torsion $\aleph(M, t)$ and curvature $\Re(M, t)$ tensor fields, for any point M at any time t instead of the affine connection field $\nabla(M, t)$. The main reason is that the affine connection is not a geometric variable. Fundamental theorems on non-Riemaniann differentiable geometry, nevertheless, ensure the reconstruction of $\nabla(M, t)$ thanks to Frobenius' theorem. Particularly, it should be stressed that in the general case, both the torsion and the curvature, together with their higher-order covariant derivatives are required to reconstruct the affine connection locally, e.g., [55], [159], [59].

4. The infinitesimal transformation is geometrically holonomic if it does not change the internal structure $\zeta_\aleph = 0$ and $\zeta_\Re = 0$. Two questions arise on: (a) the threshold of augmentation of singularity distribution and (b) the evolution rule of these singularity distributions when exceeding the threshold (i.e., on the form of ψ_i). The next section deals with the search for such a potential dissipation.

   Example. Let us consider an example of dissipation potential and, for the sake of simplicity, postulate a fictitious rate of dislocations and disclinations in the space R^n. Let C be a nonempty closed set of the rate space R^n. A generic quadratic dissipation potential is defined by:

   $$\psi(\zeta) = \frac{1}{2}\|\zeta\|^2, \zeta \in C \qquad \psi(\zeta) = \infty, \zeta \notin C.$$

   Then the conjugate (Legendre–Fenchel transform) of the quadratic potential is:

   $$\psi^*(J) = \frac{1}{2}[\|J\|^2 - \delta_C^2(J)]$$

   where $\delta_C^2(J)$ denotes the distance function of the dual variable to the set C.

5. **Phase transition.** The equations (also named inclusions) in (4.45) describe the nucleation and proliferation of translation dislocations and rotation dislocations (disclinations). Historically, the concept of dislocation found its roots in the work of Volterra and Somogliana on the singularities of equation solutions of linear elasticity. These works were mainly devoted to translation dislocations in solid mechanics. In addition to solids, dislocation theory extends its

applications to various materials as ordered media, liquid crystals, biological tissues, and most types of fluids. The first of relations (4.45) is mostly applied to quantify nucleation and propagation of dislocations in media with different length scales such as crystalline solids or earthquakes. Besides these usual applications, the theory of continuous distribution of defects, and namely the field discontinuity distribution, may also be of unexpected importance in the theory of phase transitions of some materials. Indeed, solid-liquid phase transitions may be assumed to be driven by the instantaneous nucleation and proliferation of a very large number of dislocations and disclinations (equations (4.45)). In addition, findings of intrinsic formulations of conservation laws, independently of the nature of the material (solid phase or fluid phase), may greatly help to rigorously derive the set of equations governing balance laws together with phase transition.

4.4.3 Homogeneous potential of dissipation

Consider the situation where the creation and evolution of singularity are accompanied by only dry friction. In such a case, the friction force depends only on the direction and sense of contact points and not on the intensity of the relative velocity [140]. For this purpose, consider the space of the rates of internal variables and the space of dual variables:

$$E_\zeta = \{\zeta_\aleph, \zeta_\Re\} \qquad E_J = \{\mathbf{J}_\aleph, \mathbf{J}_\Re\}.$$

Assume now an uncoupled dissipation. This hypothesis makes further development easier without loss of generality. We therefore introduce the three additive potentials $\psi = \psi_q + \psi_v + \psi_i$:

$$
\begin{aligned}
\psi_q &= \tilde{\psi}_q(\zeta_q) \\
\psi_v &= \tilde{\psi}_v(\zeta_g) \\
\psi_i &= \tilde{\psi}_i(\zeta_\aleph, \zeta_\Re).
\end{aligned}
\tag{4.46}
$$

For the sake of simplicity, let us focus on the singularity dissipation ψ_i. Assuming a dry friction, the constitutive functions of dissipative mechanisms are homogeneous of degree zero with respect to internal variable rates. We deduce a dissipation potential of degree one:

$$\tilde{\psi}_i(\lambda\zeta_\aleph, \lambda\zeta_\Re) \equiv \lambda\tilde{\psi}_i(\zeta_\aleph, \zeta_\Re) \qquad \forall\zeta_\aleph, \zeta_\Re \in E_\zeta, \forall\lambda > 0.$$

Theorem 4.4.1 *If the dissipation potential ψ is positive and homogeneous of degree one, then the conjugate potential necessarily satisfies the relation $\forall\lambda > 0$:*

$$\lambda\tilde{\psi}_i^*(\mathbf{J}_\aleph, \mathbf{J}_\Re) = \tilde{\psi}_i^*(\mathbf{J}_\aleph, \mathbf{J}_\Re) \qquad \forall\mathbf{J}_\aleph, \mathbf{J}_\Re \in E_J. \tag{4.47}$$

Proof. Using the positive homogeneity of degree one of ψ_i, we have

$$
\begin{aligned}
\lambda \psi_i^* &= \lambda \, \text{Sup}_{\zeta_\aleph, \zeta_\Re} [\mathbf{J}_\aleph : \zeta_\aleph + \mathbf{J}_\Re : \zeta_\Re - \tilde{\psi}(\zeta_\aleph, \zeta_\Re)] \\
&= \text{Sup}_{\zeta_\aleph, \zeta_\Re} [\mathbf{J}_\aleph : \lambda \zeta_\aleph + \mathbf{J}_\Re : \lambda \zeta_\Re - \tilde{\psi}(\lambda \zeta_\aleph, \lambda \zeta_\Re)] \\
&= \tilde{\psi}_i^*(\mathbf{J}_\aleph, \mathbf{J}_\Re).
\end{aligned}
$$

Since the variables $\zeta_\aleph$, $\zeta_\Re$ span the set E_ζ, we can introduce a variable transform $\zeta_\aleph' = \lambda \, \zeta_\aleph$, $\zeta_\Re' = \lambda \, \zeta_\Re$ with $\lambda \neq 0$ and then show the result.

Theorem 4.4.2 *The conjugate dissipation potential ψ_i^* is necessarily the indicator function of a convex set C in the space of E_J:*

$$
\tilde{\psi}_i^*(\mathbf{J}_\aleph, \mathbf{J}_\Re) = \begin{cases} 0 & \mathbf{J}_\aleph, \mathbf{J}_\Re \in C \\ \infty & \mathbf{J}_\aleph, \mathbf{J}_\Re \notin C. \end{cases} \tag{4.48}
$$

Proof. First, we observe that $\tilde{\psi}_i^*$ can only take the values: $-\infty$, 0 and ∞.

1. If for the particular values of $\mathbf{J}_\aleph$ and $\mathbf{J}_\Re$, ψ_i^* is equal to $-\infty$, then it remains equal to $-\infty$ for any other values of $\mathbf{J}_\aleph$ and $\mathbf{J}_\Re$. $\psi_i^* = -\infty$.

2. In the case where ψ_i^* is never equal to $-\infty$, by using the property (4.47), since λ is arbitrary, then ψ_i^* is equal to 0 or ∞. Being convex, such a conjugate potential is necessarily the indicator function of a closed convex set C of the space E_J.

It is now understandable that the dissipation potential admits the volume form, the metric, the torsion, and the curvature as parameters. Indeed, the shape of the convex set C may depend on the values of these parameters. For the sake of clarity, let C be the closure of the domain where there is no evolution of the singularity density in the dual space E_J. The convex set C may be defined by means of a yield (convex) function $Y(\mathbf{J}_\aleph, \mathbf{J}_\Re)$:

$$
C \equiv \{(\mathbf{J}_\aleph, \mathbf{J}_\Re) \in E_J | Y(\mathbf{J}_\aleph, \mathbf{J}_\Re) \leq 0\}. \tag{4.49}
$$

Let us determine now the form of the potential. Starting from the conjugation (4.41), and introducing the conjugate (4.48), indicator of C, we can write:

$$
\begin{aligned}
\tilde{\psi}_i(\zeta_g, \zeta_\aleph, \zeta_\Re, \zeta_q) &\equiv \text{Sup}_{\mathbf{J}_\aleph, \mathbf{J}_\Re \in E_J} [\mathbf{J}_\aleph : \zeta_\aleph + \mathbf{J}_\Re : \zeta_\Re - \tilde{\psi}^*(\mathbf{J}_\aleph, \mathbf{J}_\Re)] \\
\tilde{\psi}_i(\zeta_g, \zeta_\aleph, \zeta_\Re, \zeta_q) &= \text{Max} \{\text{Sup}_{\mathbf{J}_\aleph, \mathbf{J}_\Re \in C} [\mathbf{J}_\aleph : \zeta_\aleph + \mathbf{J}_\Re : \zeta_\Re], -\infty\}.
\end{aligned}
$$

We deduce, by writing explicitly the internal variables, the dissipation potential, positive homogeneous of degree one with respect to $(\zeta_\aleph, \zeta_\Re)$:

$$
\tilde{\psi}_i(\zeta_\aleph, \zeta_\Re) = \text{Sup}_{\mathbf{J}_\aleph, \mathbf{J}_\Re \in C} (\mathbf{J}_\aleph : \zeta_\aleph + \mathbf{J}_\Re : \zeta_\Re). \tag{4.50}
$$

Terms in brackets are positive and represent the internal dissipation due to the singularity density change. Hence, we deduce the principle of maximal dissipation for continuum with singularity, the analog of the "Hill–Mandel Principle," e.g., [39].

Theorem 4.4.3 *(Hill–Mandel Principle) Let there be two states of solicitation* $(\mathbf{J}_\aleph, \mathbf{J}_\Re)$ *(actual) and* $(\mathbf{J}_\aleph^*, \mathbf{J}_\Re^*)$ *(virtual) belonging to the domain C of nonevolution of singularity associated to the same rates* $(\zeta_\aleph, \zeta_\Re)$. *Then from (4.50), the principle of maximal power of internal singularity holds:*

$$(\mathbf{J}_\aleph - \mathbf{J}_\aleph^*) : \zeta_\aleph + (\mathbf{J}_\Re - \mathbf{J}_\Re^*) : \zeta_\Re \geq 0. \tag{4.51}$$

An analogous principle was proposed in the theory of elastic-plastic metal, e.g., [126], showing that, given an evolution rate of singularity, associated dual variables are those that maximize the dissipation due to singularity. In fact, relation (4.51) extends the principle of maximum plastic dissipation, credited to Von Mises, but attributed in the modern literature to Hill (1950) and Mandel (1964).

The Hill–Mandel Principle (maximum singularity dissipation) plays a central role in the mathematical formulation and computational resolution of continuum with singularity distribution. Physically, the principle states that, for given singularity rates $(\zeta_\aleph, \zeta_\Re)$ among all possible dual variables $(\mathbf{J}_\aleph, \mathbf{J}_\Re)$ (Eshelby's tensors) satisfying the yield criterion (4.49), the singularity dissipation $(\mathbf{J}_\aleph : \zeta_\aleph + \mathbf{J}_\Re : \zeta_\Re)$ attains its maximum for the actual values of the dual tensors $(\mathbf{J}_\aleph, \mathbf{J}_\Re)$. As for the classical rate-independent elastic-plastic theory, the converse is also true: The Hill–Mandel Principle implies an associative flow rule (for singularity rates) in the dual space and the convexity of the domain of nonevolution of singularity.

Theorem 4.4.4 *(Generalized Standard Material) For normal dissipative materials, constitutive laws of the continuum with singularity (4.1) may be entirely reconstructed from the free energy* ϕ *and the dissipation potential* ψ, *e.g., [64], [113], [209]:*

$$\phi = \tilde{\phi}(\omega_0, \mathbf{g}, \aleph, \Re, \theta) \qquad \psi = \tilde{\psi}(\zeta_g, \zeta_\aleph, \zeta_\Re, \zeta_q). \tag{4.52}$$

Example. Let us consider a unidimensional problem with a dissipation potential $\psi(\zeta)$ defined on the space of the real numbers R where a quadratic part and a function homogeneous of degree one are simultaneously involved:

$$\psi = \frac{1}{2}\zeta^2 + |\zeta|. \tag{4.53}$$

A direct calculation gives the conjugate dissipation potential (4.41) as follows:

$$\tilde{\psi}_i^*(J) = \begin{cases} \frac{1}{2}(|J| - 1)^2 & |J| > 1 \\ 0 & 1 \geq J \geq -1 \end{cases}. \tag{4.54}$$

For better understanding, the graphics are reported in Figure 4.2. It is quite striking that analogous potentials also appear during the adaptation process of most biological

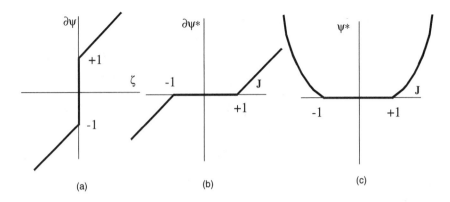

Figure 4.2. The graphics give in (a) the subgradient of the potential (4.53) from which the subgradient of the conjugate potential (b) can be calculated. The Legendre–Fenchel transform of the potential (c) is reconstructed from the conjugate subgradient (b).

tissue. As an illustration, the bone remodeling process is provoked by the alteration of mechanical stress within the tissue, e.g., [188].

From the conjugate potential, the value of the rate can be calculated according to the rate evolution laws $\zeta \in \partial \psi^*(J)$. There is no evolution of singularity distribution in the interval $]-1, +1[$. When the dual variable reaches the border $(-1, +1)$, the rate no longer vanishes. This may be visualized in Figure 4.2 (middle graphic). In other words, the internal structure of the material changes in passing the critical stress surface, here defined by the frontier -1 and $+1$. On both sides of this critical stress frontier, the material behaves in different manners. Such a problem is usual in the domain of biological tissue undergoing adaptation under mechanical stress, as for bone tissue remodeling, e.g., [188].

5
Thermoviscous Fluids

Intuitively, a fluid is a material that has neither a preferred configuration nor a "natural state." The nonexistence of such a natural configuration implies that every simple fluid, in the sense Noll gives it, has an isotropic symmetry. In the framework of a nonsimple material, the definition of fluid should be revisited. For the sake of clarity, let us give an overview of the different fluid-like continua with or without singularity. The basic method for deriving the governing equations lies within the choice of bases and affine connections. This allows us to propose a classification of the main possible models.

5.1   Fluids without singularity distribution

Classical fluid mechanics is almost exclusively based on the usual concept of continuum theory without introducing the singularity previously developed. The first basic assumption underlying the usual concept of continuum implies that matter is distributed continuously in space. The second basic assumption is that fluid motion is described by a homeomorphism which ensures the existence of a continuous inverse of the placement. Historically, the postulation form of viscous fluids draws back to Stokes, e.g., [179] when he proposed the basic hypotheses:

1. Stress σ is a continuous function of the strain rate ζ_g and is independent of all other kinematic quantities.

2. Stress σ does not depend explicitly on the point location (spatial homogeneity).

3. There is no preferred direction in space (isotropy).

4. When $\zeta_g = 0$, then the stress reduces to the hydrostatic pressure.

Starting with the above assumptions, Stokes (1845) laid down a fluid constitutive law in stress as follows:

$$\sigma = -p(\rho, \theta)\mathbf{i} + \mathbf{J}_g(\rho, \theta, \zeta_g) \tag{5.1}$$

where p was the hydrostatic pressure appropriate to the equilibrium at density ρ and temperature θ and where ζ_g was the strain rate tensor, also called the stretching tensor. Assuming an isotropic linear law, Stokes proposed the linear relationship:

$$\sigma = -p(\rho, \theta)\mathbf{i} + \lambda'(\rho, \theta)\text{tr }(\zeta_g)\mathbf{i} + 2\mu'(\rho, \theta)\zeta_g.$$

In order to extend the basic definition of Stokesian fluid to thermomechanics, Truesdell proposed constitutive laws for the stress tensor and the vector heat flux, e.g., [194]:

$$\sigma = \sigma(\rho, \theta, \zeta_g) \qquad \mathbf{J}_g = \mathbf{J}_g(\rho, \theta, \zeta_g, \zeta_q). \tag{5.2}$$

Originally, these laws have been developed to model some particular properties of rarefied gases at low densities which may explain the constitutive dependence on the density.

5.1.1 Constitutive laws

In the present section, consider a thermoviscoelastic material the elastic part of stress of which is boiled down to a hydrostatic pressure. Constitutive functions depend on the deformation through the volume change via the volume form ω_0. Dependence on the volume form ω_0 is imposed by the theorem of Cauchy–Weyl, in the Appendix E.

Definition 5.1.1 *(Classical fluids) In the class of thermoviscoelastic materials (4.37), a material is called a classical thermoviscous fluid (thermoviscous fluid without singularity distribution) if its constitutive functions are defined by*

$$\Im = \tilde{\Im}(\omega_0, \aleph \equiv 0, \Re \equiv 0, \theta, \zeta_g, \zeta_q) \qquad \mathbf{J}_g = \mathbf{J}_g(\rho, \theta, \zeta_g, \zeta_q). \tag{5.3}$$

The relation (5.3) assumes that the dual constitutive variables such as the stress tensor at a point M at a given time t are entirely determined by the volume form ω_0, the strain rate tensor ζ_g, and the temperature gradient ζ_q at that point M and at that time t. The entire histories of the primal variables are not useful. The absence of a metric tensor in the list of arguments means that the body does not react to change of shape. It is a consequence of the nonexistence of a natural configuration that every simple (simple in the sense Noll gives it) classical fluid is isotropic. The relation (5.3) implicitly assumes that no discontinuity of either scalar field or vector field occurs within the bulk fluid. The response of the fluid material to volume change is described by the volume form. Accordingly, the frame indifference of the constitutive laws with respect to referential body rotation (space isotropy) is observed when regarding the list of arguments of the fluid's free energy $\phi(\omega_0, \theta)$. Specifically for fluids, the hypothesis of isotropy of referential body (space isotropy) can be directly exploited[1] and representation theorems of tensor functions are thus applied (scalar, vectorial, tensor of second order):

1. The particular form the entropy inequality holds:

$$\mathbf{J}_q \cdot \zeta_q + \left(\sigma - \rho \frac{\partial \phi}{\partial \omega_0} : \omega_0 \mathbf{i} \right) : \zeta_g \geq 0. \tag{5.4}$$

The stress may then be split into a hydrostatic pressure p and a dissipative (viscous) stress:

$$p(\omega_0, \theta) \equiv -\rho \frac{\partial \phi}{\partial \omega_0} : \omega_0 \qquad \sigma = -p(\omega_0, \theta)\mathbf{i} + \mathbf{J}_g(\omega_0, \theta, \zeta_g, \zeta_q). \tag{5.5}$$

The viscous stress (here represented by a 2-covariant for convenience) may be expressed as a function of the strain rate, e.g., [125], by using the theorem of the representation of tensor functions:

$$
\begin{aligned}
\mathbf{J}_g &= \varpi_0 \mathbf{i} + \varpi_1 \zeta_g + \varpi_2 \zeta_g^2 + \varpi_3 \zeta_q \otimes \zeta_q \\
&+ \frac{\varpi_4}{2} [\zeta_q \otimes \zeta_g(\zeta_q) + \zeta_g(\zeta_q) \otimes \zeta_q] \\
&+ \frac{\varpi_5}{2} [\zeta_q \otimes \zeta_g^2(\zeta_q) + \zeta_g^2(\zeta_q) \otimes \zeta_q]
\end{aligned} \tag{5.6}
$$

where the ϖ_i are scalar-valued functions of the following arguments, in addition to ω_0 and θ:

$$
\begin{aligned}
I_1 &= \operatorname{tr} \zeta_g \qquad I_2 = \operatorname{tr} \zeta_g^2 \qquad I_3 = \operatorname{tr} \zeta_g^3 \qquad I_4 = \operatorname{tr} (\zeta_q \otimes \zeta_q) \\
I_5 &= \operatorname{tr} (\zeta_q \otimes \zeta_g^2(\zeta_q)) \qquad I_6 = \operatorname{tr} (\zeta_g^2(\zeta_q) \otimes \zeta_q).
\end{aligned} \tag{5.7}
$$

2. Free energy and entropy are only functions of the volume-form and the temperature:

$$\phi = \tilde{\phi}(\omega_0, \theta) \qquad s = \tilde{s}(\omega_0, \theta).$$

---

[1] Among Noll's simple materials, e.g., [146], a simple fluid is distinguished by the invariance condition of stress for any unimodular transformation of the referential configuration.

3. The heat flux vector can also be formulated by means of the representation theorem of tensor functions, e.g., [194]:

$$\mathbf{J}_q = (\kappa_0\mathbf{i} + \kappa_1\zeta_g + \kappa_2\zeta_g^2)\zeta_q \qquad \kappa_i = \tilde{\kappa}_i(\omega_0, \theta, I_1, \ldots, I_6). \tag{5.8}$$

Remark. The decomposition of the stress tensor into a pressure and a viscous stress must not be confused with the usual decomposition into a hydrostatic pressure and deviatoric stress:

$$\sigma = -p_H\mathbf{i} + \sigma' \tag{5.9}$$

where p_H is the hydrostatic pressure and σ' the deviatoric stress. Comparison of the two decompositions (5.5) and (5.9) gives the relationships between the variables $-p_H$, p, $\mathbf{J}_g$, and σ':

$$\sigma' - \mathbf{J}_g = (p_H - p)\mathbf{i}. \tag{5.10}$$

Since tr $\sigma' \equiv 0$, we obtain the following relationships:

$$\mathrm{tr}\,\mathbf{J}_g = 3(p - p_H) \qquad \sigma' = \mathbf{J}_g - \frac{1}{3}\mathrm{tr}\,\mathbf{J}_g\mathbf{i}. \tag{5.11}$$

For the normal dissipative fluid model, which is a particular class of thermo-viscous fluids, constitutive laws are summarized in the Helmholtz free energy function and the dissipation potential in terms of the constitutive primal variables (intrinsic formulation):

$$\varphi = \tilde{\varphi}(\omega_0, \theta) \tag{5.12}$$
$$\psi = \tilde{\psi}(\omega_0, \theta, \zeta_g, \zeta_q). \tag{5.13}$$

Their derivatives with respect to primal variables provide the constitutive dual variables, the total stress and the viscous stress:

$$\sigma = \rho\frac{\partial\phi}{\partial\omega_0} : \omega_0\mathbf{i} + \mathbf{J}_g \qquad \mathbf{J}_g = \frac{\partial\psi}{\partial\zeta_g},$$

and the heat flux and the entropy:

$$\mathbf{J}_q = \frac{\partial\psi}{\partial\zeta_q} \qquad s = -\frac{\partial\phi}{\partial\theta}.$$

1. **Example 1.** The constitutive law (5.5) together with (5.6), the hypothesis of linearity in ζ_g and the dropping of all thermal aspects lead to the particular Cauchy–Poisson–Stokes law:

$$\sigma = -p(\omega_0, \theta)\mathbf{i} + \varpi_1(\omega_0, \theta, \zeta_g)\mathbf{i} + \varpi_2(\omega_0, \theta)\zeta_g. \tag{5.14}$$

2. **Example 2.** Reiner–Rivlin fluids. Neglecting the thermal effects, the Reiner–Rivlin fluid model is defined by the particular constitutive law:

$$
\begin{aligned}
\sigma &= -p(\omega_0)\mathbf{i} + \mathbf{J}_g(\omega_0, \zeta_g) \\
\mathbf{J}_g &= \varpi_0(\omega_0, I_1, I_2, I_3)\mathbf{I} + \varpi_1(\omega_0, I_1, I_2, I_3)\zeta_g \\
&\quad + \varpi_2(\omega_0, I_1, I_2, I_3)\zeta_g^2.
\end{aligned}
\tag{5.15}
$$

For incompressible fluid, this law is reduced to (tr $\zeta_g = 0$):

$$
\begin{aligned}
\sigma &= -p(\omega_0)\mathbf{i} + \mathbf{J}_g(\omega_0, \zeta_g) \\
\mathbf{J}_g &= \varpi_0(\omega_0, I_1, I_2)\mathbf{I} + \varpi_1(\omega_0, I_1, I_2)\zeta_g \\
&\quad + \varpi_2(\omega_0, I_1, I_2)\zeta_g^2
\end{aligned}
\tag{5.16}
$$

where the tensor scalar invariants I_i are defined by (5.7).

Various classes of classical fluids can be recovered in the framework of normal dissipative material by choosing special forms of free energy and dissipation potential. These classes are summarized in Table 5.1.

5.1.2 Decomposition of constitutive laws on vector basis

For solving particular problems it is necessary to project the intrinsic form of constitutive and of conservation laws onto an appropriate vector basis. In classical fluid mechanics, conservation and constitutive laws are first projected on a spatial base $(\mathbf{e}_1, \mathbf{e}_2, \mathbf{e}_3)$ (see Figure 5.1). (The Lie–Jacobi bracket vanishes at each instant t since the body can be parametrized with a portion of referential Euclidean body occupied at each time t):

$$
[\mathbf{e}_i, \mathbf{e}_j] = 0 \qquad \mathbf{e}_i \times \mathbf{e}_j = \omega_0(\mathbf{e}_i, \mathbf{e}_j, \mathbf{e}_k)\mathbf{e}^k \qquad \forall i, j = 1, 2, 3.
$$

The volume form decomposed in the spatial basis and its derivative with respect to the fluid particle read:[2]

$$
J_e = \omega_0(\mathbf{e}_1, \mathbf{e}_2, \mathbf{e}_3) \qquad \frac{d^B J_e}{dt} = J_e \operatorname{tr} \zeta_g.
\tag{5.17}
$$

Without loss of generality, we can always choose a spatial basis such that the volume form component is equal $J_e = \omega_0(\mathbf{e}_1, \mathbf{e}_2, \mathbf{e}_3) = 1$ at any time. However, it is not

---

[2]Projection of a volume form on an orthonormal basis is equal to unity. However its objective time derivative is not null. The basis is in fact embedded in the continuum during only an infinitesimal lapse of time to remain parallel to the spatial axes at every instant (such a method goes back to Euler, e.g., [67]).

Free energy and dissipation potential	Stress	Viscous stress and heat flux
$\phi = \hat{\phi}(\omega_0, \aleph)$ $\psi \equiv 0$	$\sigma = \rho \frac{\partial \phi}{\partial \omega_0} : \omega_0 \mathbf{i}$	$\mathbf{J}_g \equiv 0$ $\mathbf{J}_q \equiv 0$
$\phi = \hat{\phi}(\omega_0)$ $\psi = \hat{\psi}(\zeta_g)$	$\sigma = \rho \frac{\partial \phi}{\partial \omega_0} : \omega_0 \mathbf{i} + \mathbf{J}_g$	$\mathbf{J}_g = \frac{\partial \psi}{\partial \zeta_g}$ $\mathbf{J}_q \equiv 0$
$\phi = \hat{\phi}(\omega_0, \theta)$ $\psi = \hat{\psi}(\zeta_q)$	$\sigma = \rho \frac{\partial \phi}{\partial \omega_0} : \omega_0 \mathbf{i}$	$\mathbf{J}_g \equiv 0$ $\mathbf{J}_q = \frac{\partial \psi}{\partial \zeta_q}$
$\phi = \hat{\phi}(\omega_0, \theta)$ $\psi = \hat{\psi}(\zeta_g, \zeta_q)$	$\sigma = \rho \frac{\partial \phi}{\partial \omega_0} : \omega_0 \mathbf{i} + \mathbf{J}_g$	$\mathbf{J}_g = \frac{\partial \psi}{\partial \zeta_g}$ $\mathbf{J}_q = \frac{\partial \psi}{\partial \zeta_q}$

Table 5.1. Particular laws for classical fluids. (Row 1) Perfect elastic fluid (with no stress reaction to shape change), (Row 2) Viscoelastic fluid, (Row 3) Heat conducting thermoelastic fluid, (Row 4) Heat conducting viscoelastic fluid.

really a constant since its time derivative with respect to the fluid does not vanish according to (5.17). Therefore, whenever necessary, we mention in the list of primal constitutive variables the component of the volume form $J_e = 1$. Physically, the reason is that the considered volume does not contain at any two different times the same fluid particles flowing. Mathematically, the number one cannot be considered as an absolute constant. It is no more than a component of the volume form on the 3-form base. On the other hand, classical fluid mechanics chooses to use the Euclidean connection of the referential body or at least a metric connection (Figure 5.1).

The time derivatives of the metric tensor $\mathbf{g}$ and of the temperature θ with respect to B, e.g., [162] and the temperature gradient take the following form, where the variables are the spatial coordinates $\mathbf{y}$ and the time t:

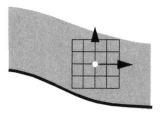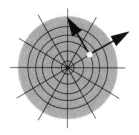

Figure 5.1. The (coordinate) spatial basis is directed by a grid attached to the referential body independent of the fluid flow. The volume form decomposed on this basis is usually chosen as equal to one. The unit vectors of the basis are orthogonal and constitute the grid represented in dashed lines. Both constitutive laws and conservation laws are projected on this spatial basis. In the case of specific symmetry (for example an axisymmetric flow), an appropriate (here a cylindrical) metric connection is used instead of the basic Euclidean connection.

$$\zeta_g = \zeta_{gij}\mathbf{e}^i \otimes \mathbf{e}^j = \frac{1}{2}(\nabla_{\mathbf{e}_i} v_j + \nabla_{\mathbf{e}_j} v_i)\mathbf{e}^i \otimes \mathbf{e}^j$$

$$\zeta_\theta = \frac{\partial\theta}{\partial t} + v^i \nabla_{\mathbf{e}_i}\theta \qquad \zeta_q = \theta\nabla_{\mathbf{e}_i}\left(\frac{1}{\theta}\right)\mathbf{e}^i.$$

Components of the total stress, the viscous stress, and the heat flux on the spatial basis are respectively given by

$$\frac{\sigma}{\rho} = \frac{\sigma^{ij}}{\rho}\mathbf{e}_i \otimes \mathbf{e}_j \qquad \mathbf{J}_g = J_g^{ij}\mathbf{e}_i \otimes \mathbf{e}_j \qquad \mathbf{J}_q = J_q^i\mathbf{e}_i. \tag{5.18}$$

The use of spatial parametrization (i.e., spatial basis, Euclidean or metric connection) provides the practical constitutive laws of (normal dissipative) thermoviscous fluids, e.g., [66]:

1. The free energy function and the dissipation potential (scalar) are:

$$\phi = \tilde{\phi}(J_e, \theta) \qquad \psi = \tilde{\psi}(J_e, \theta, \zeta_{gij}, \zeta_{qi}) \tag{5.19}$$

   in which $J_e \equiv \omega_0(\mathbf{e}_1, \mathbf{e}_2, \mathbf{e}_3)$ but remains a constitutive variable since its time derivative with respect to the continuum is not null.

2. The heat flux (vector) is given by the following relation which describes the nonlinear heat conduction in the fluid volume:

$$\mathbf{J}_q = \frac{\partial\psi}{\partial\zeta_q} \qquad \Longrightarrow \qquad \mathbf{J}_q = \tilde{\mathbf{J}}_q(J_e, \theta, \zeta_{gij}, \zeta_{qi}). \tag{5.20}$$

3. For the stress (tensor), we can project the hydrostatic pressure and the viscous stress on the basis $(\mathbf{e}_1, \mathbf{e}_2, \mathbf{e}_3)$:

$$p \equiv -\rho \frac{\partial \phi}{\partial \omega_0} : \omega_0 \qquad \Longrightarrow \qquad \frac{p}{\rho}\mathbf{i} = \frac{p}{\rho}g^{ij}\mathbf{e}_i \otimes \mathbf{e}_j$$

$$\mathbf{J}_g = \frac{\partial \psi}{\partial \zeta_g} \qquad \Longrightarrow \qquad \mathbf{J}_g = 2\frac{\partial \psi}{\partial \zeta_{gij}}\mathbf{e}_i \otimes \mathbf{e}_j.$$

Then, the stress components in the basis $(\mathbf{e}_1, \mathbf{e}_2, \mathbf{e}_3)$ take the form:

$$\sigma^{ij} = -pg^{ij} + 2\frac{\partial \psi}{\partial \zeta_{gij}} \tag{5.21}$$

in which we can notice the splitting of the stress into the hydrostatic pressure (derived from the free energy function) and the viscous stress (derived from the dissipation potential). The fluid is called non Newtonian when the viscous stress function is not linear in terms of the strain rate tensor which corresponds to a quadratic potential.

5.1.3   Conservation laws

The intrinsic form of conservation laws has been derived in Chapter 3 for a broad class of continua with singularity. The aim of this paragraph is to derive the component form of these intrinsic laws in the spatial (coordinate) basis with the help of the metric connection.

1. **Mass.** Use of spatial basis and a metric connection in the mass conservation equation (4.33) gives the classical continuity equation

$$\frac{\partial \rho}{\partial t} + \overline{\mathrm{div}}(\rho\mathbf{v}) = 0 \tag{5.22}$$

or more familiarly in component form:

$$\frac{\partial \rho}{\partial t} + \nabla_i(\rho v^i) = 0 \tag{5.23}$$

which is the classical D'Alembert–Euler equation for the density ρ.

2. **Linear momentum.** The conservation law for linear momentum (4.34) provides the Cauchy equation of motion

$$\rho\left(\frac{\partial \mathbf{v}}{\partial t} + \nabla_{\mathbf{v}}\mathbf{v}\right) = \overline{\mathrm{div}}\,\sigma + \rho\mathbf{b} \tag{5.24}$$

where stress is expressed as $\sigma = -p\mathbf{i} + \mathbf{J}_g$. Furthermore, use of the relationship gives the extended version of Euler–Navier–Stokes equation to non-Newtonian fluids

$$\rho\left(\frac{\partial \mathbf{v}}{\partial t} + \nabla_\mathbf{v}\mathbf{v}\right) = -\nabla p + \overline{\mathrm{div}}\mathbf{J}_g + \rho\mathbf{b} \tag{5.25}$$

or in component form

$$\rho\left(\frac{\partial v^i}{\partial t} + v^j\nabla_j v^i\right) = -g^{ij}\nabla_j p + \nabla_j\left(J_g^{ij}\right) + \rho b^i. \tag{5.26}$$

3. **Energy.** The energy conservation law provides the heat propagation equation in a thermoviscous fluid which may then be deduced from equation (4.35):

$$\rho C\left(\frac{\partial \theta}{\partial t} + \nabla_\mathbf{v}\theta\right) = -\overline{\mathrm{div}}\,\mathbf{J}_q - \rho\theta\frac{\partial}{\partial\theta}\left(\frac{p}{\rho}\right)\mathbf{i} : \zeta_g + \mathbf{J}_g : \zeta_g + r \tag{5.27}$$

where the heat capacity C and the entropy function s are obtained by (4.7). In components form, the equation of heat propagation within a classical fluid reads:

$$\rho C\left(\frac{\partial \theta}{\partial t} + v^i\nabla_i\theta\right) = -\nabla_i J_q^i - \rho\theta\frac{\partial}{\partial\theta}\left(\frac{p}{\rho}\right)g^{ij}\zeta_{gij} + J_g^{ij}\zeta_{gij} + r \tag{5.28}$$

where the mass conservation equation and the linear momentum equation have been accounted for.

Whether or not the constitutive laws are linear, the equations of motion (5.26) for thermoviscous fluids contain a nonlinear term that may be at least as important as the acceleration linear term (in velocity). This nonlinear term is often the source of the chaotic behavior of the fluid flow. However, this chaotic behavior must not be confused with the turbulent behavior which is rather due to a local change in the topology and, by the way, an irreversible thermodynamic transformation. The chaotic behavior may be a part of the turbulent flow but these two types of fluid flow should not be identified.

5.1.4 Navier–Stokes equations

In the framework of normal dissipative material, the most classical model of thermoviscoelastic fluid can be obtained from two particular potentials, e.g., [125]:

$$\phi = \hat{\phi}(\omega_0, \theta)$$
$$\psi = \frac{1}{2}\hat{\kappa}(\omega_0, \theta)\|\zeta_q\|^2 + \frac{1}{2}\hat{\lambda}'(\omega_0, \theta)\,\mathrm{tr}^2\zeta_g + \frac{1}{2}\hat{\mu}'(\omega_0, \theta)\,\mathrm{tr}\,\zeta_g^2. \tag{5.29}$$

The constitutive laws for Newton's viscous stress and Fourier's law for heat conduction are obtained by derivatives:

$$\mathbf{J}_g = \hat{\lambda}'(\omega_0, \theta)\operatorname{tr} \zeta_g \mathbf{i} + 2\hat{\mu}'(\omega_0, \theta)\zeta_g \qquad \mathbf{J}_q = \hat{\kappa}(\omega_0, \theta)\zeta_q. \qquad (5.30)$$

The corresponding total stress reduces to (5.14). In this particular case, the entropy inequality and the positivity of intrinsic dissipation give the inequality requirements for material constants:

$$\hat{\lambda}'(\omega_0, \theta) + \frac{2}{3}\hat{\mu}'(\omega_0, \theta) \geq 0. \qquad 2\hat{\mu}'(\omega_0, \theta) \geq 0 \quad \hat{\kappa}(\omega_0, \theta) \geq 0 \qquad (5.31)$$

In order to compare our relations with those of classical equations of thermoviscous fluids, e.g., [65], [67], [125] we use density and temperature variables with the gradients $\nabla\rho$ and $\nabla\theta$ as constitutive primal variables. Derivation of dynamic equations requires knowledge of the following vectorial relationships:

$$\nabla_{\mathbf{e}_i}[\lambda'\operatorname{tr} \zeta_g \mathbf{e}^i] = \operatorname{tr} \zeta_g \left[\frac{\partial \lambda'}{\partial \rho}\nabla\rho + \frac{\partial \lambda'}{\partial \theta}\nabla\theta\right] + \lambda'\nabla_{\mathbf{e}_i}[\lambda'\operatorname{tr} \zeta_g \mathbf{e}^i] \qquad (5.32)$$

$$\nabla_{\mathbf{e}_i}[2\mu'\zeta_g(\mathbf{e}^i)] = 2\zeta_g \left[\frac{\partial \mu'}{\partial \rho}\nabla\rho + \frac{\partial \mu'}{\partial \theta}\nabla\theta\right] + 2\mu'\nabla_{\mathbf{e}_i}[\zeta_g(\mathbf{e}^i)]$$

$$\nabla_{\mathbf{e}_i}[\zeta_g(\mathbf{e}^i)] = \frac{1}{2}\Delta\mathbf{v} + \frac{1}{2}\nabla(\operatorname{div} \mathbf{v}) \qquad \nabla_{\mathbf{e}_i}[p(\mathbf{e}^i)\mathbf{i}] = \nabla p. \qquad (5.33)$$

Introducing these relations in (5.25) leads to the extended version of Navier–Stokes equations including the thermic aspects, e.g., [125]:

$$\rho\left(\frac{\partial \mathbf{v}}{\partial t} + \nabla_{\mathbf{v}}\mathbf{v}\right) = -\nabla p + (\lambda' + \mu')\nabla(\operatorname{tr} \zeta_g) + \mu'\Delta\mathbf{v}$$

$$+ \operatorname{tr} \zeta_g \left(\frac{\partial \lambda'}{\partial \rho}\nabla\rho + \frac{\partial \lambda'}{\partial \theta}\nabla\theta\right)$$

$$+ 2\zeta_g \left(\frac{\partial \mu'}{\partial \rho}\nabla\rho + \frac{\partial \mu'}{\partial \theta}\nabla\theta\right). \qquad (5.34)$$

Remark. Historically, Euler (1741) first derived the mathematical equations of fluid dynamics (5.34). They were improved by Navier (1827) to account for the viscous friction within the bulk fluid and were put in their final form by Stokes (1845). The balance of linear momentum (5.34) describes many aspects of the fluid flow:

1. an advective effect due to the transport $\nabla_{\mathbf{v}}\mathbf{v}$ that renders the Navier–Stokes equations nonlinear whatever the case;

2. a pressure effect captured by the first line of the r.h.s. of the previous motion equation;

3. a diffusion process defined by the second term of the l.h.s. and the third term of the r.h.s. which is controlled by the kinematic viscosity. One should notice that the pressure effects of the fluid flow can lead to different velocity profiles.

By assuming a quadratic dissipation potential for viscosity, we deduce from (5.28) the heat propagation equation for Navier–Stokes fluids, e.g., [124], [125]:

$$\rho \left(\frac{\partial \theta}{\partial t} + v^i \nabla_{e_i} \theta \right) = -\nabla_{e_i} J_q^i - \rho \theta \frac{\partial}{\partial \theta} \left(\frac{p}{\rho} \right) g^{ij} \zeta_{gij}$$
$$+ \quad \lambda' \zeta_{gi}^{i2} + 2\mu' \zeta_g^{ij} \zeta_{gij} + r. \tag{5.35}$$

The thermomechanics of classical fluids are governed by the equations (5.25) and (5.27) and in a more particular case by (5.34) and (5.35). In classical theory, the equations (5.34) and (5.35) are generally accepted as describing the nonturbulent state of the fluid flow and some authors suggested that the turbulent state is not among the solutions of these two equations. However, such is the case for turbulent flow and an instability of free convection (Raylegh–Benard problem), e.g., [197]. Conversely, another point of view is the extension of the solutions set to include the discontinuous solutions of equations (5.34) and (5.35). For specific initial conditions of density, temperature, and velocity, the existence of discontinuous solutions of the Navier–Stokes equations of heat-conducting and compressible fluids has been proved, e.g., [80]. The search for these weak solutions is deemed necessary for a better understanding and description of irreversibility within the fluids. Some fundamental problems appear when modeling fluid of high grade (see below: Viscous fluids of high grade and fluids with singularity model).

Example. In the particular case where the fluid is incompressible tr $\zeta_g \equiv 0$, of uniform density $\nabla \rho \equiv 0$, and isothermal $\nabla \theta \equiv 0$, the continuity equation (conservation of mass) holds:

$$\frac{\partial \rho}{\partial t} + \nabla_{e_i} (\rho v^i) = 0 \Longrightarrow \left(\frac{\partial \rho}{\partial t} + \nabla_v \rho \right) + \rho \text{tr} (\nabla v) = 0. \tag{5.36}$$

Then for incompressible fluid, the following continuity equation is obtained:

$$\text{tr} (\nabla v) = \text{div } v = 0. \tag{5.37}$$

The Navier–Stokes equations (5.34) reduce to

$$\rho \left(\frac{\partial v}{\partial t} + \nabla_v v \right) = -\nabla p + \mu' \Delta v + \rho b. \tag{5.38}$$

The energy conservation equation (5.35) also reduces to

$$\rho C \left(\frac{\partial \theta}{\partial t} + v^i \nabla_{e_i} \theta \right) = -\nabla_{e_i} (\kappa \zeta_q^i) + 2\mu' \zeta_g^{ij} \zeta_{gij} + r. \tag{5.39}$$

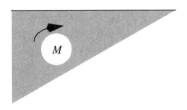

Figure 5.2. During the flow of classical fluid, vorticity may be identified as proportional to the angular momentum of a fluid particle the inertia tensor of which has spherical symmetry (material represented in dark grey color).

Further simplification of the frictional aspects $\mu' = 0$ leads to the fluid model of Stokes:

$$\rho \left(\frac{\partial \mathbf{v}}{\partial t} + \nabla_{\mathbf{v}} \mathbf{v} \right) = -\nabla p + \rho \mathbf{b}. \tag{5.40}$$

Furthermore, when no viscous dissipation occurs, the energy conservation of Stokes' fluid reads:

$$\rho C \left(\frac{\partial \theta}{\partial t} + v^i \nabla_{\mathbf{e}_i} \theta \right) = -\nabla_{\mathbf{e}_i} (\kappa \zeta_q^i) + r. \tag{5.41}$$

5.1.5 Vorticity distribution within classical fluids

An alternative way to describe the fluid flow lies on the use of vorticity theory, e.g., [8], [174] rather than directly using the velocity field. Recall that a physical interpretation of vorticity in fluid flow is provided by analyzing the relative motion near a point and by calculating the angular momentum density about its mass center of an infinitesimal fluid particle.

In classical fluid mechanics, one can obtain interesting qualitative and even quantitative information about specific flows by analyzing the distribution of vorticity. Consider a fluid flow with the velocity field $\mathbf{v}(M, t)$. By using the relations (4.19) and (4.26), the vorticity may be calculated as the exterior differential of the dual velocity field:

$$d\mathbf{v}^*(\mathbf{e}_i, \mathbf{e}_j) = \omega_0 (\text{rot } \mathbf{v}, \mathbf{e}_i, \mathbf{e}_j) = (\nabla_{\mathbf{e}_i} \mathbf{v}^*)(\mathbf{e}_j) - (\nabla_{\mathbf{e}_j} \mathbf{v}^*)(\mathbf{e}_i)$$
$$\mathbf{v}^*(\mathbf{e}_i) \equiv \mathbf{g}(\mathbf{v}, \mathbf{e}_i) \quad \forall \mathbf{e}_i \in T_M B \quad \forall \mathbf{e}_i, \mathbf{e}_j \in T_M B \tag{5.42}$$

which is the usual form of the curl vector since the torsion field equals zero within the fluid bulk. Therefore, it is easy to identify the differential $d\mathbf{v}^*$ with the field rot $\mathbf{v}$ in the classical theory of fluid mechanics, e.g., [174].

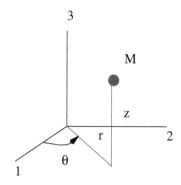

Figure 5.3. Cylindrical coordinates are defined by a mapping from an open set of R^3 to the continuum $(0 < r < \infty, 0 < \theta < 2\pi, -\infty < z < \infty)$.

Example. Vorticity in cylindrical coordinates. The vorticity can be expressed in a cylindrical coordinate system by choosing the cylindrical coordinates:

$$x^1 = r \qquad x^2 = \theta \qquad x^3 = z \tag{5.43}$$

with the associated local basis

$$\mathbf{e}_r = \frac{\partial \mathbf{OM}}{\partial r} \qquad \mathbf{e}_\theta = \frac{\partial \mathbf{OM}}{\partial \theta} \qquad \mathbf{e}_z = \frac{\partial \mathbf{OM}}{\partial z}. \tag{5.44}$$

The matrix of the metric tensor decomposed in the local basis is expressed as:

$$\mathbf{g} = \begin{pmatrix} g_{rr} = 1 & 0 & 0 \\ 0 & g_{\theta\theta} = r^2 & 0 \\ 0 & 0 & g_{zz} = 1 \end{pmatrix}. \tag{5.45}$$

The unique component of the volume form projected onto the local base equals

$$\omega_0(\mathbf{e}_r, \mathbf{e}_\theta, \mathbf{e}_z) = r. \tag{5.46}$$

The Christoffel symbols associated to the related metric connection are

$$\Gamma^\theta_{r\theta} = \Gamma^\theta_{\theta r} = \frac{1}{r} \qquad \Gamma^r_{\theta\theta} = -r \tag{5.47}$$

whereas the other symbols equal zero. Consider now the flow of a classical fluid with the velocity field:

$$\mathbf{v} = v^r \mathbf{e}_r + v^\theta \mathbf{e}_\theta + v^z \mathbf{e}_z. \tag{5.48}$$

Note that the basis vector $\mathbf{e}_z$ is constant while the vectors $\mathbf{e}_r$ and $\mathbf{e}_\theta$ depend on the angular coordinate θ. Thus, the decomposition of the $\mathbf{g}$-isomorph of the velocity field onto the local basis may calculated as follows, thanks the linearity of $\mathbf{g}$:

$$
\begin{aligned}
\mathbf{v}^*(\mathbf{e}_r) &= \mathbf{g}(\mathbf{v}, \mathbf{e}_r) = v^r \\
\mathbf{v}^*(\mathbf{e}_\theta) &= \mathbf{g}(\mathbf{v}, \mathbf{e}_\theta) = r^2 v^\theta \\
\mathbf{v}^*(\mathbf{e}_z) &= \mathbf{g}(\mathbf{v}, \mathbf{e}_z) = v^z.
\end{aligned}
$$

The components of the vorticity along this direction is then given by

$$
\omega_0(\operatorname{rot}\mathbf{v}, \mathbf{e}_r, \mathbf{e}_\theta) = r\Omega_{r\theta} = \nabla_r v_\theta^* - \nabla_\theta v_r^* = \frac{\partial(r^2 v^\theta)}{\partial r} - \frac{\partial(v^r)}{\partial \theta}. \tag{5.49}
$$

Practically, most authors prefer to handle the so-called physical components of the tensors and thus use the unit vector basis $(\hat{\mathbf{e}}_r, \hat{\mathbf{e}}_\theta, \hat{\mathbf{e}}_z)$ such that

$$
\hat{\mathbf{e}}_r = \mathbf{e}_r \qquad \hat{\mathbf{e}}_\theta = \frac{\mathbf{e}_\theta}{\|\mathbf{e}_\theta\|} \qquad \hat{\mathbf{e}}_z = \mathbf{e}_z.
$$

By rewriting the velocity components we have $\mathbf{v} = \hat{v}^r \hat{\mathbf{e}}_r + \hat{v}^\theta \hat{\mathbf{e}}_\theta + \hat{v}^z \hat{\mathbf{e}}_z$. It is straightforward to recover the usual formulation of the vorticity component as

$$
\Omega_{r\theta} = \frac{1}{r}\left[\frac{\partial(r^2 v^\theta)}{\partial r} - \frac{\partial(v^r)}{\partial \theta}\right] = \frac{\partial \hat{v}^\theta}{\partial r} - \frac{\partial \hat{v}^r}{\partial \theta} + \frac{\hat{v}^\theta}{r}. \tag{5.50}
$$

The other components are obtained according to the same method:

$$
\Omega_{\theta z} = \frac{1}{r}\frac{\partial \hat{v}^z}{\partial \theta} - \frac{\hat{v}^\theta}{\partial z} \tag{5.51}
$$

$$
\Omega_{zr} = \frac{\partial \hat{v}^r}{\partial z} - \frac{\partial \hat{v}^z}{\partial r}. \tag{5.52}
$$

The associated vorticity vector is finally defined as

$$
\vec{\Omega} = \left(\frac{1}{r}\frac{\partial \hat{v}^z}{\partial \theta} - \frac{\hat{v}^\theta}{\partial z}\right)\hat{\mathbf{e}}_r + \left(\frac{\partial \hat{v}^r}{\partial z} - \frac{\partial \hat{v}^z}{\partial r}\right)\hat{\mathbf{e}}_\theta + \left(\frac{\partial \hat{v}^\theta}{\partial r} - \frac{\partial \hat{v}^r}{\partial \theta} + \frac{\hat{v}^\theta}{r}\right)\hat{\mathbf{e}}_z. \tag{5.53}
$$

5.1.6 *Discrete distribution of vorticity*

In the simplest case of incompressible flow of classical fluids, the governing equations reduce to the continuity equation (conservation of mass) involving only the velocity field

$$
\operatorname{div}\mathbf{v} = 0,
$$

the Euler equations involving the velocity and the pressure field

$$\rho \left(\frac{\partial \mathbf{v}}{\partial t} + \nabla_{\mathbf{v}} \mathbf{v} \right) = -\nabla p + \rho \mathbf{b},$$

and finally the heat propagation

$$\rho C \left(\frac{\partial \theta}{\partial t} + \nabla_{\mathbf{v}} \theta \right) = -\text{div} \left(\kappa \zeta_q \right) + r.$$

These equations cannot describe fully the singularity occurrences since they are based on transformations that are homeomorphisms. No volume internal slip surfaces are authorized in the bulk volume. The aim of this paragraph is to present how the distribution of singularity is handled within the framework of classical fluids. Without loss of generality, we limit the development to incompressible, frictionless, and irrotational fluid flow. The main difference of the present development with the viscous dissipation theory is the diffusion of vorticity in real viscous fluids. The description of classical fluid flows for which one of the velocity components vanishes (planar flows) and the incompressible fluid can be considerably simplified by defining the stream function. Starting with the conservation of mass div $\mathbf{v} = 0$, the stream function is defined as the component of a vector field denoted $\vec{\vartheta} = \vartheta \mathbf{k}$ where the unit vector $\mathbf{k}$ is orthogonal to the plane flow, so that the velocity field is given by $v = \text{rot} \, \vec{\vartheta}$ where the divergence of $\vec{\vartheta}$ is left arbitrary. The decomposition of the rotational vector in a cylindrical coordinate (unit vector basis) may be written as follows, from (5.53):

$$\text{rot} \, \vec{\vartheta} = \left(\frac{1}{r} \frac{\partial \vartheta_z}{\partial \theta} - \frac{\vartheta_\theta}{\partial z} \right) \hat{\mathbf{e}}_r + \left(\frac{\partial \vartheta_r}{\partial z} - \frac{\partial \vartheta_z}{\partial r} \right) \hat{\mathbf{e}}_\theta + \left(\frac{\partial \vartheta_\theta}{\partial r} - \frac{\partial \vartheta_r}{\partial \theta} + \frac{\vartheta_\theta}{r} \right) \hat{\mathbf{e}}_z.$$

A convenient choice of the Coulomb gauge div $\vec{\vartheta} = 0$ leads to the Poisson equation vector $\Delta \vec{\vartheta} = -\text{rot} \, \mathbf{v} = -\Omega$. One advantage of such a formulation is that the mass conservation is satisfied identically in a simply connected region of the space since div $(\text{rot} \, \vec{\vartheta}) = 0$. For planar flows, the vorticity has only one component and Poisson's equation reduces to

$$\Delta \vartheta = -\Omega. \tag{5.54}$$

In the case in which there is no vorticity distribution and there are no internal boundaries within the fluid, meaning that the fluid flows in a simply connected region, the flow being irrotational, the Poisson equation further reduces to the Laplace equation:

$$\Delta \vartheta = 0. \tag{5.55}$$

When the fluid contains some discrete or continuous distribution of vorticity, solving the problem consists in finding the stream function ϑ from (5.54). Naturally, the flow induced by a free vortex has a tendency to have an axisymmetric symmetry, therefore

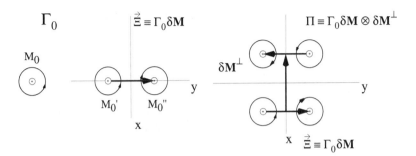

Figure 5.4. Vortex monopole, dipole, and quadrupole for incompressible fluid flow. The vortex dipole intensity and the vortex quadrupole intensity are calculated from the circulation of a free vortex and the locations of the vortex centers. The intensity of a vortex dipole is a vector quantity whereas the intensity of a vortex quadrupole is a second-order tensor.

we limit our development to a cylindrical coordinate formulation. The cylindrical form of the Poisson equation (and the Laplace equation if the vorticity vanishes) may be written as follows, e.g., [199]:

$$\Delta\,\vartheta(r,\theta) = \frac{1}{r}\frac{\partial}{\partial r}\left(r\frac{\partial\vartheta}{\partial r}\right) - \frac{1}{r^2}\frac{\partial^2\vartheta}{\partial\theta^2} = -\Omega(r,\theta).$$

Mathematically, fundamental solutions of the Laplace equation, in a simply connected region (infinite region) are obtained as high order derivatives of logarithmic functions, e.g., [8], [174], [199]:

$$\vartheta_n = \nabla^{(n)}\cdots\nabla^{(1)}(\log r).$$

Performing the differentiations, we obtain

$$\vartheta_0 = \log r \qquad \cdots \qquad \vartheta_n \propto \frac{1}{r^n}\left\{\ \cos(n\theta)\ \sin(n\theta)\ \right. \tag{5.56}$$

Physically, by considering an irrotational flow outside a circle of finite radius around the origin, the stream function of most classes of fluid flow can be represented by the superposition of independent solutions (5.56). The flow inside the circle is described by the functions $r^n\,\vartheta_n$ (n positive integer) which also satisfy the Laplace equation. Various kinds of discrete distributions of vortex are obtained by using the above solution of the Laplace equation, based on the logarithmic potential.

1. **Vortex monopole.** Consider a very small region of vortices, which could be practically idealized by a concentrated vortex with intensity Γ_0, at point M_0

(cf. Figure 5.4a). The steady irrotational flow in the region surrounding M_0, except in the origin M_0, is described by the stream function

$$\vartheta_0 = -\frac{\Gamma_0}{2\pi} \log \|\mathbf{M_0M}\| = -\frac{\Gamma_0}{2\pi} \log r. \tag{5.57}$$

The corresponding velocity field is described by showing that fluid particles move along circular paths:

$$v_r = \frac{1}{r}\frac{\partial \vartheta_0}{\partial \theta} = 0 \qquad v_\theta = -\frac{\partial \vartheta_0}{\partial r} = \frac{\Gamma_0}{2\pi}\frac{1}{r}.$$

It should be noticed that the point M_0 is a singular point, the vorticity vanishes everywhere except at M_0. The flow is steady in this case.

2. **Vortex dipole.** Consider two point vortices of respective intensities Γ_0 and $-\Gamma_0$ at the points M_0' and M_0'' with M_0 as the center - for which the distance between them is denoted $\delta\mathbf{M}$ (cf. Figure 5.4b). The irrotational fluid flow in the region surrounding these two vortices can be represented by the superposition of the previous result on the vortex monopole, taken separately. The stream function may be derived as follows:

$$\vartheta_2 = -\frac{\Gamma_0}{2\pi} \log \|\mathbf{M_0'M}\| + \frac{\Gamma_0}{2\pi} \log \|\mathbf{M_0''M}\|.$$

Now let us shorten the distance $\|\delta\mathbf{M}\|$ between the two points M_0' and M_0'' and conversely increase the vortex intensity Γ_0 so that the product $\Gamma_0\,\delta\mathbf{M}$ remains constant. The quantity

$$\vec{\Xi} = \lim_{\Gamma_0 \to \infty, \|\delta\mathbf{M}\| \to 0} \Gamma_0\,\delta\mathbf{M}$$

is defined as the intensity of the vortex dipole at the center point M_0. This vortex dipole induces an irrotational flow. The resulting stream function may be rewritten as

$$\begin{aligned}
\vartheta_2 &= \lim_{\|\delta\mathbf{M}\| \to 0} -\frac{\Gamma_0}{2\pi} \log \|\mathbf{M_0'M}\| + \frac{\Gamma_0}{2\pi} \log \|\mathbf{M_0''M}\| \\
&= \frac{\vec{\Xi} \cdot \mathbf{M_0M}}{2\pi \|\mathbf{M_0M}\|^2}.
\end{aligned}$$

The decomposition in a cylindrical coordinate with the origin at the center of the dipole and the y-axis directed by $\vec{\Xi}$ furnishes the stream function

$$\vartheta_2 = \frac{\vec{\Xi}.\mathbf{M_0M}}{2\pi \|\mathbf{M_0M}\|^2} = \frac{\Xi}{2\pi r} \sin\theta.$$

and the corresponding velocity field

$$v_r = \frac{1}{r}\frac{\partial \vartheta_2}{\partial \theta} = \frac{\Xi}{2\pi r^2}\cos\theta$$

$$v_\theta = -\frac{\partial \vartheta_2}{\partial r} = \frac{\Xi}{2\pi r^2}\sin\theta.$$

3. **Vortex quadrupole.** Consider now a vortex dipole of intensity $\vec{\Xi}$ located at the point M_0' and another vortex dipole of intensity $-\vec{\Xi}$ at M_0''. By analogy with the previous situation, let us shorten the distance $\|\delta\mathbf{M}^\perp\| = \|\mathbf{M}_0'\mathbf{M}_0''\|$ and conversely increase the intensity $\|\vec{\Xi}\|$ to maintain a finite and constant product. In fact each component of the defined nonsymmetric second-order tensor is finite and constant:

$$\lim_{\|\vec{\Xi}\|\to\infty,\,\|\delta\mathbf{M}^\perp\|\to 0} \vec{\Xi}\otimes\delta\mathbf{M}^\perp = \Pi. \tag{5.58}$$

The finiteness and constancy of the tensor are sought for every direction defined by any basis $(\mathbf{e}_1, \mathbf{e}_2, \mathbf{e}_3)$. We obtain a vortex quadrupole such that the irrotational fluid flow induced by the vortex quadrupole in the surrounding region is defined by the stream function

$$\vartheta_4 = -\frac{1}{2\pi}\nabla^2(\log r) : \Pi.$$

This stream function is represented by a tensor scalar product of the Hessian of the function $\log r$ and the above vortex quadrupole intensity. This stream function may be rewritten as

$$\vartheta_4 = \frac{\Pi : (\mathbf{n}\otimes\mathbf{n})}{2\pi r^2} \tag{5.59}$$

where $\mathbf{n}$ is the unit radius vector of M from M_0, midpoint between M_0' and M_0'' (cf. Figure 5.4c). Analytically, the stream function of the quadrupole vortex takes the form, provided the quadrupole intensity tensor is directed by the x-axis and the y-axis:

$$\vartheta_4 = \frac{1}{4\pi r^2}\left[(\Pi_{11} - \Pi_{22})\cos 2\theta + (\Pi_{12} + \Pi_{21})\sin 2\theta\right] \tag{5.60}$$

where the relation $\Pi_{11} + \Pi_{22} = 0$ has been accounted for. The corresponding velocity field have the radial and azimuthal components:

$$v_r = \frac{1}{r}\frac{\partial\vartheta_4}{\partial\theta} = -\frac{1}{2\pi r^3}[-(\Pi_{11} - \Pi_{22})\sin 2\theta + (\Pi_{12} + \Pi_{21})\cos 2\theta]$$

$$v_\theta = -\frac{\partial\vartheta_4}{\partial r} = -\frac{1}{2\pi r^3}[(\Pi_{11} - \Pi_{22})\cos 2\theta + (\Pi_{12} + \Pi_{21})\sin 2\theta]. \tag{5.61}$$

Physically, the diffusion in a real viscous fluid smooths the discontinuity of shear stresses and the vorticity (monopole, dipole, quadrupole) induced at the initial time. This fluid flow becomes unsteady in the course of time.

4. **Multipole expansion.** By considering now a vorticity distribution occupying a finite region in fluid B_v, the boundary of which extends to infinity, the stream function may be expressed as an integral of (5.57) since the vortex source point spans the vorticity region B_v:

$$\vartheta(M, t) = -\frac{1}{2\pi} \int_{B_v} \omega(M', t) \log \|\mathbf{M'M}\| \, dB_v \qquad (5.62)$$

where we have assumed a linear superposition of all distributed vortices within this vorticity region. For a point M located far from the vortices region, one can expand the logarithmic function, for any space origin O so that $\mathbf{M'M} = \mathbf{OM} - \mathbf{OM'}$:

$$\begin{aligned}
\log \|\mathbf{M'M}\| &= \log \|\mathbf{OM}\| - \nabla_{\mathbf{OM'}} (\log \|\mathbf{OM}\|) \\
&+ \frac{1}{2!} \nabla_{\mathbf{OM'}} [\nabla_{\mathbf{OM'}} (\log \|\mathbf{OM}\|)] + \cdots .
\end{aligned}$$

By introducing the orientation unit vector $\mathbf{n} = \frac{\mathbf{OM}}{\|\mathbf{OM}\|}$, expansion of the logarithmic potential implies

$$\begin{aligned}
\log \|\mathbf{M'M}\| &= \log \|\mathbf{OM}\| - \frac{\mathbf{n}.\mathbf{OM'}}{\|\mathbf{OM}\|} \\
&+ \frac{\mathbf{n} \otimes \mathbf{n} : (2\mathbf{OM'} \otimes \mathbf{OM'} - \|\mathbf{OM'}\|^2 I)}{2!\|\mathbf{OM}\|^2} + \cdots
\end{aligned}$$

where $\mathbf{I} = \delta_{ij} \, \mathbf{e}^i \otimes \mathbf{e}^j$ recalls the compact form for Kronecker symbols. By using this expanded potential, we obtain the multiple expansion of the stream function after integrating (5.62):

$$\vartheta = -\frac{\Gamma_0}{2\pi} \log r + \frac{\mathbf{n}.\vec{\Xi}}{2\pi} \frac{1}{r} + \frac{\Pi : (\mathbf{n} \otimes \mathbf{n})}{2\pi} \frac{1}{r^2} + O\left(\frac{1}{r^2}\right)$$

with $r \equiv \|\mathbf{OM}\|$ and in which we observe the following integrals equivalent of a vortex monopole intensity, a vortex dipole intensity, and a vortex quadrupole intensity respectively:

$$\Gamma_0 = \int_{B_v} \Omega(M', t) \, dB_v$$

$$\vec{\Xi} \equiv \int_{B_v} \mathbf{OM'} \, \Omega(M', t) \, dB_v$$

$$\Pi \equiv \int_{B_v} (2\mathbf{OM'} \otimes \mathbf{OM'} - \|\mathbf{OM'}\|^2 \mathbf{I}) \, \Omega(M', t) \, dB_v.$$

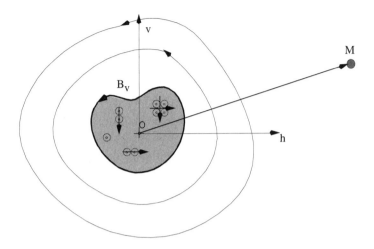

Figure 5.5. Continuous distribution of vorticity in a limited region B_v. The effect of this vortex region at any point M located at a very large distance from B_v is equivalent to a superposition of discrete distribution of the vortex monopole Γ_0, the vortex dipoles Ξ_h and Ξ_v, and the vortex quadrupole Π.

To obtain a tractable formulation in cylindrical coordinates, for the large distance $r = \|\mathbf{OM}\|$ infinitely greater than $r' = \|\mathbf{OM'}\|$ for any vortex point M' of B_v, we can rewrite the stream function:

$$
\begin{aligned}
\vartheta \;=\; & -\frac{\Gamma_0}{2\pi}\log r + \frac{A_1.\cos\theta + B_1\sin\theta}{2\pi}\frac{1}{r} \\
& + \frac{A_2.\cos 2\theta + B_2\sin 2\theta}{2\pi}\frac{1}{r^2} + O\left(\frac{1}{r^2}\right)
\end{aligned}
\tag{5.63}
$$

where the various coefficients may be calculated from the previous characteristics of the vortices zone. The corresponding terms in the case of a discrete distribution of vortices should be noticed. For a point situated at a large distance from the vortices region, the singularity is equivalently reduced to the center O.

Example. Consider the flow of an irrotational and incompressible fluid around a rigid circular cylinder of radius a. The flow is assumed uniform far from the cylinder with a constant velocity $\mathbf{v}_0$. First, let us assume that the intensity of the vorticity is equal to zero $\Gamma_0 = 0$. From direct calculation, the velocity field surrounding the circular cylinder is

$$
v_r = v_0\,\cos\theta\left(\frac{a^2}{r^2} - 1\right) \qquad v_\theta = v_0\,\sin\theta\left(\frac{a^2}{r^2} + 1\right).
$$

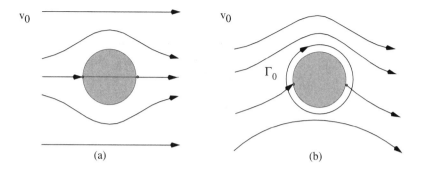

Figure 5.6. Flow around a circular cylinder of radius a (a) without and (b) with circulation of a vortex monopole. In the first case, there are no forces occurring on the cylinder. The introduction of a vortex monopole centered on the cylinder axis allows for the presence of drag and lift forces. Depending on the value of the vortex monopole intensity Γ_0, there are two $\Gamma_0 < 4\pi a v_0$, one $\Gamma_0 = 4\pi a v_0$, or no $\Gamma_0 > 4\pi a v_0$ stagnation points on the cylinder surface. For this later case, the stagnation point is outside the surface.

Second, consider an additional circulation around the cylinder, this could be assimilated as a free vortex (vortex monopole centered at the cylinder axis and of intensity Γ_0). The introduction of an additional fluid flow induced by a free vortex the stream function of which is given by (5.57), the velocity becomes

$$v_r = v_0 \, \cos\theta \, \left(\frac{a^2}{r^2} - 1\right) \qquad v_\theta = v_0 \, \sin\theta \, \left(\frac{a^2}{r^2} + 1\right) + \frac{\Gamma_0}{2\pi r}.$$

This example may highlight the role of singularity in the flight theory of object. Consider a 2-dimensional airfoil at incidence in a steady state flow of incompressible fluid. If no circulation occurs around the airfoil, we face a paradox. Actually, for a finite body of arbitrary shape in a steady, frictionless, and incompressible flow completely free from vorticity, it can be shown that the body will experience no net force, no drag, and no lift. This is what the hydrodynamicists called the D'Alembert paradox. This is the case for the flow around (finite length case) an airfoil (e.g., circular cylinder) without circulation.

To obviate this problem when considering the theory of flight, the hypothesis of Kutta–Joukowski proposes a net circulation imposed around the object immersed in the fluid. They assume that the circulation in inviscid flow is such as to ensure smooth separation of the flow at the trailing edge of the airfoil. It can be considered in the light of a multi-connected flow that we face here a basic problem of singularity distribution. The lift and drag forces can only be explained by the presence of vorticity circulation, i.e., by considering the airfoil as a flow singularity.

5.1.7 Theorems on vorticity

Definition 5.1.2 *(Circulation) Consider now a closed contour ∂C_0 in the fluid at time $t = 0$. The circulation of the velocity field around the deformed contour ∂C at any other time t during the fluid flow is defined by the integral*

$$\Gamma \equiv \int_{\partial C} \mathbf{v}^*. \tag{5.64}$$

By using the Stokes theorem, this circulation may be rewritten as

$$\Gamma \equiv \int_{\partial C} \mathbf{v}^* = \int_C d\mathbf{v}^* \tag{5.65}$$

which is the vorticity flux across the (open) surface C encircled by the closed curve ∂C. From (5.42), the practical calculus of the flux holds:

$$\Gamma = \int_C d\mathbf{v}^*(\mathbf{e}_i, \mathbf{e}_j) = \int_C \omega_0(\text{rot } \mathbf{v}, \mathbf{e}_i, \mathbf{e}_j). \tag{5.66}$$

Remark. If the circulation vanishes for all closed curves, the vorticity distribution is zero and the fluid flow is called irrotational, i.e., rot $\mathbf{v} \equiv 0$. The converse is true if the fluid is contained in a simply connected region, e.g., [195].

Theorem 5.1.1 *(Kelvin, 1869) Let B be a fluid (manifold) and ∂C a smooth closed loop (1-dimensional compact manifold) that deforms with the flow. Let $\mathbf{v}$ be the velocity field that is the solution of the governing equations (5.34) of fluid in which the pressure p is a single-valued function of the density ρ. Viscosity and heat conduction are neglected and body forces are conservative. Then the circulation of the velocity field is constant along ∂C*

$$\frac{d^B}{dt} \oint_{\partial C} \mathbf{v}^* = 0. \tag{5.67}$$

Proof. Let there be a fluid flow with a velocity field $\mathbf{v}$. The time derivative with respect to the continuum of the circulation is given by

$$\left(\frac{d^B}{dt} \oint_{\partial C} \mathbf{v}^* \right) (\mathbf{u}) = \frac{d}{dt} \left(\oint_{\partial C} \mathbf{v}^*(\mathbf{u}) \right) \qquad \mathbf{u} \in T_M \partial C$$

and then

$$\frac{d}{dt} \left(\oint_{\partial C} \mathbf{v}^*(\mathbf{u}) \right) = \oint_{\partial C} \frac{d}{dt} [\mathbf{v}^*(\mathbf{u})] = \oint_{\partial C} \left(\frac{d^B}{dt} \mathbf{v}^* \right) (\mathbf{u}).$$

If we utilize the metric of the ambient space, we do not deal with the tensor order of the velocity (5.40) and obtain

$$\frac{d^B \mathbf{v}}{dt} = \left(\frac{\partial \mathbf{v}}{\partial t} + \nabla_{\mathbf{v}} \mathbf{v} \right) = -\nabla p + \mathbf{b}.$$

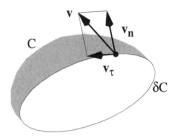

Figure 5.7. At each point of the contour, the velocity can be decomposed into a tangential and a normal component. The circulation about the contour is the summation of the tangential component along the closed curve. The flux of the vorticity is called the vorticity strength across the surface C.

In a perfect barotropic fluid (for which the density ρ is a single-valued function of the pressure p), upon which conservative forces act with a single-valued potential energy

$$\mathbf{b} = -\nabla \varpi,$$

it can be shown in a simply connected manifold B ("without handles and without holes") that:

$$\oint_{\partial C} \left(\frac{d^B}{dt} \mathbf{v}^* \right)(\mathbf{u}) = -\oint_{\partial C} \left(\frac{1}{\rho} \nabla p + \nabla \varpi \right)(\mathbf{u}).$$

Remark. A consequence of Kelvin's theorem is that if the fluid flow is irrotational once (under the conditions required by the theorem), it remains irrotational.

Theorem 5.1.2 *(Helmholtz) Let B be a fluid (manifold) and ∂C a smooth closed loop (1-dimensional compact manifold) that deforms with the flow. Let $\mathbf{v}$ the velocity field, under the same conditions as those admitted for Kelvin's theorem. Then the flux of the vorticity across a material surface C (i.e., surface moving with the flow) is constant in time:*

$$\frac{d^B}{dt} \int_C d\mathbf{v}^* = 0. \tag{5.68}$$

Proof. Starting with the curl vector, we first apply relations (5.66) and (5.65). It is then straightforward to write, by means of the Stokes theorem and the previous Kelvin theorem:

$$\frac{d^B}{dt} \int_C d\mathbf{v}^* = \frac{d^B}{dt} \oint_{\partial C} \mathbf{v}^* = 0.$$

Remark. Historically, the original Helmholtz theorem contained three parts. For that purpose, let us define first the vortex line as a line everywhere tangential to the

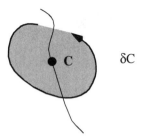

Figure 5.8. The circulation of the velocity field along ∂C is given by $\Gamma = \int_C d\mathbf{v}^*$.

local vorticity. A vortex tube is defined as the surface swept out by vortex lines passing through a given closed curve, say ∂C. Let ∂C_1 and ∂C_2 be any two circuits drawn on the vortex tube, each encircling the tube in the same direction. The Helmholtz theorem then states that the circulation about ∂C_1 is the same as the circulation about ∂C_2. Let us make a few comments on the nucleation of the vortex within the bulk fluid. As pointed out in the fundamental literature, vortex nucleation results from a change of topology. Indeed, use of differential topology for describing the fluid flow draws back to Helmholtz's 1858 paper on vortex motion. Later, Kelvin (W. Thomson) attempted to describe fundamental physics through topological properties (1875). Despite the fact that he was not successful in his explanation of atoms as knotted vortex rings in a fluid ether, Kelvin's work was seminal in the development of topological approach to fluid mechanics. An explicit attempt to include topology in the description of fluid flows was proposed by Lichtenstein in 1929 [117] and this topological approach has triggered a new interest, particularly in the domain of Cartan's differential calculus, e.g., [91].

To complete the equations of motion (5.34) and the heat conduction (5.35), boundary conditions should be considered. From such a point of view, it is theoretically impossible for irrotational flow to exist, the reason being that viscous fluids adhere to boundary surfaces while irrotational flows cannot adhere. In classical fluid dynamics, the following statement holds: Vortices cannot be generated in the bulk of a viscous incompressible fluid, but they are necessarily diffused inward from the boundary surfaces. Such a framework should be revisited.

5.1.8   Vortex dynamics in classical fluids

In the previous paragraph, the vortex distribution was assumed constant in the course of time. In a real viscous fluid, diffusion smooths the discontinuity of the vorticity (monopole, dipole, quadrupole) induced at the initial time. The fluid flow becomes unsteady in the course of time. The aim of this paragraph is to provide some basic theory on vortex diffusion. For this purpose, consider the set of equations governing the Navier–Stokes fluid. Mass conservation is written as:

$$\frac{\partial \rho}{\partial t} + \text{div } (\rho \mathbf{v}) = 0. \tag{5.69}$$

The dynamic equations reduce to:

$$\rho \left(\frac{\partial \mathbf{v}}{\partial t} + \nabla_{\mathbf{v}} \mathbf{v} \right) = -\nabla p + \mu' \, \Delta \mathbf{v} + \rho \mathbf{b}. \tag{5.70}$$

To begin, consider the particular case of a conservative body force (ρ is assumed to be a single-valued function of the pressure p) with a single-valued potential energy:

$$\mathbf{b} = -\nabla \varpi.$$

Then, taking into account the general relation (4.29), we can replace the vector $\mathbf{u}_a = \mathbf{v}$ to obtain for classical fluids, $\forall \mathbf{u}_b \in T_M B$:

$$(\nabla_{\mathbf{v}} \mathbf{v}^*)(\mathbf{u}_b) = \frac{1}{2} \left[(\nabla_{\mathbf{v}} \mathbf{v}^*)(\mathbf{u}_b) + (\nabla_{\mathbf{u}_b} \mathbf{v}^*)(\mathbf{v}) \right] + \frac{1}{2} \omega_0 (\text{rot } \mathbf{v}, \mathbf{v}, \mathbf{u}_b). \tag{5.71}$$

Since the 1-form $\mathbf{v}^* \equiv \mathbf{g}(\mathbf{v}, .)$ represents the velocity field and since we have assumed a connection compatible with the metric tensor $\nabla \mathbf{g} = 0$, we can "divide" the relation (5.71) by $\mathbf{g}$. The classical identity follows:

$$\nabla_{\mathbf{v}} \mathbf{v} = \nabla \left(\frac{\|\mathbf{v}\|^2}{2} \right) + \frac{1}{2} \text{rot } \mathbf{v} \times \mathbf{v} = \nabla \left(\frac{\|\mathbf{v}\|^2}{2} \right) + \vec{\omega} \times \mathbf{v}. \tag{5.72}$$

Therefore, the Navier–Stokes equations (5.70) become

$$\frac{\partial \mathbf{v}}{\partial t} + \nabla \left(\frac{\|\mathbf{v}\|^2}{2} \right) + \vec{\omega} \times \mathbf{v} = -\frac{\nabla p}{\rho} - \nabla \varpi + \frac{\mu'}{\rho} \Delta \mathbf{v}.$$

For barotropic fluid where ρ is a single-valued function of p, the density can be included in the gradient operator without loss of generality so as to obtain

$$\frac{\partial \mathbf{v}}{\partial t} + \vec{\omega} \times \mathbf{v} = -\nabla \left(\frac{p}{\rho} + \frac{\|\mathbf{v}\|^2}{2} + \varpi \right) + \frac{\mu'}{\rho} \Delta \mathbf{v}. \tag{5.73}$$

Equation (5.73) is the starting point of many practical results in classical hydrodynamic theory. Let us briefly recall some of them.

1. **Bernoulli's theorem and Crocco's theorem**. For perfect (nonviscous) fluid undergoing steady irrotational flow, equation (5.73) reduces to the simplest form of Bernoulli's equation:

$$\nabla \left(\frac{p}{\rho} + \frac{\|\mathbf{v}\|^2}{2} + \varpi \right) = 0.$$

Theorem 5.1.3 *(Bernoulli) Consider the steady flow of a perfect (nonviscous) barotropic fluid. If the velocity is parallel to the vorticity $\vec{\omega} \times \mathbf{v} = 0$ in a simply connected flow region, then the following quantity is constant everywhere in the flow:*

$$\frac{p}{\rho} + \frac{\|\mathbf{v}\|^2}{2} + \varpi = c^{st}. \tag{5.74}$$

Again if the flow is steady, the product of equation (5.73) multiplied by the velocity gives

$$(\vec{\omega} \times \mathbf{v})(\mathbf{v}) = -\nabla \left(\frac{p}{\rho} + \frac{\|\mathbf{v}\|^2}{2} + \varpi \right)(\mathbf{v}) + \frac{\mu'}{\rho} \Delta \mathbf{v}(\mathbf{v}) = 0$$

which implies the decrease of the term in between brackets along the streamline in the presence of viscosity and if such viscosity decelerates the fluid particles:

$$\nabla_{\mathbf{v}} \left(\frac{p}{\rho} + \frac{\|\mathbf{v}\|^2}{2} + \varpi \right) = \frac{\mu'}{\rho} \Delta \mathbf{v}(\mathbf{v}) < 0. \tag{5.75}$$

Theorem 5.1.4 *(Crocco) For a perfect barotropic fluid in a steady state flow, the term in between brackets is constant over streamlines (and path lines) and can be written as*

$$\nabla_{\mathbf{v}} \left(\frac{p}{\rho} + \frac{\|\mathbf{v}\|^2}{2} + \varpi \right) = 0. \tag{5.76}$$

Proof. Assume $\mu' = 0$ in relation (5.75).

2. **Convection and diffusion of vorticity.** By applying the rotational operator to the equation of motion (5.73), we get

$$\operatorname{rot} \left(\frac{\partial \mathbf{v}}{\partial t} + \vec{\omega} \times \mathbf{v} \right) = \operatorname{rot} \left[-\nabla_{\mathbf{v}} \left(\frac{p}{\rho} + \frac{\|\mathbf{v}\|^2}{2} + \varpi \right) \right] + \operatorname{rot} \left(\frac{\mu'}{\rho} \Delta \mathbf{v} \right).$$

The rotational operator and the derivative with respect to time may be interchanged to obtain, in a simply connected region,

$$\frac{\partial}{\partial t} (\operatorname{rot} \mathbf{v}) + \operatorname{rot} (\vec{\omega} \times \mathbf{v}) = \frac{\mu'}{\rho} \Delta (\operatorname{rot} \mathbf{v}).$$

By introducing the vectorial relation

$$\operatorname{rot} (\vec{\omega} \times \mathbf{v}) = \nabla_{\mathbf{v}} \vec{\omega} - (\operatorname{div} \vec{\omega}) \mathbf{v} - \nabla_{\vec{\omega}} \mathbf{v} + (\operatorname{div} \mathbf{v}) \vec{\omega} \tag{5.77}$$

we can further simplify the equation, thanks to the relation div $\vec{\omega} = 0$,

$$\frac{\partial}{\partial t}\vec{\omega} + \nabla_{\mathbf{v}}\vec{\omega} - \nabla_{\vec{\omega}}\mathbf{v} + (\text{div } \mathbf{v})\vec{\omega} = \frac{\mu'}{\rho}\Delta\vec{\omega} \qquad (5.78)$$

which is the classical equation that describes the diffusion of vorticity in the viscous fluid flow. First, the final form of this equation holds when introducing the total derivative:

$$\frac{d\vec{\omega}}{dt} = \frac{\partial}{\partial t}\vec{\omega} + \nabla_{\mathbf{v}}\vec{\omega} = \nabla_{\vec{\omega}}\mathbf{v} - (\text{div } \mathbf{v})\vec{\omega} + \frac{\mu'}{\rho}\Delta\vec{\omega}. \qquad (5.79)$$

Relation (5.79) means that the rate of change of the vorticity following a fluid particle is equal to the interaction between the velocity gradient and the vorticity plus the diffusion of the vorticity. Second, by observing the derivative with respect to the continuum, it is quite simple to write the relation (5.79) as

$$\frac{d^B\vec{\omega}}{dt} = \frac{d\vec{\omega}}{dt} - \nabla_{\vec{\omega}}\mathbf{v} = -(\text{div } \mathbf{v})\vec{\omega} + \frac{\mu'}{\rho}\Delta\vec{\omega}. \qquad (5.80)$$

Relation (5.80) describes the derivative of the vorticity with respect to the fluid and the diffusion of the vorticity in the flow. The problem is to gain some information on how vorticity diffuses in a fluid flow. Starting with the basic equation of continuity (5.69), we can replace div $\mathbf{v}$ in terms of density and of its time derivative:

$$\frac{1}{\rho}\frac{d\vec{\omega}}{dt} - \frac{1}{\rho^2}\frac{d\rho}{dt}\vec{\omega} = \frac{d\frac{\vec{\omega}}{\rho}}{dt} = \nabla_{\frac{\vec{\omega}}{\rho}}\mathbf{v} + \frac{\mu'}{\rho^2}\Delta\vec{\omega}. \qquad (5.81)$$

Theorem 5.1.5 (*Nanson, vorticity convection, 1874*) *For a nonviscous fluid undergoing a barotropic flow, the convection of the vortices weighted by the inverse of the density is governed by the first-order differential equation:*

$$\frac{d}{dt}\frac{\vec{\omega}}{\rho} = \nabla_{\frac{\vec{\omega}}{\rho}}\mathbf{v}. \qquad (5.82)$$

Proof. Starting from (5.81) for the particular case of nonviscous fluid undergoing barotropic flow, we have the following evidence:

$$\frac{d}{dt}\frac{\vec{\omega}}{\rho} - \nabla_{\frac{\vec{\omega}}{\rho}}\mathbf{v} = 0.$$

Remark. This equation governs the convection of vorticity within fluid and recalls the very celebrated Helmholtz theorem.

Theorem 5.1.6 *(Helmholtz) The vortex curves in a nonviscous baro-tropic fluid flow are material curves, constituted by the same material particles all along the flow, in the course of time:*

$$\frac{d^B}{dt}\frac{\vec{\omega}}{\rho} = 0. \tag{5.83}$$

Proof. We can introduce the derivative with respect to the continuum together with Nanson's theorem to obtain the diffusion of vorticity equation, e.g., [8], [125], [151]:

$$\frac{d^B}{dt}\frac{\vec{\omega}}{\rho} = \frac{d}{dt}\frac{\vec{\omega}}{\rho} - \nabla_{\frac{\vec{\omega}}{\rho}}\mathbf{v} = 0.$$

This theorem implicitly states that there is only a convection of vorticity during the flow of a nonviscous fluid.

Theorem 5.1.7 *(Vorticity diffusion) For viscous fluids, the more general relation expressing the rate of the vortex with respect to the fluid is given by the convection-diffusion equation, which illustrates that the diffusion of vorticity occurs in addition to convection:*

$$\frac{d^B}{dt}\frac{\vec{\omega}}{\rho} = \frac{\mu'}{\rho^2}\Delta\vec{\omega}. \tag{5.84}$$

Proof. This equation describes the diffusion of vorticity in viscous fluids during the course of time. The proof is directly drawn from relation (5.81).

5.1.9   Turbulence and Reynolds number

Even in the smoothest and most symmetric boundary conditions, such as a water flow within an infinitely long pipe, experimental measurements of fluid flows highlight the irregular state of turbulence. Various theories have been suggested and among them, the occurrence of dynamic chaos has been one of the most attractive theories. Some authors have defined turbulent flow as chaotic flow, e.g., [151]. Nevertheless, the change of local topology in the region of flow suggests the need for a more complicated theory which takes into account the turbulence irreversibility. Subsequently, the goal of the present chapter is to point out the role of stream flow singularity, such as the creation of vortices, on the irreversible flow of fluids. To estimate the relative importance of each term in the Navier–Stokes equations (5.38), a set of dimensionless variables are commonly used as follows, e.g., [149]:

$$\mathbf{v}' \equiv \frac{\mathbf{v}}{v_0} \qquad \mathbf{OM}' \equiv \frac{\mathbf{OM}}{l_0} \qquad t' \equiv t\,\frac{l_0}{v_0} \tag{5.85}$$

where the two values prescribed by the problem are respectively the characteristic velocity v_0 (example: velocity at the infinite) and the characteristic length l_0 (example: a pipe diameter). In some situations, their handling is not so evident. The dimensionless pressure definition can be inspired by the Bernoulli relation (5.74):

$$p' \equiv \frac{p}{\rho v_0^2}. \tag{5.86}$$

By introducing these variables in the dynamic equation (5.70), we obtain the dimensionless Navier–Stokes equations in which the primes are dropped for notation convenience and in which the body force is neglected without loss of generality:

$$\left(\frac{\partial \mathbf{v}}{\partial t} + \nabla_{\mathbf{v}} \mathbf{v}\right) = -\nabla p + \frac{1}{R_e} \Delta \mathbf{v} \tag{5.87}$$

in which the Reynolds[3] number is classically defined as:

$$R_e \equiv \rho \frac{v_0 l_0}{\mu'}. \tag{5.88}$$

Remark. This number is the most familiar control parameter of the flow. The definition of the dimensionless pressure by scaling with the kinetic energy (5.86) is quite correct when $R_e \gg 1$. In this case the dimensionless equation (5.87) is used in this range of Reynolds number. In the case where $R_e \ll 1$, the viscous term is overweighted by the inverse of R_e and may artificially dominate the equation of motion (5.87). Another set of dimensionless variables is therefore proposed in that case. The Reynolds number has gained a great interest in the analysis of fluid flow due to the fact that it has been "possible" to distinguish two classes of fluid flows, laminar flow and turbulent flow, according to the values of Reynolds number. Fluid mechanics exploits to a great extent the fact that the R_e is often either large or small. Various situations may occur according to the values taken on by the Reynolds number. Some examples are sketched below for incompressible fluid div $\mathbf{v} = 0$.

1. **Incompressible flows.** Consider a steady incompressible fluid flow where $R_e \gg 1$. The dimensionless equation is obtained from (5.87) by dropping the unsteady component of the acceleration and by introducing the vorticity vector (5.72):

$$\nabla_{\mathbf{v}} \mathbf{v} = \nabla \left(\frac{\|\mathbf{v}\|^2}{2}\right) + \vec{\omega} \times \mathbf{v} = -\nabla p + \frac{1}{R_e} \Delta \mathbf{v}.$$

---

[3]Reynolds in 1883–1894 identified laminar and turbulent flows by using the criterion for the onset of turbulence in terms of the Reynolds number. Furthermore, he decomposed the flow into a mean part and fluctuating parts and therefore identified additional stresses due to turbulence: "Reynolds stresses."

It is well known that the resolution of a 2-dimensional problem leads to the classical Prandtl boundary layer and celebrated Blasius equation around a heated infinitely thin plate (see, e.g., [149] for a concise but complete reminder). Application of the rotational operator to the previous equation gives, by using (5.77),

$$\text{rot } (\vec{\omega} \times \mathbf{v}) = \nabla_{\mathbf{v}} \vec{\omega} - (\text{div } \vec{\omega}) \mathbf{v} - \nabla_{\vec{\omega}} \mathbf{v} + (\text{div } \mathbf{v}) \vec{\omega} = \frac{1}{R_e} \Delta \vec{\omega}.$$

Since the two divergences vanish and the torsion tensor vanishes everywhere for classical fluids, we deduce the following relation:

$$[\vec{\omega}, \mathbf{v}] = \frac{1}{R_e} \Delta \vec{\omega}. \tag{5.89}$$

Unsteady slow flow where the nonlinear term can be neglected leads to the dimensionless equation

$$\frac{\partial \mathbf{v}}{\partial t} = -\nabla p + \frac{1}{R_e} \Delta \mathbf{v}.$$

Again, the application of the rotational operator gives the evolution law of the vorticity within the fluid:

$$\frac{\partial \vec{\omega}}{\partial t} = \frac{1}{R_e} \Delta \vec{\omega}. \tag{5.90}$$

2. **Flows with low Reynolds number.** Consider a fluid flow with a very low Reynolds number $R_e \ll 1$. Another way of nondimensionalization should be adopted in this case by defining the dimensionless pressure as, e.g., [149]:

$$p' \equiv \frac{1}{R_e} \frac{p}{\rho v_0^2}.$$

The dynamic equation (5.87) becomes

$$R_e \left(\frac{\partial \mathbf{v}}{\partial t} + \nabla_{\mathbf{v}} \mathbf{v} \right) = -\nabla p + \Delta \mathbf{v} \tag{5.91}$$

which is another dimensionless version of the Navier–Stokes equations. As previously, after the application of the rotational operator and of the vectorial relation already exploited before, we arrive at the vorticity equation

$$R_e \left(\frac{\partial \vec{\omega}}{\partial t} + [\vec{\omega}, \mathbf{v}] \right) = \Delta \vec{\omega}. \tag{5.92}$$

For a creeping flow, which is a very slow flow, the acceleration may be neglected, the motion equation reduces to

$$\Delta \mathbf{v} = \nabla p.$$

Remark. In principle, solving the Navier–Stokes equations (5.87) allows us to fully understand both laminar and turbulent flows. Nevertheless, the turbulence problem remains an unsolved problem in continuum mechanics due to, among other things, the lack of a precise definition of the turbulence state itself. Curiously, the nonturbulent state could be easily related to the smooth solutions of the Navier–Stokes equations but the turbulent state is not clearly defined. Thus, various approaches have been proposed over the years.

The Leray definition of turbulence is related to the exclusion of Navier–Stokes equations solutions. This definition pictures the turbulent flows as not leading to solutions of Navier–Stokes solutions. This may lead to confusion since there are discontinuous solutions of these classical equations in thermomechanics, e.g., [80].

Komolgorov's 1941 definition takes on the statistical approach, stating that turbulence consists of vortices of all scales with random intensities. Komolgorov's essential idea was that the small scale motions in developed turbulence are universal, in the sense that they are independent of the statistical properties of small eddies in the nature of fluid (water, air, interstellar dust or other), independent of the mechanism stirring the flow, and finally independent of the particular geometry of the fluid container. This approach is not explicitly related to the Navier–Stokes equations.

On the other hand, the turbulence theory of Hopf–Landau is based on local instability and suggests that the turbulence is a quasi-periodic phenomenon in which there are infinitely many periods sequentially generated to give the appearance of randomness, e.g., [28]. These infinite eigenmotions are excited at the same time. Conversely, the theory of Ruelle–Takens claims that the transition to turbulence is due only to a few transitions to the chaotic state. This means that the transition is due only to few eigenmotions defined by a "strange attractor." The following question holds: Is the chaotic motion of flow the same as turbulent flow, e.g., [91]? In 1922, Richardson has proposed a cascade picture of turbulence in the sense that turbulence is initiated by the creation of instabilities of the main stream flow. For high Reynolds numbers, eddies exist at various scales and this situation is denoted as developed turbulence.

Inspired by Cartan's work, Kiehn proposed an extended version of the turbulence theory by distinguishing the two flows: Turbulent flow may be chaotic, e.g., [151] but turbulence includes more features such as irreversibility [91]. Chaotic flow (at least the deterministic chaotic) can be reversible but not turbulent flow. Beginning with this remark, the turbulence state is here defined as a change of topology within the continuum. In the present work, the creation of vorticity in the bulk is defined as nucleation of singularity. A transition from a connected topology (simply connected region of flow) to a disconnected topology (multiply connected region of topology) is merely a consequence of the creation of singularity distribution. Therefore, the aim of the following section is to present an outline of the basic theory of an extended version of fluid mechanics which may account for the creation of discontinuity of scalar fields and vector fields [163]. This announces the next section devoted to non classical fluid flows. The transition from laminar to turbulent flow is then believed to involve a change of topology (torsion and curvature tensors). The intention of this monograph is far from developing any turbulence theory of fluid flow. Much more work should be undertaken for this purpose. I would like to suggest only a possible way of modeling the unavoidable local change of topology within the fluid continuum during a turbulent flow.

5.2   Fluids with singularity distribution

When the circulation is zero for all closed curves, the vorticity field is zero and the fluid flow remains irrotational. The converse is true if the fluid is contained in a simply connected region, e.g., [195]. In a perfect barotropic fluid (ρ is a single-valued function p), acted upon by conservative forces with a single-valued potential energy, the following have been shown: (a) the first Helmholtz law, e.g., [190], [198], stating that a fluid particle originally free of vortices remains free of vortices; (b) the Kelvin circulation theorem stating that, e.g., [174], [195] the circulation around a material curve remains zero if initially zero.

The basic question then arises whether, in the same conditions, vortices can be created within the fluid volume without violating these two theorems. In fact, the ability of a classical fluid model (5.3) to capture the source of vortices, which is in fact a local change of the continuum topology, e.g., [174], is questionable. At this stage, we face the problems of vortices and in creation of internal slip surfaces. Indeed, the occurrence of internal slip surfaces implies that the rates (4.45) do not necessarily vanish, at least not in some regions of the fluid. To extend the fluid model of Stokes, one of the first authors to propose a law containing the vorticity tensor as a constitutive variable was Boussinesq (1868) when he wrote:

$$\sigma = -p(\rho, \theta)\mathbf{i} + \mathbf{J}_g(\rho, \theta, \zeta_g, \Omega) \tag{5.93}$$

where Ω is the vorticity tensor and where $\mathbf{J}_g(\rho, \theta, \zeta_g = 0, \Omega) = 0$. Later, in the framework of simple material theory, Noll showed that the principle of frame indifference reduces the Boussinesq definition to the common Stokes definition and that, furthermore, it enforces the viscous stress as an isotropic tensor function. Nevertheless, the extension of Boussinesq's definition beyond the theory of simple material is not so simple and certainly requires further additional investigation.

As another way to extend Noll's simple material theory [146], higher gradients of the metric tensor and temperature were introduced into constitutive equations (viscous fluids of high grade). Specifically, some authors e.g. [73] implicitly admitted the density gradient as a primal variable to capture the long range spatial dependence in constitutive laws. Various problems related to the consistency of the theory of high grade fluids have been pointed out and discussed in the literature. We will discuss it later after having introduced the extended fluid model capable of capturing the occurrence of bulk singularity.

5.2.1 Constitutive laws

By extending the definition of classical thermoviscous fluids, we propose the following definition which accounts for the bulk nucleation of vortices and for the increase of vortex density within the fluid volume.

Definition 5.2.1 *(Fluids with singularity distribution) Among the materials defined by the constitutive functions (4.1), a thermoviscous fluid with singularity is defined by the particular class of constitutive functions:*

$$\Im = \hat{\Im}(\omega_0, \aleph, \Re, \theta, \zeta_g, \zeta_\aleph, \zeta_\Re, \zeta_q) \qquad \Im = \rho, \sigma, s, \phi, \mathbf{J}_g, \mathbf{J}_\aleph, \mathbf{J}_\Re, \mathbf{J}_q. \qquad (5.94)$$

It is worthwhile to notice that the metric tensor **g** does not appear as an explicit argument of tensor functions (5.94). This allows the fluid model not to be dependent on the initial configuration's response, except by its volume change. A second interesting point is that the torsion tensor (skew symmetric with respect to the lower indices) appears as a more appropriate variable than the classic vorticity tensor (5.93) to capture the nucleation of internal slip surface. Regarding the results of the free energy formulation in Chapter 4, the particular class of normal dissipative fluids admits the Helmholtz free energy and the dissipation potential in terms of primal constitutive variables:

$$\phi = \hat{\phi}(\omega_0, \aleph, \Re, \theta) \qquad \psi = \hat{\psi}(\zeta_g, \zeta_\aleph, \zeta_\Re) \qquad (5.95)$$

where, again, ω_0, $\aleph$, $\Re$, and θ may be parameters for the dissipation potential. A first question to ask ourselves is whether such a fluid model remains isotropic due to the presence of the torsion tensor and the curvature tensor in the list of primal variables. To answer it, the tensor representation of constitutive laws, valid for classical fluids, at the beginning of this chapter should be revisited. The constitutive dual variables are derived from these potentials, the total stress, the hydrostatic pressure, and the viscous stress:

$$\sigma = -p\mathbf{i} + \mathbf{J}_g \qquad p \equiv -\rho \frac{\partial \phi}{\partial \omega_0} : \omega_0 \qquad \mathbf{J}_g = \frac{\partial \psi}{\partial \zeta_g}$$

the Eshelbian-like tensors:

$$\mathbf{J}_\aleph = \frac{\partial \psi}{\partial \zeta_\aleph} \qquad \mathbf{J}_\Re = \frac{\partial \psi}{\partial \zeta_\Re}$$

the heat flux and the entropy:

$$\mathbf{J}_q = \frac{\partial \psi}{\partial \zeta_q} \qquad s = -\frac{\partial \psi}{\partial \theta}.$$

To describe the flow of a fluid with singularity distribution, it is also worthwhile to adopt the same spatial base $(\mathbf{e}_1, \mathbf{e}_2, \mathbf{e}_3)$, directed by the referential body Σ, as previously. Nevertheless, the amount of singularity generated within the fluid bulk will be captured by the affine connection ∇ with torsion and curvature tensors $\aleph$ and $\mathfrak{R}$. The governing equations of a fluid with singularity (constitutive laws and conservation laws) are the same as for any continuum with singularity model we developed previously in Chapter 3. Again we can decompose the equations of both conservation laws and constitutive laws onto a spatial vector base such that $J_e = 1$ without loss of generality. The component form of constitutive laws are similar to those of classical fluids by introducing additionally the torsion and the curvature variables and their conjugate Eshelbian stress tensors.

Example. Consider a closed convex set C of E_J that contains the null stress $\mathbf{J}_\alpha = 0$. For the stresses within C, no evolution of the singularity density occurs; for those on the boundary of C, yielding occurs. As an extended formulation of the free energy function (5.29), we can propose the following model:

$$\phi = \hat{\phi}(\omega_0, \aleph, \mathfrak{R}, \theta) \tag{5.96}$$

and for the quadratic dissipation potential, including a homogeneous function of degree one in terms of singularity rates:

$$
\begin{aligned}
\psi &= \frac{1}{2}\hat{k}(\omega_0, \theta)\|\zeta_q\|^2 + \frac{1}{2}\hat{\lambda}'(\omega_0, \theta)\mathrm{tr}^2\zeta_g + \frac{1}{2}\hat{\mu}'(\omega_0, \theta)\mathrm{tr}\,(\zeta_g^2) \\
&+ \mathrm{Sup}_{\mathbf{J}_\aleph, \mathbf{J}_\mathfrak{R} \in C}(\mathbf{J}_\aleph : \zeta_\aleph + \mathbf{J}_\mathfrak{R} : \zeta_\mathfrak{R})
\end{aligned} \tag{5.97}
$$

in which C is the convex set defined by

$$C = \{\mathbf{J}_\aleph, \mathbf{J}_\mathfrak{R}, Y(\mathbf{J}_\aleph, \mathbf{J}_\mathfrak{R}) < k\}.$$

This model example indeed captures the existence of a threshold phenomenon defined by the convex set C, a domain where there is neither nucleation of singularity nor increase of existing singularity density. Again, most classes of nonclassical fluids are recovered by the choice of special forms of free energy and dissipation potential, as reported in the Table 5.2 below.

5.2.2   Source of vortices in a fluid

As with classical fluids, the vorticity field may be directly calculated with the exterior derivative and projected onto the spatial basis $(\mathbf{e}_1, \mathbf{e}_2, \mathbf{e}_3)$:

$$\omega_0(\mathrm{rot}\,\mathbf{v}, \mathbf{e}_i, \mathbf{e}_j) = (\nabla_{\mathbf{e}_i}\mathbf{v}^*)(\mathbf{e}_j) - (\nabla_{\mathbf{e}_j}\mathbf{v}^*)(\mathbf{e}_i) + \aleph(\mathbf{e}_i, \mathbf{e}_j)$$
$$\forall \mathbf{e}_i, \mathbf{e}_j \in T_M B \tag{5.98}$$

Free energy and dissipation potential	Stress	Viscous stress, heat flux, and Eshelbian stresses
$\phi = \hat{\phi}(\omega_0, \aleph, \Re, \theta)$ $\psi \equiv 0$	$\sigma = \rho \frac{\partial \phi}{\partial \omega_0} : \omega_0 \mathbf{i}$	$\mathbf{J}_g \equiv 0$ $\mathbf{J}_q \equiv 0$
$\phi = \hat{\phi}(\omega_0, \aleph, \Re, \theta)$ $\psi = \hat{\psi}(\zeta_g, \zeta_q)$	$\sigma = \rho \frac{\partial \phi}{\partial \omega_0} : \omega_0 \mathbf{i} + \mathbf{J}_g$	$\mathbf{J}_g = \frac{\partial \psi}{\partial \zeta_g}$ $\mathbf{J}_q = \frac{\partial \psi}{\partial \zeta_q}$
$\phi = \hat{\phi}(\omega_0, \aleph, \Re, \theta)$ $\psi = \hat{\psi}(\zeta_g, \zeta_q)$	$\sigma = \rho \frac{\partial \phi}{\partial \omega_0} : \omega_0 \mathbf{i} + \mathbf{J}_g$	$\mathbf{J}_g = \frac{\partial \psi}{\partial \zeta_g}$ $\mathbf{J}_q = \frac{\partial \psi}{\partial \zeta_q}$ $\mathbf{J}_\aleph = \frac{\partial \psi}{\partial \zeta_\aleph}$ $\mathbf{J}_\Re = \frac{\partial \psi}{\partial \zeta_\Re}$

Table 5.2. Particular constitutive laws for nonclassical fluids. (Row 1) Perfect elastic fluid with frozen distribution of singularity: $\zeta_\aleph \equiv 0$ and $\zeta_\Re \equiv 0$. (Row 2) Heat-conducting viscoelastic fluid with frozen distribution of singularity: $\zeta_\aleph \equiv 0$ and $\zeta_\Re \equiv 0$. (Row 3) Heat-conducting viscoelastic fluid with nucleation and evolution of singularity distribution: $\zeta_\aleph \in \partial \hat{\psi}^*_{\mathbf{J}_\aleph}(\mathbf{J}_\aleph, \mathbf{J}_\Re)$ and $\zeta_\Re \in \partial \hat{\psi}^*_{\mathbf{J}_\Re}(\mathbf{J}_\aleph, \mathbf{J}_\Re)$.

where the additional terms represent the distribution of singularity. As with classical fluids, it should be noticed that there are rotational motions where each fluid particle moves along a straight line and there are irrotational flows where fluid particles move along curve paths. Moreover, in relation (5.98) the presence of the torsion field (distribution of scalar discontinuity) emphasizes the participation of singularity in "rotating" the fluid particle. This extends to a continuous volume distribution the notion of isolated internal slipping surfaces (2.5) in Chapter 2.

Example. As for a classical fluid model, the vorticity components are derived on the local base of the cylindrical coordinate system. The matrix of the metric tensor and the component of the volume form remain the same as previously. Further, the Christoffel symbols associated to the related metric part of connection remain unchanged:

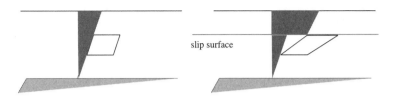

Figure 5.9. For the first flow of fluid without singularity (left) the vorticity is fully obtained from the continuous distribution of velocity (here a triangular shape distribution) whereas for the second flow (right) with singularity (internal surface slipping), the vorticity is enhanced by the velocity discontinuity (in other words the distribution of torsion tensor).

$$\bar{\Gamma}^\theta_{r\theta} = \bar{\Gamma}^\theta_{\theta r} = \frac{1}{r} \qquad \bar{\Gamma}^r_{\theta\theta} = -r. \tag{5.99}$$

The other coefficients of the metric tensor are zero. The velocity field of the flow is:

$$\mathbf{v} = v^r \mathbf{e}_r + v^\theta \mathbf{e}_\theta + v^z \mathbf{e}_z. \tag{5.100}$$

During the flow of a fluid with singularity, the components of the vorticity field along the basis are then given by

$$r\Omega_{r\theta} = \frac{\partial}{\partial r} r^2 v^\theta - \frac{\partial}{\partial \theta} v^r + \aleph^r_{r\theta} v^r + \aleph^\theta_{r\theta} r^2 v^\theta + \aleph^z_{r\theta} v^z \tag{5.101}$$

$$r\Omega_{\theta z} = \frac{\partial}{\partial \theta} v^z - \frac{\partial}{\partial z} r^2 v^\theta + \aleph^r_{\theta z} v^r + \aleph^\theta_{\theta z} r^2 v^\theta + \aleph^z_{\theta z} v^z \tag{5.102}$$

$$r\Omega_{zr} = \frac{\partial}{\partial z} v^r - \frac{\partial}{\partial r} v^z + \aleph^r_{zr} v^r + \aleph^\theta_{zr} r^2 v^\theta + \aleph^z_{zr} v^z. \tag{5.103}$$

For the particular problem of axisymmetry, only a limited number of the torsion components are retained. Generally, this type of symmetry permits us to keep only the following coefficients $\Gamma^r_{\theta\theta}, \Gamma^\theta_{r\theta}, \Gamma^\theta_{\theta r}$ of the connection. All components not explicitly listed and not connected by a symmetrical relation to a listed one are equal to zero. Thus, the torsion components that do not vanish are restricted to $\aleph^\theta_{r\theta}$. In this particular case, the vorticity reduces to

$$r\Omega_{r\theta} = \frac{\partial}{\partial r} r^2 v^\theta + \aleph^\theta_{r\theta} r^2 v^\theta$$

$$r\Omega_{\theta z} = -\frac{\partial}{\partial z} r^2 v^\theta$$

$$r\Omega_{zr} = \frac{\partial}{\partial z} v^r - \frac{\partial}{\partial r} v^z.$$

The component $\aleph^\theta_{r\theta}$ captures an axial singularity.

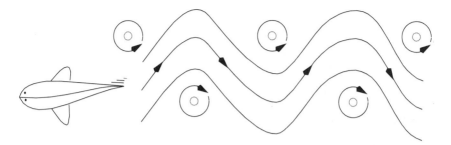

Figure 5.10. During the locomotion of some aquatic animal, vortex wakes are left behind it. The figure illustrates the vortices' distribution produced by the fish's motion and the oscillation from left to right of the caudal fin edge. Also pictured is the streamline pattern induced by these free vortices (adapted from Karman and Burgers (1934)) [84].

5.2.3 Theorems on vorticity

Let us consider now a closed curve ∂C within the fluid. In presence of internal slip surfaces, the circulation of the velocity field along ∂C is also expressed by the integral

$$\Gamma \equiv \oint_{\partial C} \mathbf{v}^* = \int_C d\mathbf{v}^*. \tag{5.104}$$

The practical form projected onto the spatial basis is

$$\Gamma = \int_C [(\nabla_{\mathbf{e}_i} \mathbf{v}^*)(\mathbf{e}_j) - (\nabla_{\mathbf{e}_j} \mathbf{v}^*)(\mathbf{e}_i) + \aleph(\mathbf{e}_i, \mathbf{e}_j, \mathbf{v}^*)]. \tag{5.105}$$

Compared to (5.66), the circulation (5.105) includes an additional term due to the nucleation of singularity. The additional term represents some "permanent" circulation that occurs in the fluid flow.

From (5.105), it can be stated that if the circulation Γ vanishes for all closed curves C then it implies that the vorticity is zero and that the flow is irrotational. Regarding the integrand in (5.105), we notice that the converse is true if the fluid is contained in a simply connected region (which is the case when no singularity occurs, "no holes and no handles"). However, if the velocity field satisfies only the usual condition, which is sometimes considered as the curl of the velocity:

$$(\nabla_{\mathbf{e}_i} \mathbf{v}^*)(\mathbf{e}_j) - (\nabla_{\mathbf{e}_j} \mathbf{v}^*)(\mathbf{e}_i) \equiv 0,$$

the converse may be not true when singularity such as scalar and vector discontinuity occur since we can observe a "residual" circulation:

$$\Gamma_r = \int_C \aleph(\mathbf{e}_i, \mathbf{e}_j, \mathbf{v}^*) \neq 0.$$

Indeed, the fluid is multi-connected in this case (see Figure 5.10). Let us propose an extended version of isothermic barotropic fluid defined by defining the free energy and dissipation potential functions (we drop the thermal aspects for the sake of simplicity, without loss of generality):

$$\phi = \hat{\phi}(\omega_0, \aleph, \Re) \qquad \psi = \hat{\psi}(\zeta_g, \zeta_\aleph, \zeta_\Re). \tag{5.106}$$

Given the extended model (5.106), the vorticity is expressed by the relation

$$\omega_0(\text{rot } \mathbf{v}, \mathbf{e}_i, \mathbf{e}_j) = (\nabla_{\mathbf{e}_i}\mathbf{v}^*)(\mathbf{e}_j) - (\nabla_{\mathbf{e}_j}\mathbf{v}^*)(\mathbf{e}_i) + \aleph(\mathbf{e}_i, \mathbf{e}_j, \mathbf{v}^*). \tag{5.107}$$

Basically, this relation clearly points out that the fluid motion, here combined with rotation, is influenced by the singularity distribution, not solely by the velocity distribution. In the framework of normal dissipative mechanism, the presence of torsion fields allows the model to capture the vortices' source once the threshold of vortices creation is surpassed in view of (4.45):

$$\zeta_\aleph \in \partial\hat{\psi}_{\mathbf{J}_\aleph}(\mathbf{J}_\aleph, \mathbf{J}_\Re) \neq 0 \tag{5.108}$$

$$\zeta_\Re \in \partial\hat{\psi}_{\mathbf{J}_\Re}(\mathbf{J}_\aleph, \mathbf{J}_\Re) \neq 0. \tag{5.109}$$

Therefore, the continuous distribution of vortices may appear within the fluid bulk, not necessarily at the boundary. We can notice not only the irreversible aspect of the vortices but also the ability of the continuum model with singularity to capture the vortices' source in a fluid flow. The inclusions (5.108) and (5.109) (they become equations whenever the potentials are differentiable as quadratic functions) describe the creation of vortices which occur through discontinuous transformations (Figure 5.10). The creation of vortices and therefore the turbulence would be, in principle, forced by a discontinuity of slope (brusque variation) in (5.108) and (5.109). Conversely, the decay of vortex density, and turbulence, may be described by a continuously smooth phenomenon in the course of time.

For the constitutive functions (5.94) or the more particularized form (5.106), we have to rewrite the flux of the vortices field across any closed surface ∂U:

$$\Phi = \oint_{\partial U} h_0(\Omega) = \oint_{\partial U} h_0(\text{rot } \mathbf{v}) \tag{5.110}$$

in which we use the ω_0-isomorphism (3.3) of the vorticity. The application of Stokes' theorem implies, provided relation (3.6),

$$\Phi = \oint_{\partial U} h_0(\Omega) = \int_U dh_0(\Omega) = \int_U \text{div } \Omega\omega_0. \tag{5.111}$$

In the case in which the flux is zero, we obtain again the intrinsic condition, assessing that the vorticity field is solenoidal, e.g., [195]:

$$\text{div } \Omega = 0. \tag{5.112}$$

This equation has often been used for classical models of fluids without singularity. In the present study, projecting relation (5.112) on the spatial vectorial basis $(\mathbf{e}_1, \mathbf{e}_2, \mathbf{e}_3)$, and provided (4.15), gives

$$\overline{\nabla}_{\mathbf{e}_i} \Omega^i + \aleph^j_{ij} \, \Omega^i = 0. \tag{5.113}$$

Relation (5.113) constitutes the local version of the more usual form of vortex circulation in a multiply connected region, e.g., [195]. This is illustrated in Figure 5.10. As for the mass and linear momentum conservation equations and the heat propagation equation, the appearance of a source term can be observed but again it is implicitly included in the nonsymmetric part of the affine connection. A question arises about the existence of this vortex' source following a change of local topology, e.g., [174] and which plays a fundamental role in the vortex theory of fluids in the same way as in the dislocation theory of solids.

Remark. As an extension, the transition from an initial laminar state (no singularity) to a final turbulent state is due (at least in part) to a change of topology in the fluid bulk. In other words, as a consequence of the tangential discontinuity of the velocity field, the transition to turbulence involves transformations that are not of class C^2. From another point of view, change of topology may be considered as a route to turbulent flow. Beyond the threshold of vortex creation, the motion so generated cannot necessarily be described by homeomorphism. The fluid topology in the deformed state is no longer the same as in the initial state, due to the creation of singularity.

5.2.4 Conservation laws

The intrinsic form of conservation laws has been developed in Chapter 3. The present section is devoted to the development of the component form of these laws for fluids. For this purpose we choose a spatial (coordinate) basis such that $\aleph^k_{0ij} \equiv 0$ in the course of time and a (sym)-metric connection $\overline{\nabla}$. In the following equations, the resulting affine connection ∇ is not symmetric with respect to the lower indices because the torsion tensor is not null, due to nucleation and increasing of singularity within the bulk fluid. Therefore, the antisymmetric part of the connection reduces to the torsion tensor such that $\Gamma^k_{ij} = \overline{\Gamma}^k_{ij} + \aleph^k_{ij}$. The spatial connection $\overline{\Gamma}^k_{ij}$ is chosen initially as being symmetric (and metric) for convenience, depending on the symmetry of the problem, while the antisymmetric part $\aleph^k_{ij}$ has to be calculated according to the creation and evolution of singularity. The conservation laws in this particular case may be written directly in component forms as follows since the intrinsic formulation is given in the previous chapter.

1. The conservation of mass is obtained from (4.33):

$$\frac{\partial \rho}{\partial t} + \overline{\nabla}_{\mathbf{e}_i}(\rho v^i) = -\rho \aleph^i_{ij} \, v^j. \tag{5.114}$$

It should be noticed that the presence of a term in the r.h.s. of the equation (5.114) does not imply a nonconservation of the mass locally.

2. The conservation of linear momentum is derived from (4.34) to give

$$\rho\left(\frac{\partial v^i}{\partial t} + v^j \overline{\nabla}_{e_j} v^i\right) = \overline{\nabla}_{e_i}\sigma^{ij} + \rho b^i + \aleph^j_{jl}\sigma^{il} - \rho\aleph^k_{jk}v^j v^i \qquad (5.115)$$

provided that $\aleph^i_{jl}\sigma^{lj} = 0$, due to the skew symmetry of $\aleph^i_{jl}$ for lower indices and to the symmetry of the stress for upper indices. The stress σ may be decomposed as a hydrostatic pressure and a viscous stress:

$$\sigma^{ij} = -pg^{ij} + J_g^{ij}.$$

We recover, of course, the classical equations of motion in the absence of scalar and vector singularity fields. Moreover, we must point out that some additional forces occur in the dynamic equation (5.115). They are generated by the singularity distribution multiplied by the stress and by the singularity distribution multiplied by the velocity field (a kind of drag force).

3. The energy conservation is deduced from the equation (4.35) to give the heat propagation equation:

$$
\begin{aligned}
\rho C\left(\frac{\partial\theta}{\partial t} + v^i\overline{\nabla}_{e_i}\theta\right) &= -\overline{\nabla}_{e_i}J_q^i - \rho\theta\frac{\partial}{\partial\theta}\left(\frac{p}{\rho}\right)g^{ij}\zeta_{gij} + J_g^{ij}\zeta_{gij} \\
&+ \quad r + \left[J_{\aleph k}^{ij} - \rho\theta\frac{\partial}{\partial\theta}\left(\frac{J_{\aleph k}^{ij}}{\rho}\right)\right]\zeta_{\aleph ij}^k \\
&+ \quad \left[J_{\Re k}^{ij} - \rho\theta\frac{\partial}{\partial\theta}\left(\frac{J_{\Re k}^{ij}}{\rho}\right)\right]\zeta_{\Re ij}^k \\
&- \quad \rho C\theta v^i\aleph_{ij}^j - \aleph_{ij}^j J_q^i .
\end{aligned}
\qquad (5.116)
$$

Again, various additional sources of heat may be noted when compared to the heat propagation equation within a classical fluid. These volume sources are on the one hand due to the rates of singularity density (second and third lines of (5.116)) and on the other hand due to the singularity distributions (fourth line of (5.116)).

Remarks. Direct comparison of the present nonclassical fluid models and the complex fluid flow such as turbulence needs more development and is beyond the scope of this book. However, it is tempting to give some remarks, which should be considered as suggestions rather as conjectures. Consider the very basic model of turbulent fluid flow, the $K - \varepsilon$ model. First, this model assumed the superposition of the disturbing velocity field $\mathbf{v}'$ to the basic velocity field $\mathbf{v}$. Conservation laws for the $K - \varepsilon$ model are resumed to the mass and momentum balance, by assuming that the perturbing density $\rho' \sim 0$ and also its partial time derivative $\frac{\partial \rho'}{\partial t} \sim 0$ (these assumptions are not important for the purposes of this remark section):

$$\frac{\partial \rho}{\partial t} + \overline{\mathrm{div}}\,(\rho \mathbf{v}) = -\overline{\mathrm{div}}\,(\rho \mathbf{v}')$$

$$\rho \left(\frac{\partial \mathbf{v}}{\partial t} + \overline{\nabla}_{\mathbf{v}} \mathbf{v} \right) = \overline{\mathrm{div}}\,(\sigma - \rho \mathbf{v}' \otimes \mathbf{v}') + \rho \mathbf{b}.$$

In addition to the basic constitutive laws relating linearly the stress tensor σ and the rate of deformation $\mathbf{D} = \frac{1}{2}\left(\nabla \mathbf{v} + \nabla \mathbf{v}^T\right)$, say $\sigma = \lambda' \mathrm{tr}\,\mathbf{D}\mathbf{I} + 2\mu'\mathbf{D}$, a $K - \varepsilon$ model defines a supplementary constitutive law between the local disturbing kinetic energy and the disturbing strain rate:

$$K \equiv \frac{1}{2}\rho \mathrm{tr}\,(\mathbf{v}' \otimes \mathbf{v}') \qquad \varepsilon \equiv 2\mu \mathrm{tr}\,(\mathbf{D}'^2) \qquad K = f(\varepsilon).$$

For more details on thermomechanic theory of turbulent flow, readers should refer to specialized lectures, e.g., [28]. We just like to recall the conservation laws for the nonclassical fluid models here developed as

$$\frac{\partial \rho}{\partial t} + \overline{\mathrm{div}}(\rho \mathbf{v}) = -\rho \tilde{\aleph}(\mathbf{v})$$

$$\rho \left(\frac{\partial \mathbf{v}}{\partial t} + \overline{\nabla}_{\mathbf{v}} \mathbf{v} \right) = \overline{\mathrm{div}}\sigma + \sigma(\tilde{\aleph}) - \rho \mathbf{v} \otimes \mathbf{v}(\tilde{\aleph}) + \rho \mathbf{b}$$

in which we have defined the 1-form $\tilde{\aleph} \equiv \aleph^j_{ij}\mathbf{u}^i$.

1. First, we can observe the influence of discontinuity of the velocity field $\rho \mathbf{v} \otimes \mathbf{v}(\tilde{\aleph})$ in the previous equation (right-hand side equation) which has nearly the same role as the turbulent Reynolds stress tensor $\overline{\mathrm{div}}(\rho \mathbf{v}' \otimes \mathbf{v}')$. The presence of the 1-form $\tilde{\aleph}$ reflects no more than a discrete gradient due to singularity distribution. It should be noted that the nonclassical fluid models include a second term due to the discontinuity of velocity field, $\sigma(\tilde{\aleph})$. Physical interpretation of this term remains difficult in fluid but is much clearer in the framework of dislocation theory in deformable solids, e.g., [152].

2. Second, constitutive laws associated to the singularity distribution is derived from a dissipation potential:

$$\zeta_{\aleph} \in \partial \psi^*_{\mathbf{J}_{\aleph}}(\mathbf{J}_{\aleph}, \mathbf{J}_{\Re}).$$

Such a model may be designed to include the threshold behavior of fluid flow (change of flow state when "surpassing the Reynolds number") although this task is not so easy in practice. Another point is that these nonclassical fluid models then fullfill the entropy inequality after choosing an appropriate potential of dissipation. Indeed, by applying the usual method of Coleman and Noll, it has been shown that some anisotropic and nonlinear turbulence models based on $K - \varepsilon$ are thermodynamically not admissible, depending on the shape of the closure function $K = f(\varepsilon)$, e.g., [173]. It could be suggested that future works may gain clarity in comparing more systematically the present nonclassical fluid models to some previous turbulent models, since the discontinuity of velocity in the bulk fluid is in turn closely tight to nucleation of vortices in the fluid volume.

5.2.5 High grade fluids and fluid with singularity

Besides Noll's simple material theory [146], continuum models, in which higher gradients of the metric tensor and temperature entered into constitutive equations, were proposed long ago. The fluids are called viscous fluids of high grade. Different authors, e.g., [73] implicitly admitted the density gradient as primal variables to capture the long range spatial dependence in constitutive laws. A classical example is given by a constitutive law conceived by Korteweg, e.g., [99], [194]:

$$
\begin{aligned}
\sigma &= [\lambda \text{tr } \zeta_g - p + \lambda_1 \text{tr } (\nabla^2 \rho) + \lambda_2 \|\nabla \rho\|^2] \mathbf{i} + 2\mu \zeta_g \\
&+ \lambda_3 (\nabla \rho \otimes \nabla \rho) + \lambda_4 \nabla^2 \rho.
\end{aligned} \tag{5.117}
$$

The density ρ has been chosen as primal variable instead of the volume form ω_0 projected on the spatial vector base.

Such material can model the capillary effects of fluids and more generally the liquid-vapor transition phase where the density $\rho \propto \sqrt{\det(\mathbf{g})}$ has strong discontinuity. The Korteweg constitutive model depended not only on temperature and on density but also depended on the first and second gradient of the density. Combining the ideas of Korteweg with the theory of Maxwell on the thermal transpiration (i.e., production of gaseous motion and nonequilibrated stresses), Truesdell proposed general constitutive laws [194] where the gradients (first and second order) of density and temperature are also considered as primary variables.

Another problem arises when considering fluids in which abrupt change of density can occur. The incompatibility between the formulation of constitutive laws for fluid-like materials such as "Maxwellian fluid" [194] and classical continuum thermodynamics is a matter of fact. Previous studies, e.g., [45] pointed out the incompatibility of (5.117) with classical formulation of conservation laws (except if all coefficients associated to the density gradient are identically null).

The problem is not in fact specific to fluid mechanics. In the domain of solid mechanics, the dependence of energy, heat flux, and stress on the metric and temperature and on a certain number of their higher gradients was proposed by Toupin [189] and

Green and Rivlin [70]. The concept of an elastic simple material was extended systematically by Eringen [52] and latter by Suhubi [186] to include the current value of the metric tensor $\mathbf{g}$ (projected on the deformed tangent space $g_{ab} = \mathbf{g}(\mathbf{u}_a, \mathbf{u}_b)$) and its first p-gradients as well as the current value of the temperature θ and its q-first gradients, as constitutive variables. For these constitutive laws to be admitted, the Clausius–Duhem inequality together with Coleman and Noll's method should be verified [33]. They proposed a rigorous procedure by which the laws of thermodynamics could be used to obtain constitutive laws restrictions on a large variety of materials.

In this way, Gurtin has fundamentally established [73] that, even if the constitutive functions (energy, entropy, heat flux, and stress in his paper) were permitted to depend on the current temperature and metric fields throughout the continuum B, the application of Coleman and Noll's method allowed only a restricted dependence on the current values of temperature θ and metric tensor $\mathbf{g}$ (and additionally of the volume form ω_0 in the present approach). This very important result was confirmed by the investigation of Eringen stating that the dependence on higher gradients of temperature and the metric tensor was not permitted [52].

All of the constitutive laws previously quoted, and therefore the specified theories deduced from them, have been shown to be incompatible with the usual form of thermomechanic laws [45] when the restrictions imposed by the Coleman and Noll's method are applied. In, e.g., [45], Dunn and Serrin proposed a nonclassical thermomechanics theory. They introduced ad hoc an interstitial power in the equation of energy conservation and in the entropy inequality. The approach has been extended in [46]. They proposed constitutive laws in which the internal energy, the entropy, the stress, and the heat flux are given by smooth functions of the type[4]

$$\Im = \hat{\Im}(\mathbf{g}, \theta, \nabla\mathbf{g}, \nabla^2\mathbf{g}, \zeta_g, \zeta_q). \tag{5.118}$$

In their original work, Dunn and Serrin [45] and Dunn [46] utilized the deformation gradient instead of the metric tensor. When applying the Coleman and Noll method, the material of Korteweg reduces to

$$\Im = \hat{\Im}(\mathbf{g}, \theta, \zeta_g, \zeta_q) \tag{5.119}$$

if the usual form of thermodynamic laws are used. In their paper [45] and [46], a nonclassical form of continuum thermodynamics is proposed by postulating that there

---

[4]For another point of view, e.g., Batra [9] has suggested a further dependence of the constitutive dual variables on the rate of temperature and has defined an equivalent temperature variable $\theta^*(\theta, \zeta_\theta)$, in addition to the metric, the temperature, and their gradients. A theory of nonsimple material thermomechanics has thus been developed based on the constitutive laws $\Im = \hat{\Im}(\mathbf{g}, \theta, \nabla\mathbf{g}, \nabla\theta, \zeta_\theta)$. Without going into details, it has been shown mainly that either the derivative $\frac{\partial\theta^*}{\partial\zeta_\theta} = 0$ and then thermal disturbances propagate with infinite speed, and constitutive laws writes $\Im = \hat{\Im}(\mathbf{g}, \theta, \nabla\mathbf{g}, \nabla\theta, \zeta_\theta)$ (nonsimple materials), or $\frac{\partial\theta^*}{\partial\zeta_\theta} \neq 0$ and thus finite speed of thermal disturbances occur. But the Coleman and Noll method imposes independence from higher metric gradients to give $\Im = \hat{\Im}(\mathbf{g}, \theta, \nabla\theta, \zeta_\theta)$. This result may bring new insight for the analysis.

exists an interstitial working flux W_i such that the balance of energy is modified. The interstitial working (rate of W_i) is assumed to be induced (and therefore measures) by the long range spatial interactions. The conservation laws (motion law, entropy inequality) therefore become

$$\operatorname{div} \sigma + \rho \mathbf{b} = \rho \mathbf{a}$$
$$-\rho(\dot{\psi} + s\dot{\theta}) + \sigma : \zeta_g + \operatorname{div} \mathbf{u} - \mathbf{J}_q : \zeta_q \geq 0 \qquad (5.120)$$

where the divergence operators are the usual spatial divergence defined from a metric connection. Additionally, the interstitial work flux becomes a dual variable:

$$W_i = \hat{W}_i(\mathbf{g}, \theta, \nabla \mathbf{g}, \nabla^2 \mathbf{g}, \zeta_g, \zeta_q). \qquad (5.121)$$

They solved the incompatibility problem by introducing a frame indifferent work quantity. It should be noticed that a point that remains unclear is the use of the metric gradient $\nabla \mathbf{g}$ and eventually of other higher metric gradients, as constitutive variables. Indeed, the connection used in all of these theories is implicitly the Levi-Civita connection based on the fundamental metric compatibility:

$$\nabla \mathbf{g} \equiv 0. \qquad (5.122)$$

This is a fact. Of course, this may be not the case for the gradients of deformations at first sight. But the rigorous analysis of the combination of the frame indifference axiom and the use of higher gradients of deformation should be addressed before asserting that these higher gradients are admissible constitutive variables within the chosen spacetime structure, where the continuum evolves. In this book, instead of using a second grade material, e.g., [52], [63], [114], [194] (gradient of metric tensor) to capture the effects of the gradient of gradient, we prefer to introduce geometrical variables such as a tensor metric and separately an affine connection (together with torsion and curvature) which captures the field singularity. In addition, by extending the geometrical internal structure of the continuum, we observe that the divergence operator must be redefined accordingly. The present study shows that when using intrinsic divergence and rotational operators, the existence of such interstitial working may be obtained by deduction. The use of basic geometrical variables (metric $\mathbf{g}$, volume form ω_0, and affine connection ∇) enables us, on the one hand, to capture the long range spatial dependence (up to a certain limit and at least on the first gradient of metric tensor and temperature) and on the other hand, to capture field singularity. The material model defined by (5.34) may be considered as a viscoelastic material of order three which permits us to describe complex phenomena such as jumps of tensor fields (scalar and vectorial).

5.3   Overview of fluid-like models

For fluid-like continua, it first seems obvious to use a spatial coordinate base $(\mathbf{e}_1, \mathbf{e}_2, \mathbf{e}_3)$ (directed by the referential Euclidean body Σ). Consequently, the constants of structure vanish identically at any time t and at any point M. Second, it is possible to choose different affine connections according to Table 5.3. The different cases are reported in the following table.

Vector Base	Constants of Structure	Connection	Torsion	Type of Material
$(\mathbf{e}_1, \mathbf{e}_2, \mathbf{e}_3)$	$\aleph^k_{0ij} \equiv 0$	$\overline{\Gamma}^k_{ij} \equiv \overline{\Gamma}^k_{ji}$	$\aleph^k_{ij} \equiv 0$	No singularity
$(\mathbf{e}_1, \mathbf{e}_2, \mathbf{e}_3)$	$\aleph^k_{0ij} \equiv 0$	$\overline{\Gamma}^k_{ij} \equiv \overline{\Gamma}^k_{ji}$	$\aleph^k_{ij} \neq 0$	Not coherent
$(\mathbf{e}_1, \mathbf{e}_2, \mathbf{e}_3)$	$\aleph^k_{0ij} \equiv 0$	$\Gamma^k_{ij} \neq \Gamma^k_{ji}$	$\aleph^k_{ij} \equiv 0$	Not coherent
$(\mathbf{e}_1, \mathbf{e}_2, \mathbf{e}_3)$	$\aleph^k_{0ij} \equiv 0$	$\Gamma^k_{ij} \neq \Gamma^k_{ji}$	$\aleph^k_{ij} \neq 0$	$\Gamma^k_{ij} - \Gamma^k_{ji} = \aleph^k_{ji}$

Table 5.3. Types of fluid-like continua with singularity or not.

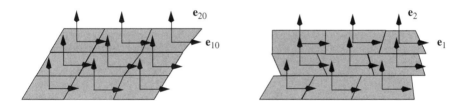

Figure 5.11. An initial coordinate base, endowed by the referential solid is chosen and, after deformation, another coordinate base directed by the referential solid is retained. During the fluid flow, the constants of structure associated to these bases vanish, even in the presence of vortex and other singularity. The model presented here can capture the flow of a granular material assumed to remain a continuous medium.

1. **No singularity distribution.** When no internal slip surfaces (continuous distribution) exist within the fluid, then the cases 1 (row 1) and 3 (row 3) hold. Model 1 (row 1) with a choice of symmetrical connections reduces to a classical fluid model for which there is no nucleation of singularity (no internal slip surfaces). Model 3 (row 3) is not geometrically coherent when regarding the definition and components of the torsion tensor.

2. **With singularity distribution.** In the presence of singularity distribution, the choice of symmetric connection is not suitable since incoherent (Model 2 (row 2)) and it is necessary to choose a nonsymmetric affine connection such that $\Gamma^k_{ij} - \Gamma^k_{ji} = \aleph^k_{ij}$. For the practical problem solving, integration of constitutive equations (4.45) must be done to obtain $\aleph$ and $\Re$. Model 4 (row 4) is the only one suitable for fluid with singularity among the four presented for which the torsion tensor and the curvature tensor become additional constitutive primal variables.

To sum up, the two situations that are coherent with the torsion definition show that either the model does not allow internal slip to occur (Model 1) or that models usually proposed in existing works are not complete since we have to account for the nonsymmetric part of the connection. Model 4 is then suitable to capture the fluid with singularity distribution. This aspect is crucial and has been discussed in the section concerning the viscous fluid of high grade, e.g., [179]. Once again, we face here the incoherence of some fluid flow models in the presence of internal slips. As a possible extension of the present model, since the turbulence in fluid flows is a statistical distribution of vortices, an approach of the turbulence through the continuum with singularity ("weakly continuous medium") could bring new insight for understanding and hopefully better modeling and classifying dissipation in turbulent flow.

6
Thermoviscous Solids

In everyday language, materials are considered as solid if they have different responses when they are deformed from different initial configurations. This rough definition will be refined all along this chapter to include nonholonomic deformations. For clarity's sake, let us give an overview of the different models of deformable solids with singularity. A choice of bases and affine connections allows us to classify all possible models.

6.1   Solids without singularity distribution

6.1.1   Constitutive laws

Consider continua of the rate type defined by the constitutive functions (4.1) for which any deformation remains geometrically holonomic $\aleph \equiv 0$ and $\Re \equiv 0$, e.g., [71], [130], [183].

Definition 6.1.1 *(Classical solids) In the class of thermoviscoelastic materials of rate type (4.37), a material is said to be a strongly continuous solid, without singularity distribution, if its constitutive functions are defined by*

$$\Im = \hat{\Im}(\omega_0, \mathbf{g}, \aleph \equiv 0, \Re \equiv 0, \theta, \zeta_g, \zeta_q) \qquad \Im = \rho, \sigma, s, \phi, \mathbf{J}_g, \mathbf{J}_q. \tag{6.1}$$

The presence of the metric tensor $\mathbf{g}$ in the list of arguments indicates a stress reaction of the continuum to change of shape or size, which is not the case for fluid model. Each constitutive function has, additionally, as arguments the components of the metric tensor in a basis embedded in the referential body $\mathbf{i} = \delta_{ab} \, \mathbf{e}^a \otimes \mathbf{e}^b$ [194]. For a normal dissipative solid model, which is a particular class of thermoviscous solids, the constitutive laws are summarized in the Helmholtz free energy function and in the dissipation potential in terms of the constitutive primal variables (intrinsic formulation):

$$\phi = \hat{\phi}(\omega_0, \mathbf{g}, \aleph \equiv 0, \Re \equiv 0, \theta) \qquad \psi = \hat{\psi}(\zeta_g, \zeta_q). \tag{6.2}$$

When not necessary, we do not mention hereafter the arguments of the functions ϕ and ψ to avoid cumbersome notation. According to the classical scheme, their derivatives with respect to primal variables provide the constitutive dual variables:

1. for mechanics, the total stress and the viscous stress:

$$\sigma = \rho \frac{\partial \phi}{\partial \omega_0} : \omega_0 \mathbf{i} + 2\rho \frac{\partial \phi}{\partial \mathbf{g}} + \mathbf{J}_g \qquad \mathbf{J}_g = \frac{\partial \psi}{\partial \zeta_g}$$

2. for thermics, the heat flux vector and the entropy:

$$\mathbf{J}_q = \frac{\partial \psi}{\partial \zeta_q} \qquad s = -\frac{\partial \phi}{\partial \theta}.$$

The main difference with the fluid model is the presence of the metric tensor in the list of arguments, which implies a response of the solid under the change of shape and size.

Example. The simplest solid model is derived from the two quadratic potentials with respect to the variables $\mathbf{g}$ and ζ_g:

$$
\begin{aligned}
\phi &= \frac{1}{2}\hat{\lambda}(\omega_0, \theta)\mathrm{tr}^2\left(\frac{\mathbf{g} - \mathbf{I}}{2}\right) + \frac{1}{2}\hat{\mu}(\omega_0, \theta)\mathrm{tr}\left(\frac{\mathbf{g} - \mathbf{I}}{2}\right)^2 \\
\psi &= \frac{1}{2}\hat{\kappa}(\omega_0, \mathbf{g}, \theta)\|\zeta_q\|^2 + \frac{1}{2}\hat{\lambda}'(\omega_0, \mathbf{g}, \theta)\mathrm{tr}^2\zeta_g + \frac{1}{2}\hat{\mu}'(\omega_0, \mathbf{g}, \theta)\mathrm{tr}\zeta_g^2 \quad (6.3)
\end{aligned}
$$

which define a Kelvin–Voigt material with an elastic behavior of Kirchhoff–St-Venant, a heat conduction law of Fourier and finally a dissipative viscosity of Newton.

Various forms of free energy and dissipative potentials have been successfully applied to model the mechanical behavior of viscoelastic biological tissues undergoing large deformations, e.g., [157]. For solids undergoing large strains, this does not imply linear behavior of responses. Various classes of classical solids are recovered through the choice of special forms for the free energy and the dissipation potential. Most of existing solid models are obtained in this framework of normal dissipative continuum.

Free energy and dissipation potential	Stress	Viscous stress and heat flux
$\phi = \hat{\phi}(\omega_0, \mathbf{g})$ $\psi \equiv 0$	$\sigma = \rho \frac{\partial \phi}{\partial \omega_0} : \omega_0 \mathbf{i} + 2\rho \frac{\partial \phi}{\partial \mathbf{g}}$	$\mathbf{J}_g \equiv 0$ $\mathbf{J}_q \equiv 0$
$\phi = \hat{\phi}(\omega_0, \mathbf{g})$ $\psi = \hat{\psi}(\zeta_g)$	$\sigma = \rho \frac{\partial \phi}{\partial \omega_0} : \omega_0 \mathbf{i} + 2\rho \frac{\partial \phi}{\partial \mathbf{g}} + \mathbf{J}_g$	$\mathbf{J}_g = \frac{\partial \psi}{\partial \zeta_g}$ $\mathbf{J}_q \equiv 0$
$\phi = \hat{\phi}(\omega_0, \mathbf{g}, \theta)$ $\psi = \hat{\psi}(\zeta_q)$	$\sigma = \rho \frac{\partial \phi}{\partial \omega_0} : \omega_0 \mathbf{i} + 2\rho \frac{\partial \phi}{\partial \mathbf{g}}$	$\mathbf{J}_g \equiv 0$ $\mathbf{J}_q = \frac{\partial \psi}{\partial \zeta_q}$
$\phi = \hat{\phi}(\omega_0, \mathbf{g}, \theta)$ $\psi = \hat{\psi}(\zeta_g, \zeta_q)$	$\sigma = \rho \frac{\partial \phi}{\partial \omega_0} : \omega_0 \mathbf{i} + 2\rho \frac{\partial \phi}{\partial \mathbf{g}} + \mathbf{J}_g$	$\mathbf{J}_g = \frac{\partial \psi}{\partial \zeta_g}$ $\mathbf{J}_q = \frac{\partial \psi}{\partial \zeta_q}$

Table 6.1. Particular laws for classical solids. (Row 1) Elastic solid with a stress reaction to shape change. (Row 2) Viscoelastic solid with a stress reaction to shape change and to velocity gradient. (Row 3) Heat conducting thermoelastic solid. (Row 4) Heat conducting viscoelastic solid.

6.1.2 Decomposition of constitutive laws on a vector basis

To solve solid mechanics problems it is necessary to project the intrinsic equations on an appropriate vector basis and to use an appropriate affine connection. In classical solid mechanics, on the one hand, the decomposition of conservation and constitutive laws requires a choice of a vector basis in the deformed configuration, say $(\mathbf{f}_1, \mathbf{f}_2, \mathbf{f}_3)$, which may be embedded in B. The corresponding initial basis is chosen in such a way that the Lie–Jacobi bracket is null:

$$[\mathbf{f}_{a0}, \mathbf{f}_{b0}] = 0 \qquad \forall a, b = 1, 2, 3.$$

It is always possible to use an embedded basis $(\mathbf{f}_1, \mathbf{f}_2, \mathbf{f}_3)$ such that $\mathbf{f}_a = d\varphi(\mathbf{f}_{a0})$ to project the basic equations because the placement of B is a homeomorphism from

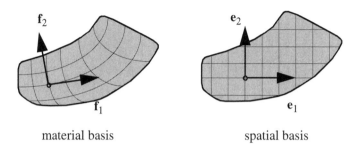

material basis   spatial basis

Figure 6.1. Vectorial basis (material and spatial). At least two vector basis fields can be defined on the continuum: the material basis, which is the embedded basis in the solid during its deformation, and the spatial basis, which is directed by some spatial frame.

B_0 to B and may be used as a parametrization of, e.g., [160] (Figure 6.1). This choice is worthwhile although not compulsory. In such a case, the deformed vector basis has the following properties, at any time t for any particle M:

$$[\mathbf{f}_a, \mathbf{f}_b] = 0 \qquad \mathbf{f}_a \times \mathbf{f}_b = \omega_0(\mathbf{f}_a, \mathbf{f}_b, \mathbf{f}_c)\, \mathbf{f}^c \qquad \forall a, b = 1, 2, 3. \qquad (6.4)$$

The vector basis embedded in the continuum B is called the material basis. When the body B undergoes only strongly continuous transformations, i.e., without nucleation of singularity, the affine connection ∇ can be reduced to the Euclidean or at least to a metric connection of the referential body. The coefficients of these connections in the embedded vector basis Γ_{ab}^c vanish identically. The tangent basis $(\mathbf{f}_1, \mathbf{f}_2, \mathbf{f}_3)$ is directed initially by a Cartesian frame. This is merely a convenient way of proceeding not the rule. For constitutive primal variables, the component of the volume form and of its time derivative with respect to the continuum along the material basis is, the variables being x and t,

$$J_f = \omega_0(\mathbf{f}_1, \mathbf{f}_2, \mathbf{f}_3) \qquad \frac{d^B}{dt} J_f = J_f \operatorname{tr} \zeta_g. \qquad (6.5)$$

On the other hand, since only strongly continuous transformations can occur, the continuum B remains always affinely equivalent to the referential Euclidean body. Therefore, we can choose a Euclidean connection. The time derivative of the metric $\mathbf{g}$ and its time derivative with respect to B are given by

$$\mathbf{g} = g_{ab}\mathbf{f}^a \otimes \mathbf{f}^b \qquad \zeta_g = \frac{1}{2}\dot{g}_{ab}\mathbf{f}^a \otimes \mathbf{f}^b. \qquad (6.6)$$

The time derivative of the temperature θ with respect to B and the gradient of temperature are

$$\zeta_\theta = \frac{\partial \theta}{\partial t} \qquad \zeta_q = \theta \, \nabla_{\mathbf{f}_a} \left(\frac{1}{\theta} \right) \mathbf{f}^a. \tag{6.7}$$

For constitutive dual variables, the components of the stress tensor and those of the 1 "material" heat flux vector on a material basis are written, in which $\rho_0 \equiv \rho J_f$,

$$\frac{\sigma}{\rho} = \frac{\sigma^{ab}}{\rho_0} \mathbf{f}_a \otimes \mathbf{f}_b \qquad Q \equiv J_f \mathbf{J}_q = Q^a \mathbf{f}_a. \tag{6.8}$$

Remark. Components of the stress tensor projected onto the material basis (6.8) are strictly equal to those of the so-called second Piola–Kirchhoff stress which is commonly used in solid mechanics, e.g., [191]. This aspect was extensively investigated in an earlier work devoted to the theory of generalized tensor strain and stress measures, e.g., [36], [160]. The formulation of solid thermomechanics in term of matrices $(\sigma^{ab}, g_{ab}, Q^a)$ is called the convective description of the thermoelasticity, e.g., [183]. It is possible to visualize in a 2-dimensional case the physical interpretation of the different components of the stress tensor. For such a purpose, we consider an area element formed by two vectors of the embedded vector basis. Due to the fact that the continuum is endowed with a metric tensor, we can "raise" and "lower" the indices of the stress tensor by means of the metric tensor. However, it is essential to notice that the physical interpretation of these stress tensor components depends on the position (lower, upper) of the indices [160].

The components of the stress tensor (6.8) have been associated to the concept of generalized stress measures, in combination with strain measures, e.g., [36] although it has been shown that they may also be interpreted physically as nonorthogonal components of the so-called Cauchy stress tensor [160]. Some incompatibilities of the strain rate definition pointed out in [160] suggest to us the adoption of an approach based on the deformed state of the continuum which seems to be a more rigorous approach. Indeed, the concept of a "material" description based only on the initial configuration combined with a homeomorphism (topologically reversible) is quite restrictive from the theoretical point of view except if some other intermediate configurations are introduced, e.g., [111] to capture singularity such as dislocations and disclinations. The "material parametrization" (material basis, Euclidean or metric connection) gives the classical notation of large strain thermomechanics, e.g., [124], [130], [201]. Finally, use of material parametrization in (6.1) provides constitutive laws of (normal dissipative) thermoviscoelastic solids, as follows:

1. The free energy function and the dissipation potential (scalar) are written

$$\phi = \hat{\phi}(J_f, g_{ab}, \theta) \qquad \psi = \hat{\psi}(J_f, g_{ab}, \theta, \dot{g}_{ab}, \zeta_{qa}) \tag{6.9}$$

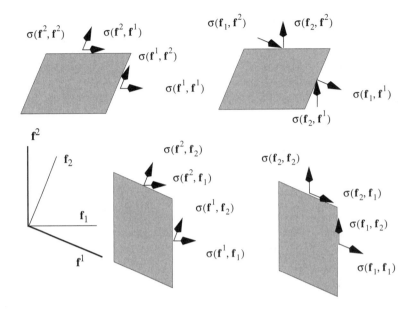

Figure 6.2. Various components of the stress tensor $(a, b = 1, 2)$. (a) $\sigma(\mathbf{f}^a, \mathbf{f}^b)$; (b) $\sigma(\mathbf{f}^a, \mathbf{f}_b)$; (c) $\sigma(\mathbf{f}_a, \mathbf{f}^b)$; (d) $\sigma(\mathbf{f}_a, \mathbf{f}_b)$. The various components have no meaning if the base onto which they have been decomposed is not specified. This may lead to a confusion and a misinterpretation as mentioned earlier, e.g., [160].

2. The heat flux (vector) is given by:

$$\mathbf{J}_q = \frac{\partial \psi}{\partial \zeta_q} \qquad \Longrightarrow \qquad Q = \hat{Q}(J_f, g_{ab}, \theta, \dot{g}_{ab}, \zeta_{qa}) \qquad (6.10)$$

3. For the stress tensor, the hydrostatic pressure and the viscous stress are decomposed on the material basis $(\mathbf{f}_1, \mathbf{f}_2, \mathbf{f}_3)$ as follows:

$$p \equiv -\rho \frac{\partial \phi}{\partial \omega_0} : \omega_0 \qquad \Longrightarrow \qquad \frac{p}{\rho} i = J_f \frac{p}{\rho_0} g^{ab} \mathbf{f}_a \otimes \mathbf{f}_b$$

$$\mathbf{J}_g = \frac{\partial \psi}{\partial \zeta_g} \qquad \Longrightarrow \qquad \mathbf{J}_g = 2 \frac{\partial \psi}{\partial \dot{g}_{ab}} \mathbf{f}_a \otimes \mathbf{f}_b.$$

Then, starting from the relation (4.10)2 the stress components on the material basis $(\mathbf{f}_1, \mathbf{f}_2, \mathbf{f}_3)$ of a Generalized Standard Material take the form:

$$\sigma^{ab} = -J_f p g^{ab} + 2\rho_0 \frac{\partial \phi}{\partial g_{ab}} + 2J_f \frac{\partial \psi}{\partial \dot{g}_{ab}}. \qquad (6.11)$$

Various models of materials can therefore be proposed by choosing two identified potentials ϕ and ψ. Both of them should satisfy convexity requirements.

6.1.3 Conservation laws

The derivation of conservation laws for solids without singularity distribution can be reduced to nonlinear elasticity approach. For solids, the mass conservation can be directly expressed from the integral invariance (Poincaré) to obtain:

$$\frac{d^B}{dt}\rho\omega_0 = 0 \qquad \rho(t)\omega_0(\mathbf{f}_1, \mathbf{f}_2, \mathbf{f}_3) = \rho(t_0)\omega_0(\mathbf{f}_{01}, \mathbf{f}_{02}, \mathbf{f}_{03}) = \rho_0. \tag{6.12}$$

The conservation of linear momentum is obtained by using the intrinsic divergence, the material acceleration (2.76), and the continuity equation (6.12) in a more concise form:

$$\rho\frac{\partial\mathbf{v}}{\partial t} = \overline{\operatorname{div}}\,\sigma + \rho\mathbf{b}. \tag{6.13}$$

Projection of (6.13) in a basis $(\mathbf{e}_1, \mathbf{e}_2, \mathbf{e}_3)$ embedded in the referential body leads to the first Piola–Kirchhoff stress matrix, labeled $\mathbf{P}$, the components of which are strictly equal to the terms under the divergence operator (6.13), e.g., [160], $\rho_0 \equiv \rho\,J_f$,:

$$J_f\sigma(\mathbf{f}^a) \equiv P^{ia}\,\mathbf{e}_i \qquad \rho_0\frac{\partial v^i}{\partial t} = \nabla_{\mathbf{f}_a}P^{ia} + \rho_0 b^i. \tag{6.14}$$

This later formulation has often been rewritten in the theory of large elastic deformation as $\rho_0\mathbf{a} = \operatorname{Div}\mathbf{P} + \rho_0\mathbf{b}$. The conservation of the moment of momentum implies the symmetry of the stress tensor σ:

$$\sigma(\mathbf{f}^a, \mathbf{f}^b) = \sigma(\mathbf{f}^b, \mathbf{f}^a) \qquad \sigma^{ab} = \sigma^{ba}. \tag{6.15}$$

The conservation of total energy may be written by applying the divergence operator and calculating the derivative of the internal energy, from (4.35),

$$\rho C\frac{\partial\theta}{\partial t} = -\overline{\operatorname{div}}\,\mathbf{J}_q + \rho\theta\frac{\partial}{\partial\theta}\left(\frac{\sigma}{\rho}\right) : \zeta_g + \left[\mathbf{J}_g - \rho\theta\frac{\partial}{\partial\theta}\left(\frac{\mathbf{J}_g}{\rho}\right)\right] : \zeta_g + r. \tag{6.16}$$

Projection of (6.16) onto $(\mathbf{e}_1, \mathbf{e}_2, \mathbf{e}_3)$ provides the heat propagation equation with the volume heat source $r_0 \equiv r J_f$, the heat capacity C, and the entropy s obtained with relation (4.36),

$$\begin{aligned}
\rho_0 C\frac{\partial\theta}{\partial t} &= -\nabla_{\mathbf{f}_a}Q^a + \rho_0\theta\frac{\partial}{\partial\theta}\left(\frac{\sigma^{ab}}{\rho_0}\right)\frac{\dot{g}_{ab}}{2} \\
&\quad + \left[J_f J_g^{ab} - \rho_0\theta\frac{\partial}{\partial\theta}\left(\frac{J_f J_g^{ab}}{\rho_0}\right)\right]\frac{\dot{g}_{ab}}{2} + r_0.
\end{aligned} \tag{6.17}$$

Remark. Extended Duhamel–Neumann hypothesis. The intrinsic theory applied to linear thermoelasticity allows us to revisit the famous Duhamel–Neumann hypothesis in linear thermoelastic solid models. Let $\phi(\mathbf{g}, \theta)$ be a free energy, a function of the metric and of the temperature. The stress tensor is given by the derivative

$$\sigma = 2\rho \frac{\partial \phi}{\partial \mathbf{g}} \qquad \tau \equiv \frac{\sigma}{\rho}.$$

The time derivative of stress (divided by the density) with respect to the continuum is written:

$$\frac{d^B \tau}{dt} = 2 \frac{\partial^2 \phi}{\partial \mathbf{g}^2} : \frac{d^B \mathbf{g}}{dt} + 2 \frac{\partial^2 \phi}{\partial \theta \partial \mathbf{g}} \frac{d^B \theta}{dt}.$$

Introducing the rate of $\mathbf{g}$, we obtain

$$\frac{d^B \tau}{dt} = 4 \frac{\partial^2 \phi}{\partial \mathbf{g}^2} : \zeta_g + 2 \frac{\partial^2 \phi}{\partial \theta \partial \mathbf{g}} \frac{d^B \theta}{dt}.$$

Let us now perform a partial Legendre transform to derive the complementary free energy (conjugate potential):

$$\phi^*(\tau, \theta) \equiv \frac{1}{2} \tau : \mathbf{g} - \phi(\mathbf{g}, \theta).$$

The derivative with respect to the tensor τ, followed by a time derivative with respect to the continuum, gives

$$\zeta_g \equiv \frac{1}{2} \frac{d^B \mathbf{g}}{dt} = \frac{\partial^2 \phi^*}{\partial \tau^2} : \frac{d^B \tau}{dt} + \frac{\partial^2 \phi^*}{\partial \theta \partial \tau} \frac{d^B \theta}{dt}.$$

This resembles the Duhamel–Neumann hypothesis which states that the total deformation rate is equal to the mechanical rate plus the thermal rate. This, therefore, extends the Duhamel–Neumann hypothesis from the infinitesimal linear thermoelasticity to nonlinear cases.

Remark. Plasticity and field discontinuity. The generic model (6.1) has been extensively used to analyze the elasticity and viscosity of solids, combined with heat production and propagation, e.g., [71], [124], [130]. This generic model is identified as the basic framework to investigate irreversible deformation such as the plastic behavior of materials, by introducing internal variables such as plastic strains to describe the evolution of internal structure, e.g., [125], [131]. An incoherence of the classical elastic-plastic continuum theory appears at this level.

In the particular case of plastic deformation without hardening, it has been shown, e.g., [131] that the velocity field, a vector field, has discontinuity across internal slip surfaces [4]. For a material obeying the von Mises criterion, it is easy to show that the jump of a velocity field [**v**] occurs if the second principal stress (middle eigenstress) is equal to half the sum of the two other principal stresses (smallest and greatest eigenvalues):

$$\sigma_{II} = \frac{1}{2}(\sigma_I + \sigma_{III}).$$

Therefore, the use of classical solid constitutive laws, even with the introduction internal variables, to model the elastic-plastic behavior is questionable. Indeed, it seems contradictory to assume on the one hand a "full" continuity of the velocity field and then of the displacement field (transformations remain homeomorphisms) and to obtain on the other hand a velocity discontinuity in these internal slip regions.

The purpose of the next paragraph is to propose an extended constitutive framework to model solids with singularity which is distributed continuously within the bulk. The proposed model is derived from a geometrical consideration in the same way as is Noll's theory on nonhomogeneous solids, e.g., [148], [200]. It has recently regained new interest by using affine connection, e.g., [163] and the concept of transplant mappings, e.g., [50].

6.2 Solids with singularity distribution

6.2.1 Compatibility conditions

The metric tensor **g** has six components that are expressible in terms of the three components of the displacement vector. According to our notation, the relationships take the form

$$g_{ab} = \mathbf{g}(\mathbf{u}_a, \mathbf{u}_b) = \mathbf{g}(d\varphi(\mathbf{u}_{0a}), d\varphi(\mathbf{u}_{0b})) \tag{6.18}$$

in which the vectorial base $(\mathbf{u}_1, \mathbf{u}_2, \mathbf{u}_3)$ is embedded in the continuum. It is simple to express the deformed embedded basis in terms of displacement components. Given the displacement field, we determine the metric tensor components by means of the relations (6.18). If on the other hand the six functions g_{ab} are given and claimed to be the components of the metric tensor, a question arises as to the existence of a single-valued continuous displacement field corresponding to these six functions. The six equations (6.18) constitute an overestimated system and must satisfy compatibility conditions for a solution to exist.

The physical interpretation of compatibility conditions may be pictured as follows. Let us cut the body into small cubes before it is strained and then give each cube a certain strain. Generally the strained cubes cannot be fitted back together to form a continuous body without further deformation (Volterra's process of dislocations). In linear elasticity, the compatibility conditions are attributed to St-Venant (1864). The proof that compatibility conditions are necessary and sufficient conditions for a single-valued displacement field in a simply connected continuum has originally been developed by Cesaro (1906) [24].

In the present work, it first should be noticed that classical compatibility conditions for nonlinear elasticity lead to the vanishing of the torsion tensor and the curvature tensor, e.g., [124], [130]. Secondly, the existence of scalar and vector singularity distribution implies that the torsion or/and the curvature may not vanish:

$$\aleph \neq 0 \qquad \Re \neq 0 \tag{6.19}$$

which are the incompatibility conditions for the displacement field.

6.2.2 Constitutive laws

Let us imagine an initial configuration for the solid without singularity distribution. After deformation, we find the presence of scalar discontinuity and vectorial discontinuity (6.19). To extend the definition of classical thermoviscous solids to defected solids, we once more propose the following definition which accounts for the bulk nucleation of dislocations and disclinations and for the increase of singularity density.

Definition 6.2.1 (*Solids with singularity distribution*) *Among the materials defined by the constitutive functions (4.1), a thermoviscous solid with singularity distribution is defined by the particular class of constitutive functions:*

$$\Im = \hat{\Im}(\omega_0, \mathbf{g}, \aleph, \Re, \theta, \zeta_g, \zeta_\aleph, \zeta_\Re, \zeta_q) \qquad \Im = \rho, \sigma, s, \phi, \mathbf{J}_g, \mathbf{J}_\aleph, \mathbf{J}_\Re, \mathbf{J}_q. \tag{6.20}$$

Derivation of conservation laws and of constitutive laws is quite similar to that of fluid with singularity, developed in previous chapter, except that we prefer to choose the embedded basis to project the basic equations. Furthermore, we adopt a Riemannian connection for the derivative operations.

Without going into the physical implications of the dislocation theory, the most fundamental consequence of the basic equations of singularity evolution is that the body loses its original topology and undergoes irreversible deformations beyond a stress threshold. In other words, the original simply connected continuum may become multiply connected. All topological properties of the continuum are invariant only with respect to processes (motion and temperature evolution) that are described by homeomorphism (continuous maps with continuous inverse images). For example, connectedness is one of the most important topological properties. The modification of topological properties we consider in the present study is governed by the usual equations of Generalized Standard Material with two potentials, the Helmholtz free energy and the dissipation potential:

$$\phi = \hat{\phi}(\omega_0, \mathbf{g}, \aleph, \Re, \theta) \qquad \psi = \hat{\psi}(\zeta_g, \zeta_\aleph, \zeta_\Re, \zeta_q). \tag{6.21}$$

The constitutive dual variables are derived from these potentials, the total stress, the hydrostatic pressure, and the viscous stress:

$$\sigma = -p\mathbf{i} + 2\rho\frac{\partial\phi}{\partial\mathbf{g}} + \mathbf{J}_g \qquad p = -\rho\frac{\partial\phi}{\partial\omega_0} : \omega_0 \qquad \mathbf{J}_g = \frac{\partial\phi}{\partial\zeta_g}$$

Remark. In dynamic processes, when irreversible deformation occurs in material, it can be postulated in view of relation (6.30) that a proportion of the mechanical power is converted into heat [4]. By the way, if the strain rate in the considered region is high, then there may not be enough time for the heat to diffuse away from the deforming zone. The effect would be a local thermal softening of the material. If the strength loss due to thermal softening becomes greater than the increase in strength due to strain or strain rate hardening, the irreversible deformation will become unstable. The term adiabatic is of course an oversimplified abbreviation of a complicated coupled (mechanical and thermodynamic) phenomenon such as we can observe in the relations (6.27), (6.28), (6.29), and (6.30). Conversely, stress concentration may be generated at the crack under uniform heat flux in any direction [78]. This illustrates the high thermal-mechanical coupling effects.

Remark. Strain-gradient plasticity. Recent literature has shown experimental evidence that materials display very strong size effects when the characteristic length scale associated with nonuniformity of plastic strain is of the order of a few micrometers. To model this size dependence, many authors have developed the so-called strain gradient plasticity theory, in which a third-order tensor:

$$\eta_{ijk} \equiv \frac{\partial u_k}{\partial x_i \partial x_j} = \frac{\partial \varepsilon_{ik}}{\partial x_j} + \frac{\partial \varepsilon_{kj}}{\partial x_i} - \frac{\partial \varepsilon_{ij}}{\partial x_k}$$

has been introduced as an additional variable to strain. Since we have the linearized relation $2\varepsilon_{ij} = g_{ij} - \delta_{ij}$, it is easy to show that:

$$\eta_{ijk} = \frac{1}{2}\left(\frac{\partial g_{ik}}{\partial x_j} + \frac{\partial g_{kj}}{\partial x_i} - \frac{\partial g_{ij}}{\partial x_k}\right) = \Gamma_{ijk}.$$

To develop constitutive laws that take the strain gradient into account, Fleck and Hutchinson, e.g., [57] have defined an equivalent strain gradient that measures the density of geometrically necessary dislocations:

$$\eta \equiv \sqrt{\frac{1}{4}\eta'_{ijk}\eta'_{ijk}} \qquad \eta'_{ijk} \equiv \eta_{ijk} - \frac{1}{4}(\delta_{ik}\eta_{jll} + \delta_{kj}\eta_{ill}).$$

It can be observed that this strain-gradient theory is a particular case of the present constitutive laws modeling the continuous distribution of singularity. Indeed, it is quite straightforward to rewrite the corresponding variables as:

$$\Gamma \equiv \sqrt{\frac{1}{4}\Gamma'_{ijk}\Gamma'_{ijk}} \qquad \Gamma'_{ijk} \equiv \Gamma_{ijk} - \frac{1}{4}(\delta_{ik}\Gamma_{jll} + \delta_{kj}\Gamma_{ill}).$$

However, since neither the torsion tensor nor the constants of structure vanish, the connection coefficients need not be symmetric with respect to the lower indices in the present theory. By analogy, on the microscale, the quantity Γ is a measure of the density of geometrically necessary dislocations. Analogous development for the curvature tensor seems to be missing in the literature. Recent publications report attempts to introduce a gradient of damage variables to develop new constitutive laws. These resemble the theory we have developed in the sense that the curvature is in a way (the geometrical way) a gradient of the torsion tensor.

6.3   Intermediate configurations

When the solid (6.20) undergoes an irreversible deformation resulting from a loss of affine equivalence with the Euclidean referential body, we recover the continuum model (4.37). Conservation laws for mass (4.33), linear momentum (4.34), and energy (4.35) still hold while constitutive equations for field singularity are obtained from both state laws (4.44) and evolution laws (4.45). When such is the case, the method based on the affine connection choice is analogous to the method based on the intermediate configuration choice: intermediate configuration and relaxed configuration.

Suppose the initial configuration B_0 is homogeneous—that is, free from singularity, both dislocations and disclinations—and without residual stresses (ideal initial configuration). In this initial configuration, the torsion field and the curvature field are null, at any point M:

$$\aleph^{(0)} = 0 \qquad \Re^{(0)} = 0. \tag{6.31}$$

Let B be the deformed configuration of the solid in which singularity occurs with a continuous distribution in the bulk. As there is nucleation of incompatibility, the torsion and curvature do not vanish anymore at the deformed configuration:

$$\aleph \neq 0 \qquad \Re \neq 0. \tag{6.32}$$

Various scenarios could be imagined to explain (and in fact to solve) the nonholonomic deformation of a solid in the course of time. Accordingly, various methods have been suggested to calculate singularity evolution step-by-step. Among them the methods of intermediate and relaxed configurations are the most usual.

6.3.1   Intermediate configurations

Physically, the intermediate configuration has been introduced to picture (and indeed corresponds to) an actual configuration of B without the distortion induced by singularity. Such a configuration is obtained by cutting the matter so as to relax all residual stress and make the pieces of matter homogeneous, e.g., [3], [95], [132]:

$$\aleph^{(i)} = 0 \qquad \Re^{(i)} = 0. \tag{6.33}$$

Integrating (6.33) provides for this intermediate configuration a certain connection $\nabla^{(e)}$, Euclidean or Riemannian (with null torsion and curvature) on B, fully obtained by a holonomic transformation from the initial configuration.

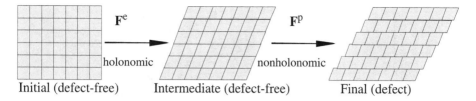

Initial (defect-free)   Intermediate (defect-free)   Final (defect)

Figure 6.3. Intermediate configurations: The linear tangent motion $d\varphi$, represented by the deformation gradient matrix $\mathbf{F}$, is decomposed into two transformations, the lattice deformation $\mathbf{F}^e$ and the dislocation deformation $\mathbf{F}^p$ such that $\mathbf{F} = \mathbf{F}^p\mathbf{F}^e$. Neither $\mathbf{F}^e$ nor $\mathbf{F}^p$ represents a real gradient.

6.3.2 Relaxed configurations

Consider again the initial configuration and the deformed configuration. Kröner [101] has proposed a relaxed configuration by first relaxing the deformed configuration B to remove the residual stresses. Then he introduced an affine connection in such a way that it is not modified when the material is deformed from the relaxed (sometimes called natural) configuration to the deformed configuration. In the natural configuration obtained by a nonholonomic transformation of the initial configuration, the torsion and the curvature tensors are not null:

$$\aleph^{(r)} \neq 0 \qquad \mathfrak{R}^{(r)} \neq 0. \tag{6.34}$$

Even when it is free from all external interactions, the body in its relaxed configuration $B^{(r)}$ may possess an energy that depends on the distribution of field singularity. Indeed, field singularity may be a source of residual stress that exists in the body in the absence of external loading. Residual stresses are in fact due to the incompatibility of the irreversible part of deformation with the Euclidean reference, e.g., [83].

In the same way, Kröner, Le, and Stumpf [111] determined the particular case of an anholonomic configuration (here labeled crystal configuration, but which they called a crystal reference since it may consist of a nonviable configuration) by means of the metric tensor (elastic strain) and the Cartan torsion tensor (density of translation dislocations). This configuration belongs to the class of relaxed configurations. The first basic assumption is the existence of a crystal configuration that is generally anholonomic and also time dependent. The linear tangent motion represented by the deformation (shape) gradient is decomposed into elastic (lattice) and plastic (dislocation) deformations as in:

$$\mathbf{F} = \mathbf{F}^e\,\mathbf{F}^p \qquad \det\mathbf{F} > 0 \qquad \det\mathbf{F}^e > 0 \qquad \det\mathbf{F}^p > 0$$

in which each map is assumed to be as many times continuously differentiable as required and also invertible.

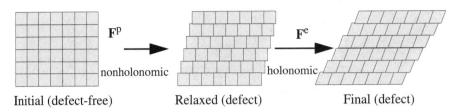

Initial (defect-free)   Relaxed (defect)   Final (defect)

Figure 6.4. Relaxed configurations: The linear tangent motion $d\varphi$, represented by the deformation gradient matrix $\mathbf{F}$, is decomposed into two transformations, the lattice deformation $\mathbf{F}^e$ and the dislocation deformation $\mathbf{F}^p$ such that $\mathbf{F} = \mathbf{F}^e\mathbf{F}^p$. Neither $\mathbf{F}^e$ nor $\mathbf{F}^p$ is a true gradient.

Example. For a crystal with a single slip system transcribed by the set of unit vectors $\mathbf{a}, \mathbf{b}$ attached to the lattice, the plastic flow $\mathbf{F}^p$ (plastic deformation) is given by the relation:

$$\mathbf{F}^p = \mathbf{I} + \gamma \, \mathbf{a} \otimes \mathbf{b}$$

in which γ is the plastic shearing on the crystallographic slip. Relative to the crystal configuration, a free energy is supposed to depend on the elastic deformation gradient and on its first derivatives with respect to the crystal configuration coordinates:

$$\phi = \hat{\phi}(\mathbf{F}^e, \mathbf{F}^e_\alpha).$$

In this energy, the introduction of the first derivative of the elastic gradient deformation $\mathbf{F}^e_\alpha$ leads to the second-order gradient material model. The frame indifference of the above free energy function imposes the appearance of the torsion tensor as an additional variable [111]:

$$\phi = \hat{\phi}(\mathbf{F}^e\mathbf{g}\mathbf{F}^{eT}, \aleph) \tag{6.35}$$

in which all arguments are relative to the crystal configuration. In this sense, the free energy per crystal volume unit depends only on the deformed configuration and the crystal configuration and it is insensitive to the change of the initial configuration (principle of initial scaling indifference [110]). It would be interesting to investigate using a similar method the plastic-elastic decomposition $\mathbf{F}^p \, \mathbf{F}^e$ so as to consider the intermediate reference [32], [132].

In practice, it is convenient to project all tensor fields (primal constitutive variables) onto a vector basis of the initial configuration labeled $(\mathbf{u}_{01}, \mathbf{u}_{02}, \mathbf{u}_{03})$ which becomes $(\mathbf{u}_{p1}, \mathbf{u}_{p2}, \mathbf{u}_{p3})$ with the dislocation transformation and finally into $(\mathbf{u}_1, \mathbf{u}_2, \mathbf{u}_3)$ in the deformed state. On the one hand, the components of the metric tensor projected onto any base vector in the crystal configuration are given by:

$$\mathbf{F}^e\mathbf{g}\mathbf{F}^{eT}(\mathbf{u}_{pa}, \mathbf{u}_{pb}).$$

On the other hand, the decomposition of the metric tensor on any deformed base vector is written as follows:

$$\mathbf{g}(\mathbf{u}_a, \mathbf{u}_b) = \mathbf{g}[\mathbf{F}^e(\mathbf{u}_{pa}), \mathbf{F}^e(\mathbf{u}_{pb})] = \mathbf{F}^e \mathbf{g} \mathbf{F}^{eT}(\mathbf{u}_{pa}, \mathbf{u}_{pb}) = g_{ab}. \tag{6.36}$$

Therefore, it should be pointed out that the components of the tensor $\mathbf{F}^e \mathbf{g} \mathbf{F}^{eT}$ on the crystal base are exactly the same as the components of the metric tensor but are projected on the actual deformed basis. Indeed, the vector base of the crystal configuration onto which the tensor $\mathbf{F}^e \mathbf{g} \mathbf{F}^{eT}$ is decomposed may be always chosen as the transformed base vector $(\mathbf{u}_{p1}, \mathbf{u}_{p2}, \mathbf{u}_{p3})$ of an initial base $(\mathbf{u}_{01}, \mathbf{u}_{02}, \mathbf{u}_{03})$. Their existence and uniqueness are ensured since both $\mathbf{F}^e$ and $\mathbf{F}^p$ are assumed continuously differentiable and invertible. The connection of the crystal configuration relative to the initial configuration is defined in component form from the plastic deformation gradient:

$$\Gamma^c_{ab} = (F^{p^{-1}})^c_\alpha (F^p)^\alpha_{b,a}.$$

The associated torsion tensor is accordingly deduced:

$$\aleph^c_{ab} = (F^{p^{-1}})^c_\alpha (F^p)^\alpha_{b,a} - (F^{p^{-1}})^c_\alpha (F^p)^\alpha_{a,b}.$$

The presence of a nonvanishing torsion tensor over the course of time reveals the change of topology within the continuum during the deformation. Finally in [111], a crystal connection for which the curvature should vanish to ensure the existence of the reference is proposed:

$$\mathfrak{R}^c_{abd} \equiv 0.$$

To sum up, the crystal configuration approach may be considered as a practical way to calculate the evolution of the singularity by means of the successive configurations. We can thus give explicitly in component form the metric and the torsion tensors. To check up the results, the linearization of the equations governing the location of the crystal reference allows us to compare the obtained crystal reference with the natural reference of Kröner [101]. The intrinsic formulation (independent from any base vector) of the free energy function (6.35) is given by:

$$\phi = \hat{\phi}(\mathbf{g}, \aleph, \mathfrak{R} \equiv 0).$$

We remark that the model of material defined by the Helmholtz free energy (6.35) constitutes a particular case of the free energy we have found in (6.21). Nevertheless, contrary to equations (6.24), which govern the nucleation and increase of dislocation and disclination densities, the requirement that the curvature should vanish seems to be a strong limitation for the concept of crystal reference to capture the vectorial singularity as disclination distribution within solids. In the face of equation (2.24) in Chapter 2, only translation dislocations and some particular kinds of rotation can be modeled with the concept of crystal configuration.

6.3.3   G-structure method

In the framework of relaxed configuration and in the geometric framework of G-structure, e.g., [114], the concept of relaxed configuration was studied by combining the notion of material uniformity and the multiplicative decomposition of the deformation gradient [51]. A second-grade elastic material is sensitive in its mechanical response (defined by the Helmholtz free energy) to the first and second gradient of deformation. In this theory, the integrability of the second-grade G-structure is equivalent to the local homogeneity of the material. Therefore, authors such as de Leon and coworkers have proposed that the Cartan torsion tensor $\aleph$ associated to the chosen crystal reference, defining the inhomogeneity, captures entirely the dislocations and the disclinations [114]. This is not the case in the present work since, first, we introduce the affine connection as an extra variable independently on the metric tensor and, second, the associated torsion and curvature must both be nonnull in presence of scalar and vectorial singularity (translation dislocations and rotational dislocations). It should be pointed out that the curvature alone does not model the disclinations when taking relation (2.20). A null curvature but nonvanishing torsion may even induce some particular classes of disclinations. Practically, for isotropic and anisotropic elastic solids, residual stress distribution can be determined by using stress functions, e.g., [17] and Green's functions, e.g., [134].

6.4   Overview of solid-like models

The method we adopt in the present book insists rather on the existence of an infinite number of intermediate configurations than on one intermediate or one relaxed configuration. Indeed, we calculate, in a continuous manner, the evolution of the torsion and of the curvature by considering their derivatives with respect to the continuum:

$$\aleph(t+dt) \;=\; \aleph(t) + \frac{d^B}{dt}\aleph(t)\,dt + \frac{d^{B2}}{dt}\aleph(t)\,\frac{dt^2}{2!} + \cdots \tag{6.37}$$

$$\Re(t+dt) \;=\; \Re(t) + \frac{d^B}{dt}\Re(t)\,dt + \frac{d^{B2}}{dt}\Re(t)\,\frac{dt^2}{2!} + \cdots \tag{6.38}$$

in which the rate of singularity density is given by inclusions (6.24). The resolution of the time derivative of the singularity density could be in practice replaced by an Euler discrete time integration as follows (implicit scheme $k = i+1$, explicit scheme $k = i$, or intermediate $k = i + \alpha$, $0 < a < 1$):

$$\aleph^{(i+1)} \;=\; \aleph^{(i)} + \zeta_\aleph^{(k)}(t_{i+1} - t_i) \qquad \aleph^{(0)} = 0 \tag{6.39}$$

$$\Re^{(i+1)} \;=\; \Re^{(i)} + \zeta_\Re^{(k)}(t_{i+1} - t_i) \qquad \Re^{(0)} = 0. \tag{6.40}$$

The most stable algorithm scheme for integrating the internal variables' evolution in the presence of a threshold stress value seems to be the implicit integration that has been extensively used in the modeling of elastic-plastic deformation of either isotropic, e.g., [145] or anisotropic material, e.g., [161]. From the integration scheme point of view, it is clear that the concept of intermediate or relaxed configuration when applied between the initial state and the final state can be considered as an algorithm to define the level and the distribution of singularity (either the dislocation or the disclination). The combination of the notion of intermediate configurations and of the broad class of integration algorithms can be found in, e.g., [181].

For the model we present in this book, the solid-like continua with singularity, we first choose a (non) coordinate base $(\mathbf{u}_1, \mathbf{u}_2, \mathbf{u}_3)$. Secondly, an affine connection has to be chosen to derive the conservation laws and constitutive laws of the solid model. Practically, when deriving the governing equations of solid thermomechanics, we have to choose an initial coordinate base $(\mathbf{u}_{01}, \mathbf{u}_{02}, \mathbf{u}_{03})$ which is transformed into a (non) coordinate base $(\mathbf{u}_1, \mathbf{u}_2, \mathbf{u}_3)$ such that $\mathbf{u}_a = d\varphi(\mathbf{u}_{0a})$. From Table 6.3, the four first cases occur if the transformation is a diffeomorphism (2.38). The four last cases occur when there is nucleation of singularity and an increase of density.

Vector base	Constants of structure	Connection	Torsion	Type of material
$(\mathbf{u}_1, \mathbf{u}_2, \mathbf{u}_3)$	$\aleph^c_{0ab} \equiv 0$	$\overline{\Gamma}^c_{ab} \equiv \overline{\Gamma}^c_{ba}$	$\aleph^c_{ab} \equiv 0$	No singularity
$(\mathbf{u}_1, \mathbf{u}_2, \mathbf{u}_3)$	$\aleph^c_{0ab} \equiv 0$	$\overline{\Gamma}^c_{ab} \equiv \overline{\Gamma}^c_{ba}$	$\aleph^c_{ab} \neq 0$	Not coherent
$(\mathbf{u}_1, \mathbf{u}_2, \mathbf{u}_3)$	$\aleph^c_{0ab} \equiv 0$	$\Gamma^c_{ab} \neq \Gamma^c_{ba}$	$\aleph^c_{ab} \equiv 0$	Not coherent
$(\mathbf{u}_1, \mathbf{u}_2, \mathbf{u}_3)$	$\aleph^c_{0ab} \equiv 0$	$\Gamma^c_{ab} \neq \Gamma^c_{ba}$	$\aleph^c_{ab} \neq 0$	$\Gamma^c_{ab} - \Gamma^c_{ba} = \aleph^c_{ab}$
$(\mathbf{u}_1, \mathbf{u}_2, \mathbf{u}_3)$	$\aleph^c_{0ab} \neq 0$	$\overline{\Gamma}^c_{ab} \equiv \overline{\Gamma}^c_{ba}$	$\aleph^c_{ab} \equiv 0$	Not coherent
$(\mathbf{u}_1, \mathbf{u}_2, \mathbf{u}_3)$	$\aleph^c_{0ab} \neq 0$	$\overline{\Gamma}^c_{ab} \equiv \overline{\Gamma}^c_{ba}$	$\aleph^c_{ab} \neq 0$	$\aleph^c_{ab} = -\aleph^c_{0ab}$
$(\mathbf{u}_1, \mathbf{u}_2, \mathbf{u}_3)$	$\aleph^c_{0ab} \neq 0$	$\Gamma^c_{ab} \neq \Gamma^c_{ba}$	$\aleph^c_{ab} \equiv 0$	$\Gamma^c_{ab} - \Gamma^c_{ba} = \aleph^c_{0ab}$
$(\mathbf{u}_1, \mathbf{u}_2, \mathbf{u}_3)$	$\aleph^c_{0ab} \neq 0$	$\Gamma^c_{ab} \neq \Gamma^c_{ba}$	$\aleph^c_{ab} \neq 0$	$\aleph^c_{ab} = (\Gamma^c_{ab} - \Gamma^c_{ba}) - \aleph^c_{0ab}$

Table 6.3. Types of solid-like continua with singularity.

Therefore, the following classification, based essentially on the values of the constants of structure and on the value of the torsion tensor, permits us to obtain the eight cases:

1. **Coordinate base.** When a coordinate base is chosen, we have the first four cases in the above table which are characterized by $\aleph_{0ab}^c \equiv 0$. When no singularity occurs, model 1 (row 1) reports the classical continuum. Models 2 and 3 (rows 2 and 3) are not geometrically coherent with respect to the torsion definition. Model 4 (row 4) must be given an affine connection such that $\Gamma_{ab}^c - \Gamma_{ba}^c = \aleph_{ab}^c$. The creation of vortex (torsion) is governed by the inclusions (6.24) combined with the integration schemes (6.39) and (6.40).

2. **Noncoordinate base.** When a noncoordinate base is chosen $\aleph_{0ab}^c \neq 0$, we recover the cases 5 to 8 (rows 5 to 8). Model 5 is not geometrically coherent. When no singularity occurs, model 5 may be based on the choice of the so-called Levi-Civita connection. By the way, a metric connection, i.e., $\nabla \mathbf{g} \equiv 0$, with a torsion identically null is called a Levi-Civita connection [144]. The coefficients of a Levi-Civita connection are given in any noncoordinate base [29]:

$$\overline{\Gamma}_{ab}^c = \frac{1}{2} g^{cd} [\mathbf{u}_b(g_{ad}) + \mathbf{u}_a(g_{db}) - \mathbf{u}_d(g_{ab})]$$
$$- \frac{1}{2}(\aleph_{0ba}^c + g^{cd} g_{ae}\aleph_{0bd}^e + g^{cd} g_{eb}\aleph_{0ad}^e) \tag{6.41}$$

in which the Christoffel symbols are classically defined as:

$$\{_{ab}^c\} = \frac{1}{2} g^{cd} [\mathbf{u}_b(g_{ad}) + \mathbf{u}_a(g_{db}) - \mathbf{u}_d(g_{ab})]. \tag{6.42}$$

Otherwise, the choice of a nonsymmetric torsion-free connection must satisfy $\Gamma_{ab}^c - \Gamma_{ba}^c = \aleph_{0ab}^c$. In the presence of singularity, it is possible to use a symmetric connection (model 6) and the torsion tensor will be the opposite of the constants of structure $\aleph_{ab}^c = -\aleph_{0ab}^c$. In this case, the first three terms of the r.h.s (6.41) are the Christoffel symbols (containing the metric property) and the last three terms represent the contortion tensor. The use of a nonsymmetric connection (last row) such that $\Gamma_{ab}^c - \Gamma_{ba}^c = \aleph_{ab}^c + \aleph_{0ab}^c$ is also possible, although complicated for practical use.

6.5   Elastic waves in nonclassical solids

Consider a continuum B endowed with a metric tensor $\mathbf{g}$, a volume form ω_0, and an affine connection ∇. The continuum is assumed to undergo small deformations and small displacements. The goal of this section is to develop the wave propagation equation in a linear elastic solid with a continuous distribution of singularity (microdefects) [162].

6.5.1 Wave propagation equations

Starting with the motion equation and the linear elastic stress-strain law, it is easy to obtain the Navier equation in the absence of body force, e.g., [124]:

$$\rho \frac{\partial^2 \mathbf{u}}{\partial t^2} = (\lambda + \mu)\nabla(\operatorname{div} \mathbf{u}) + \mu \Delta \mathbf{u}. \tag{6.43}$$

To analyze wave propagation in an isotropic elastic medium, it is usual to introduce the longitudinal and transversal velocities of sound:

$$c_L^2 = \frac{\lambda + 2\mu}{\rho} \qquad c_T^2 = \frac{\mu}{\rho}. \tag{6.44}$$

Introduction of (6.44) in (6.43) leads to the wave propagation equation:

$$\frac{\partial^2 \mathbf{u}}{\partial t^2} = (c_L^2 - c_T^2)\nabla(\operatorname{div} \mathbf{u}) + c_T^2 \Delta \mathbf{u}. \tag{6.45}$$

By analogy with the wave separation in an electromagnetic field, the displacement vector is usually represented as the sum of two terms satisfying the two following conditions, e.g., [184]:

$$\mathbf{u} = \mathbf{u}_L + \mathbf{u}_T \qquad \operatorname{div} \mathbf{u}_T = 0 \qquad \operatorname{rot} \mathbf{u}_L = 0. \tag{6.46}$$

This is the Poisson decomposition of the displacement. Other decompositions such as the Helmholtz additive decomposition can also be used to separate various waves in elastic solids, e.g., [184]. Classical development in a continuum field gives the two equations that describe the two components of the displacement vector:

$$\frac{\partial^2 \mathbf{u}_L}{\partial t^2} = c_L^2 \Delta \mathbf{u}_L \qquad \frac{\partial^2 \mathbf{u}_T}{\partial t^2} = c_T^2 \Delta \mathbf{u}_T. \tag{6.47}$$

Then, by applying the intrinsic differential operators, we can derive the basic equations of wave propagation in solids with continuous distribution of singularity. More particularly, we give a complete derivation of the Laplacian operator. However, we will start directly with equation (6.45) hereafter. The reason is that the result (6.47) obtained from (6.45) and (6.46) should be revisited in the case of a multiply connected manifold.

Remark. The classical wave equation generally holds for low frequency excitation but fails for a very high frequency wave. Indeed, for disordered crystal solids where the length scale is of the order of several lattice constants (10Å-50Å), the divergence of a vector field should also account for the material structural flaw at the mesoscopic level as for strain gradient continuum.

6.5.2   Laplacian of a scalar field

To derive the Laplacian of a scalar field, it is necessary to extend the definition of the divergence from a vector (1-contravariant) to a 1-form (1-covariant). The divergence of a vector projected onto a vector basis $(\mathbf{u}_1, \mathbf{u}_2, \mathbf{u}_3)$ has been derived previously:

$$\text{div } \mathbf{v} = \frac{1}{J_u} \nabla_{\mathbf{u}_a} [J_u \mathbf{v}(\mathbf{u}^a)] + \sum_{(abc)} \aleph_{0ab}^{d} \varepsilon_{dce} \mathbf{v}(\mathbf{u}^e) \tag{6.48}$$

where $J_u = \omega_0(\mathbf{u}_1, \mathbf{u}_2, \mathbf{u}_3)$, $\aleph_{0ab}^{c} \mathbf{u}_c = [\mathbf{u}_a, \mathbf{u}_b]$, and $\varepsilon_{dce} = J_u \omega_0(\mathbf{u}_d, \mathbf{u}_c, \mathbf{u}_e)$. A more concise formulation of (6.48) is obtained by applying the circular permutation $(abc) = \{123\,231\,312\}$ and by defining the 1-form field $\tilde{\aleph}_0 = \aleph_{0ab}^{b} \mathbf{u}^a$, the components of which are:

$$\tilde{\aleph}_{01} = \aleph_{012}^{2} + \aleph_{013}^{3} \qquad \tilde{\aleph}_{02} = \aleph_{023}^{3} + \aleph_{021}^{1} \qquad \tilde{\aleph}_{03} = \aleph_{031}^{1} + \aleph_{032}^{2}. \tag{6.49}$$

The divergence of a vector becomes:

$$\text{div } \mathbf{v} = \frac{1}{J_u} \nabla_{\mathbf{u}_a} (J_u v^a) + \tilde{\aleph}_{0a} v^a. \tag{6.50}$$

Thus, the intrinsic divergence includes a classical divergence (macroscopic) and a contribution of the singularity distribution (mesoscopic):

$$\text{div } \mathbf{v} = \overline{\text{div}} \ \mathbf{v} + \tilde{\aleph}_0(\mathbf{v}). \tag{6.51}$$

Definition 6.5.1 *(Divergence of a 1-form) Let ω be any 1-form on B and $\mathbf{w}$ a unique vector field such that $\omega(\mathbf{u}) \equiv \mathbf{g}(\mathbf{w}, \mathbf{u})$, $\forall \mathbf{u} \in T_M B$. Hence, the divergence of the 1-form ω on B is defined by the relation:*

$$\text{div } \omega \equiv \text{div } \mathbf{w}. \tag{6.52}$$

In component form, since $\omega_a = g_{ab} w^b$ and $w^a = g^{ab} \omega_b$, the intrinsic divergence of the 1-form is obtained directly from (6.48):

$$\text{div } \omega = \frac{1}{J_u} \nabla_a (J_u g^{ab} \omega_b) + \sum_{(abc)} g^{ef} \aleph_{0ab}^{d} \varepsilon_{dce} \omega_f. \tag{6.53}$$

It is easy to rewrite (6.53) in a coordinate-free form:

$$\text{div } \omega = \overline{\text{div}} \ \omega + \mathbf{g}^{-1}(\tilde{\aleph}_0, \omega). \tag{6.54}$$

Definition 6.5.2 *(Laplacian of a scalar field) Let Θ be any scalar field on the continuum B. The Laplacian of Θ is a scalar field defined as:*

$$\Delta \Theta \equiv \text{div } (\nabla \Theta). \tag{6.55}$$

In component form, the use of a covariant derivative and of the divergence of a 1-form (6.53) allows us to write first $\nabla\Theta = \nabla_a\Theta\, \mathbf{u}^a$. Then the Laplacian of a scalar field decomposed in any base gives:

$$\Delta\Theta = \frac{1}{J_u}\nabla_a(J_u g^{ab}\nabla_b\Theta) + \sum_{(abc)} g^{ef}\aleph^d_{0ab}\varepsilon_{dce}\nabla_f\Theta \tag{6.56}$$

or in a more concise form:

$$\Delta\Theta = \frac{1}{J_u}\nabla_a(J_u g^{ab}\nabla_b\Theta) + g^{ab}\tilde{\aleph}_{0a}\nabla_b\Theta. \tag{6.57}$$

Relation (6.57) gives the most general formulation of the Laplacian operator in component form. A coordinate-free formulation of the Laplacian (6.57) is:

$$\Delta\Theta = \overline{\Delta}\Theta + \mathbf{g}^{-1}(\tilde{\aleph}_0, \nabla\Theta). \tag{6.58}$$

The second term of the r.h.s is the contribution of the continuous distribution of singularity on the Laplacian of a scalar field. This relation may be directly checked:

$$\Delta\Theta = \mathrm{div}(\nabla\Theta) = \overline{\mathrm{div}}(\nabla\Theta) + \mathbf{g}^{-1}(\tilde{\aleph}_0, \nabla\Theta). \tag{6.59}$$

In a Cartesian coordinate system defined by unit and orthogonal base vectors, the Laplacian operator (6.58) takes the form:

$$\Delta\Theta = \overline{\Delta}\Theta + \sum_{a=1}^{3}\tilde{\aleph}_{0a}\frac{\partial\Theta}{\partial x^a} \tag{6.60}$$

where $\overline{\Delta}\Theta$ stands for the Laplacian operator within continuum without singularity distribution.

6.5.3 Laplacian of a vector field

The extended definition for the divergence of a vector can be generalized for any second-order tensor by "raising" and "lowering" the tensor indices with the metric tensor $\mathbf{g}$, e.g., [58]:

$$\pi^{ab} = g^{ac}\pi^b_c = g^{cb}\pi^a_c = g^{ac}g^{bd}\pi_{cd}.$$

Definition 6.5.3 (*Divergence of tensor*) *Let π be any tensor on B, of order $(1, 1)$, $(0, 2)$, or $(2, 0)$. The divergence of any second-order tensor π, denoted $\mathrm{div}\,\pi$, is a vector field equal to the divergence of the corresponding 2-contravariant tensor:*

$$\mathrm{div}\,\pi \equiv \mathrm{div}\,\pi_{2contravariant}. \tag{6.61}$$

The divergence of a 2-contravariant tensor π is written in component form as follows:

$$\text{div } \pi = \frac{1}{J_u} \nabla_{u_a} (J_u \pi^{ab} u_b) + \sum_{(abc)} \aleph^d_{0ab} \varepsilon_{dce} \pi^{fe} u_f. \tag{6.62}$$

For our purpose, the divergence of a 1-covariant, 1-contravariant tensor is transcribed:

$$\text{div } \pi = \frac{1}{J_u} \nabla_{u_a} (J_u g^{ac} \pi^b_c u_b) + \sum_{(abc)} g^{fg} \aleph^d_{0ab} \varepsilon_{dce} \pi^{fe} u_f. \tag{6.63}$$

In a more concise form, the divergence (6.63) is given in a coordinate-free formulation:

$$\text{div } \pi = \overline{\text{div }} \pi + \pi(\tilde{\aleph}_0). \tag{6.64}$$

This is possible by observing the following relations in (6.63):

$$\begin{aligned}
\pi &= \pi^c_a \, \mathbf{u}^a \otimes \mathbf{u}_c \\
\tilde{\aleph}_0 &= \tilde{\aleph}_{0d} \, \mathbf{u}^d \\
\pi(\tilde{\aleph}_0) &= \pi^c_a \, \mathbf{u}^a \otimes \mathbf{u}_c (\tilde{\aleph}_{0d} \, \mathbf{u}^d) = \tilde{\aleph}_{0c} \pi^c_a \, \mathbf{u}^a = g^{ab} \tilde{\aleph}_{0c} \pi^c_a \mathbf{u}_b.
\end{aligned}$$

Definition 6.5.4 *(Laplacian of a vector) Let* $\mathbf{v}$ *be any vector field on a continuum B, endowed with a metric tensor* $\mathbf{g}$*. The Laplacian of* $\mathbf{v}$*, denoted* $\Delta \mathbf{v}$*, is the vector field defined by:*

$$\Delta \mathbf{v} \equiv \text{div } (\nabla \mathbf{v}). \tag{6.65}$$

Replacing the mixed tensor π by the gradient of velocity $\nabla \mathbf{v}$ in the divergence formula (6.63), we deduce the general form of (6.65) projected in any coordinate or noncoordinate vector basis $(\mathbf{u}_1, \mathbf{u}_2, \mathbf{u}_3)$:

$$\Delta \mathbf{v} = \frac{1}{J_u} \nabla_{\mathbf{u}_a} (J_u g^{ac} \nabla_c v^b \mathbf{u}_b) + \sum_{(abc)} (g^{fg} \aleph^d_{0ab} \varepsilon_{dce} \nabla_g v^e) \mathbf{u}_f. \tag{6.66}$$

By analogy with the Laplacian of a scalar field, a more concise form of (6.66) is also obtained by introducing the variables (6.49):

$$\Delta \mathbf{v} = \frac{1}{J_u} \nabla_{\mathbf{u}_a} (J_u g^{ac} \nabla_c v^b \mathbf{u}_b) + (g^{ab} \tilde{\aleph}_{0c} \nabla_a v^c) \mathbf{u}_b. \tag{6.67}$$

Relations (6.66) and (6.67) give the most general formulation for the Laplacian of a vector on a continuum and may be rewritten in coordinate-free form as follows:

$$\Delta \mathbf{v} = \overline{\Delta} \mathbf{v} + \nabla \mathbf{v}(\tilde{\aleph}_0). \tag{6.68}$$

To check this, we can write:

$$\Delta \mathbf{v} = \text{div } (\nabla \mathbf{v}) = \overline{\text{div}}(\nabla \mathbf{v}) + \nabla \mathbf{v}(\tilde{\aleph}_0) = \overline{\Delta} \mathbf{v} + \nabla \mathbf{v}(\tilde{\aleph}_0).$$

6.5.4   Wave scattering in defected solids

The extended formulation of wave propagation equations in solids with singularity distribution are obtained by using intrinsic differential operators, that is by introducing (6.51) and (6.68) in equations (6.45):

$$\frac{\partial^2 \mathbf{u}}{\partial t^2} = (c_L^2 - c_T^2)\nabla(\overline{\text{div}}\,\mathbf{u}) + c_T^2\overline{\Delta}\mathbf{u} + (c_L^2 - c_T^2)\nabla[\tilde{\aleph}_0(\mathbf{u})] + c_T^2\nabla\mathbf{u}(\tilde{\aleph}_0). \quad (6.69)$$

The additional terms represent the configurational forces and cannot be eliminated by the deliberate choice of an affine connection ∇ and of a base vector $(\mathbf{u}_1, \mathbf{u}_2, \mathbf{u}_3)$. They are responsible for wave scattering within the material bulk. For the sake of simplicity, consider now a Cartesian vector basis $(\mathbf{u}_1, \mathbf{u}_2, \mathbf{u}_3)$ directed by the referential body. The differential operators simplify to:

$$\text{div}\,\mathbf{v} = \overline{\text{div}}\,\mathbf{v} + \tilde{\aleph}_{0a}v_a \qquad \Delta\mathbf{v} = \overline{\Delta}\mathbf{v} + \tilde{\aleph}_{0a}\frac{\partial v_a}{\partial x_b}\mathbf{u}_b.$$

Without loss of generality, we can in such a case lower all indices and drop the Kronecker symbols. Change the notation of $\mathbf{v}$ into $\mathbf{u}$. The projection of (6.69) onto a vector base then gives more tractable equations (summation for index a):

$$\begin{aligned}
\frac{\partial^2 u_b}{\partial t^2} &= (c_L^2 - c_T^2)\frac{\partial}{\partial x_b}\left(\frac{\partial u_a}{\partial x_a}\right) + c_T^2\frac{\partial^2 u_b}{\partial x_a^2} \\
&+ (c_L^2 - c_T^2)\frac{\partial}{\partial x_b}\left(\tilde{\aleph}_{0a}u_a\right) + c_T^2\tilde{\aleph}_{0a}\frac{\partial u_a}{\partial x_b}.
\end{aligned} \quad (6.70)$$

6.5.5   Steady-state wave in elastic solid

For the particular case in which the displacement vector $\mathbf{u}$ depends only on one coordinate $x^1 = x$ and on the time t, further simplification gives:

$$\begin{aligned}
\frac{\partial^2 u_1}{\partial t^2} &= c_L^2\frac{\partial^2 u_1}{\partial x^2} + c_L^2\tilde{\aleph}_{0a}\frac{\partial u_a}{\partial x} + (c_L^2 - c_T^2)\frac{\partial\tilde{\aleph}_{0a}}{\partial x}u_a \\
\frac{\partial^2 u_2}{\partial t^2} &= c_T^2\frac{\partial^2 u_2}{\partial x^2} \\
\frac{\partial^2 u_3}{\partial t^2} &= c_T^2\frac{\partial^2 u_3}{\partial x^2}.
\end{aligned} \quad (6.71)$$

The first equation governs the longitudinal wave propagation whereas the two last ones describe the transverse wave propagation. The existence of a continuous distribution of singularity implies a coupling of the wave propagation along the three directions. The first equation looks like a linear damped Klein–Gordon wave equation, e.g., [96].

The resolution of the two last equations gives transverse waves under some boundary conditions:

$$u_2(x,t) = (A_2 \cos \omega t + B_2 \sin \omega t)\left(C_2 \cos \frac{\omega x}{c_T} + D_2 \sin \frac{\omega x}{c_T}\right)$$

$$u_3(x,t) = (A_3 \cos \omega t + B_3 \sin \omega t)\left(C_3 \cos \frac{\omega x}{c_T} + D_3 \sin \frac{\omega x}{c_T}\right). \quad (6.72)$$

The first equation is more complicated and can be rewritten after separating the variables $u(x,t) = U(x)\kappa(t)$ (valid only under some boundary conditions):

$$
\begin{aligned}
U\kappa'' = {} & c_L^2 U'' \kappa + c_L^2 \tilde{\aleph}_{01} U' \kappa + (c_L^2 - c_T^2)\frac{\partial \tilde{\aleph}_{01}}{\partial x} U\kappa \\
& + c_L^2 \tilde{\aleph}_{02}\frac{\partial u_2}{\partial x} + c_L^2 \tilde{\aleph}_{03}\frac{\partial u_3}{\partial x} \\
& + (c_L^2 - c_T^2)\frac{\partial \tilde{\aleph}_{02}}{\partial x} u_2 + (c_L^2 - c_T^2)\frac{\partial \tilde{\aleph}_{03}}{\partial x} u_3.
\end{aligned}
$$

The equation is not homogeneous due to the second and third lines which may be considered as a known from (6.72).

Some remarks.

1. We observe that the presence of defects scatters the wave transversally. Continuous distribution of plastic strains induces a coupling between the wave propagation along the three directions ($\mathbf{e}_1, \mathbf{e}_2, \mathbf{e}_3$). Given the measurement of wave characteristics such as attenuation, the quantitative determination of the nonholonomic residual strain $\aleph_0$, engendered by plastic strains, becomes theoretically tractable.

2. The determination of $\aleph_0$ becomes a keystone to evaluate the irreversible defects in continuum. A more concise formulation is obtained by considering the Cartan 1-form as follows:

$$\tilde{\aleph}_{01} = \aleph_{012}^2 + \aleph_{013}^3 \qquad \tilde{\aleph}_{02} = \aleph_{023}^3 + \aleph_{021}^1 \qquad \tilde{\aleph}_{03} = \aleph_{031}^1 + \aleph_{032}^2.$$

This would facilitate the construction of hierarchical models in the sense that the models at one scale level would provide rational arguments to develop the models at another scale level. Namely, the consideration of physical mesoscopic mechanisms has a major consequence on the wave propagation. When the dimensions of a sample to be tested are of the same order as $\aleph_0$, the strain gradient effects are no longer negligible.

3. The 1-form $\aleph_0$ has three independent components which could be associated to three values of length scales e.g. [57]. It connects the continuum macroscopic description and the mesoscopic behaviour in strain gradient plasticity.

6.5.6 Homogeneous distribution of singularity

During the experimental measurement of material properties, it is often assumed that the characteristics are not coordinate dependent (homogeneous distribution) in the "small" piece of material to be tested. For further simplification, let us assume that the wave propagation is unidirectional $u_1 = u$ and $u_2 = u_3 = 0$ by denoting the defect distribution $\tilde{\aleph}_0 = \tilde{\aleph}_{01}$:

$$\frac{\partial^2 u}{\partial t^2} = c_L^2 \frac{\partial^2 u}{\partial x^2} + c_L^2 \tilde{\aleph}_0 \frac{\partial u}{\partial x}. \tag{6.73}$$

The shape of the steady-state solutions depends on the amount of singularity density within the solid. Three types of solutions exist according to the microdefect density values. For the convenience of physical interpretation, we define the characteristic defect length $d_\aleph$:

$$d_\aleph \equiv \frac{2}{\tilde{\aleph}_0}. \tag{6.74}$$

Let us also define the defect circular frequency $\omega_\aleph$ and calculate the discriminant:

$$\omega_\aleph \equiv \frac{c_L}{d_\aleph} \qquad \Delta_\aleph \equiv \tilde{\aleph}_0^2 c_L^4 - 4\omega^2 c_L^2 = 4c_L^2(\omega_\aleph^2 - \omega^2). \tag{6.75}$$

1. Without singularity distribution $\tilde{\aleph}_0 = 0$, we obtain the classical steady-state elastic waves:

$$u(x,t) = \left(A e^{j\frac{\omega}{c_L}x} + B e^{-j\frac{\omega}{c_L}x} \right)(C \sin \omega t + D \cos \omega t). \tag{6.76}$$

2. For high density of singularity distribution, we have from (6.75) $\omega_\aleph^2 - \omega^2 \geq 0$. It is equivalent to the situation where the excitation frequency is greater than the defect circular frequency. The corresponding solutions are:

$$u(x,t) = \left(A e^{\frac{\sqrt{\omega_\aleph^2 - \omega^2}}{c_L}x} + B e^{-\frac{\sqrt{\omega_\aleph^2 - \omega^2}}{c_L}x} \right) e^{-\frac{x}{d_\aleph}}$$

$$\times \quad (C \sin \omega t + D \cos \omega t). \tag{6.77}$$

3. For a low density of singularity distribution, $\omega_\aleph^2 - \omega^2 < 0$, we may write:

$$\omega_\aleph^2 - \omega^2 = j^2(\omega^2 - \omega_\aleph^2)$$

and the solution becomes:

$$u(x,t) = \left(A e^{j\frac{\sqrt{\omega^2 - \omega_\aleph^2}}{c_L}x} + B e^{-j\frac{\sqrt{\omega^2 - \omega_\aleph^2}}{c_L}x} \right) e^{-\frac{x}{d_\aleph}}$$

$$\times \quad (C \sin \omega t + D \cos \omega t). \tag{6.78}$$

Remark. Coefficients A, B, C, and D are dependent on the boundary conditions and on the initial conditions. It might be noted that there is no change in the frequency, no decay of the time function. Conversely, the wave amplitude is modified within the bulk material depending on the singularity distribution density $\aleph$. Thus, a disturbance of arbitrary form propagates with a decay of the amplitude without having its shape changed.

6.5.7 Example of wave propagation

The present model could find an application in the measuring of the loss in an ultrasonic signal due to the propagation of wave, through a defected sample of material (sample with residual nonholonomic strain).[1]

Suppose that a plate of elastic material of thickness L is subjected to the steady-state displacement boundary condition $u(0, t) = \bar{u} \cos \omega_0 t$ (ω_0 is not to be confused with the volume form on the manifold) at the left boundary and that the plate is bonded to a fixed support at the right boundary. Let us determine the steady-state motion of the material according to the amount of microdefects within the solid.

1. **No distribution of singularity.** Case $\tilde{\aleph}_0 = 0$. Boundary conditions to the left and to the right enable us to define the coefficients by writing:

$$u(0, t) = \bar{u} \cos \omega_0 t \qquad u(L, t) = 0. \tag{6.79}$$

   D may be set equal to 1 without loss of generality and $C = 0$. Additionally, for a steady-state wave we obtain $\omega = \omega_0$. The coefficients are given or may

---

[1] **Ultrasonic techniques for microcracking detection.** For most brittle materials the ultrasonic technique has been developed to characterize the internal degradation by measuring the attenuation of ultrasonic waves rather than the modification of its natural frequencies. Two nonlinear ultrasonic techniques are usually proposed to characterize and monitor the fatigue microcracking damage: acoustic-elastic effects (stress dependence of the attenuation) and higher harmonic generation. The first technique measured the wave attenuation on the basis of Taylor expansion of the sound velocity with respect to the pre-strain level whereas the second technique was used to capture the material degradation by assuming a nonlinear stress-strain law. These methods seemed not suitable for the case of microcracked brittle material, for which the stress-strain law remained linear although with a lower elastic modulus than intact material. For instance, these two methods were not obviously suitable to detect the presence of uniformly distributed microcracks. In the same way, the experimental analysis of elastic wave propagation in microflawed ceramics (pores 1μm) showed strong attenuation and cutoff of frequency, which could not be explained in the light of existing macroscopic models of wave propagation. The cutoff frequency phenomenon exhibited distinct frequency bands with energy transmission (pass bands) and with near-zero energy transmission (stop band with cutoff frequency). Due to the shortness of these microcracks' characteristic length (1μm to 10μm) compared to the usual wavelength used in ultrasonic techniques, the continuum wave theory has often no sufficient sensitivity to apprehend material degradation at the mesoscopic level. Theoretical models should be developed not only for improving the measurement processes as for ultrasonic inspection techniques but also and mainly for better interpretation of the measured data.

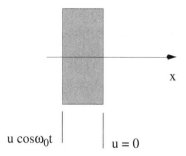

$u \cos \omega_0 t$ | | $u = 0$

Figure 6.5. Steady-state wave propagating through an infinite plate of thickness L, excited at the left side and bonded to a rigid wall at the right side. The material is elastic linear with distribution of scalar discontinuity and assumed isotropic with the Lame's coefficients λ and μ.

be calculated from the following system of equations:

$$A + B = \bar{u} \qquad (Ae^{j\frac{\omega}{c_L}L} + Be^{-j\frac{\omega}{c_L}L}) = 1. \qquad (6.80)$$

When introducing these coefficients in the general wave solution, the steady-state wave in the infinite plate is fully described by the function:

$$u(x, t) = \bar{u} \, \frac{\sin \frac{\omega}{c_L}(x - L)}{\sin \frac{\omega}{c_L}L} \, \cos \omega_0 t.$$

This solution exhibits the phenomenon of mechanical resonance. The displacement amplitude is infinite when $\sin \frac{\omega}{c_L}L = 0$, that is when the excitation to the left (imposed displacement) has a frequency:

$$\omega_n = n\pi \, \frac{c_L}{L} \qquad n = 1, \ldots, \infty. \qquad (6.81)$$

2. **High density of singularity.** $\omega_\aleph^2 - \omega^2 \geq 0$. Similarly, we obtain the steady-state wave:

$$u(x, t) = \bar{u} \, \frac{\sinh \frac{\sqrt{\omega_\aleph^2 - \omega^2}}{c_L}(x - L)}{\sinh \frac{\sqrt{\omega_\aleph^2 - \omega^2}}{c_L}L} \, e^{-\frac{x}{d_\aleph}} \cos \omega_0 t.$$

Apparently, no resonance phenomenon occurs, at least in the classical sense. However, the amplitude of the space-dependent part of the function may rise to infinity when:

$$\sinh \frac{\sqrt{\omega_\aleph^2 - \omega^2}}{c_L}L = 0 \Longrightarrow \omega_n = \omega_\aleph,$$

that is when and only when the frequency excitation equals to defect frequency.[2] It should be noted that the wave amplitude decays with respect to the space location. To illustrate this, the resonance in stainless steel material is obtained at the frequency $\omega_n \approx 0.83 \, 10^3 \aleph$ which may give a frequency up to 10^{10}Hz (domain of microwaves) for a high density of singularity. This is far beyond the upper limit of mechanical resonance.

3. **Low density of singularity.** $\omega_\aleph^2 - \omega^2 < 0$. The steady-state wave is given by:

$$u(x, t) = \bar{u} \, \frac{\sin \frac{\sqrt{\omega^2 - \omega_\aleph^2}}{c_L} (x - L)}{\sin \frac{\sqrt{\omega^2 - \omega_\aleph^2}}{c_L} L} \, e^{-\frac{x}{d_\aleph}} \cos \omega_0 t.$$

In addition to the decay of amplitude already observed in the previous case, there is a resonance phenomenon when the denominator vanishes:

$$\sin \frac{\sqrt{\omega^2 - \omega_\aleph^2}}{c_L} L = 0.$$

This happens to be when the excitation frequency is the solution of the characteristic equation. The resonance frequencies for this plate are:

$$\omega_n = \sqrt{\omega_\aleph^2 + n^2 \pi^2 \frac{c_L^2}{L^2}} \qquad n = 1, \ldots, \infty.$$

For the extreme case where the ratio $\omega/\omega_\aleph \gg 1$, the wave behavior should be comparable with reflection and refraction at locally planar interfaces. Moreover, the energy peaks in this pass-band are transmitted through the medium. For all cases, the attenuation of the wave depends on the frequency and on the amount of microplastic strains. This cut off frequency would be a starting point for experimental measurement of the plastic strain distribution. For fully 3-dimensional samples, the material is supporting compression and shear waves and the situation becomes more complicated but remains tractable.

6.5.8 Discussion

Another way of deriving the Laplacian operator consists in projecting the Laplacian operator onto a system of orthogonal vectors $(\mathbf{u}_1, \mathbf{u}_2, \mathbf{u}_3)$ such that $[\mathbf{u}_a, \mathbf{u}_b] \equiv 0$,

---

[2]This frequency equation may be used as a starting point for experimental measurements of the singularity distribution when the density is sufficiently high.

Material	c_L [m/s]	c_T [m/s]
Polyethylene	$1.95 \ 10^3$	$0.54 \ 10^3$
Tungsten	$5.20 \ 10^3$	$2.90 \ 10^3$
Stainless steel	$5.79 \ 10^3$	$3.10 \ 10^3$
Aluminium	$6.42 \ 10^3$	$3.04 \ 10^3$

Table 6.4. Longitudinal and transversal phase velocities for various solids.

that is $\aleph^c_{0ab} \equiv 0$. The Laplacian expression (6.57) apparently simplifies but this is not truly the case. Indeed, a more explicit formulation of the Laplacian operator needs a reminder of the existing relationships between affine connection coefficients, constants of structure and torsion components: $\aleph^c_{ab} = (\Gamma^c_{ab} - \Gamma^c_{ba}) - \aleph^c_{0ab}$. The choice of a vector basis with the properties mentioned above imposes the use of nonsymmetric affine connection related to the torsion tensor. Therefore, the Laplacian operator may be calculated as a covariant derivative of a vector field followed by a contraction:

$$\Delta\Theta = \frac{1}{J_u}\nabla_a(J_u g^{ab}\nabla_b\Theta) = \frac{1}{J_u}\frac{\partial}{\partial x^a}\left(J_u g^{ab}\frac{\partial\Theta}{\partial x^b}\right) + \aleph^a_{ab}g^{bc}\frac{\partial\Theta}{\partial x^c}. \tag{6.82}$$

An initial affine connection $\overline{\Gamma}^c_{ab} \equiv 0$ (which is a fortiori symmetric) is transformed into an antisymmetric connection whenever there is nucleation and increase of singularity distribution $\aleph^c_{ab} = (\Gamma^c_{ab} - \Gamma^c_{ba})$ at the deformed state. The Laplacian of a scalar field does not reduce to the second order derivatives of the scalar Θ itself but rather involves first-order derivatives coupled with changes in the local topology (torsion tensor) in the continuum. These affine connection coefficients describe the effects of singularity distribution in the Laplacian operator. Projected on a Cartesian reference frame, the Laplacian operator (6.82) can be brought down:[3]

$$\Delta\Theta = \overline{\Delta}\Theta + (\aleph^2_{21} + \aleph^3_{31})\frac{\partial\Theta}{\partial x^1} + (\aleph^1_{12} + \aleph^3_{32})\frac{\partial\Theta}{\partial x^2} + (\aleph^1_{13} + \aleph^2_{23})\frac{\partial\Theta}{\partial x^3}. \tag{6.83}$$

Once again, it should be remembered that no change of coordinate frame can eliminate the terms due to singularity. They represent the irreversible change in the continuum which from a simply connected state becomes multi-connected after deformation. It

---

[3]The dimension of the constants of structure may be defined to be $[m^{-1}]$ (inverse to a length dimension).

should be noticed that

$$\Delta \Theta \;=\; \overline{\Delta \Theta} - \aleph_a \frac{\partial \Theta}{\partial x^a}$$

$$\Delta \Theta \;=\; \overline{\Delta \Theta} + \overline{\aleph}_a \frac{\partial \Theta}{\partial x^a}$$

which implies that

$$\tilde{\aleph}_a + \aleph_a = 0.$$

Practically, any of the two variables $\tilde{\aleph}_a$ or $\aleph_a$ can be used to derive the Navier equations and consequently the wave propagation equation. This has been shown in a previous work.

Remark. The dislocation density varies in the range of $10^4/m^2$ to $10^{16}/m^2$. $\aleph$ may be estimated, as a first approximation, to range between $10^2/m$ to $10^8/m$.

7
Solids with Dry Microcracks[1]

The theories of macroscopic properties of cracked solids predict the modification of the stiffness and the change of anisotropy orientation caused by microcracking. Hence, the development of continuum theory dealing with microcrack distribution has an obvious interest for mechanics of materials. Basically, a microcrack is often associated to internal slipping and debonding in matter, e.g., [13], [25]. Following a closed path around the defect, the displacement field has a jump. When the occurrence of microcracks appears with sufficiently high density in some matter, then a continuous volume distribution is a reasonable hypothesis. For a continuous distribution of cracks, displacement and velocity are not single-valued functions of position of a material point M. At least three approaches may be proposed to construct the theory of microcrack continuous distribution: statistical models, micromechanical models, and continuum models [100]. The statistical models, based on statistical physics, are beyond the scope of this book. The micromechanical models consider a basic "microscopic cell" including the crack. In these models, the contact mechanics theory and the homogenization method are applied to each crack in order to account for the influence of crack on the global material properties, e.g., [131]. The continuum models are based on continuum mechanics accounting for discontinuity of scalar and vector fields [163].

---

[1] Part of this chapter was written with Dr. N. Ramaniraka and some of the results on contact thermomechanics may be found in [165] and in a joint paper [166].

Most models of microcrack distribution are based on the theory of effective fields [87] after homogenization of the "microscopic cells." As a keystone, a basic postulate assumes that a representative mathematical variable of inhomogeneity is embedded into the effective field [106], [141]. This is in a sense equivalent to an internal variable [113] or multi-field description [127]. Cracks have a strong orientation dependence and the choice of a density parameter to capture the distribution of cracks in solids is far from being established. To this purpose, the concept of crack density tensor was first introduced by Vakulenko and Kachanov [196] and later Kachanov developed it more fully in [86], in the framework of elastic cracked body. Basically, the crack density tensor was defined as a second-order symmetric tensor directed by the normal unit vector of surface cracks [87]. It was pointed out that for a 3-dimensional situation these tensors did not capture the crack density entirely and so fourth-order tensors were introduced although their impact on the accuracy was not so important. Many crack density tensors have been based on the invariant functions of vector arguments rather than tensorial ones [100].

In parallel, some authors have adopted the concept of a damage tensor to model the existence of crack density, e.g., [11], [142]. At a very basic level, occurrence of void in a matter is related to the occurrence of discontinuity of the normal velocity across an internal slip surface, leading to an unilateral contact phenomenon, e.g., [30], [138], [140]. At this interface, friction lies at the origin of the two fundamental principles of thermodynamics, at least for solid mechanics, e.g., [7], [56]. The mechanical power dissipated by friction is transformed into heat and is positive, e.g., [16]. Yet the thermomechanics of contact between two deformable solids is rather underdeveloped comparatively to continuum thermomechanics. Theories and models of contact thermomechanics are rare, e.g., [85], [210], [211]. They treat mechanical contact as a two-body problem but oddly introduce a third body for thermal aspects [171]. In this chapter, we adopt the theory of a two-body contact [165]. The contact thermomechanics underlying the behavior of a basic cell is a two part theory for the present study. The conservation laws of thermomechanical contact problems are derived from thermodynamic principles. The constitutive laws of unilateral contact, friction, and heat conduction across the interface, are developed following a nonsmooth analysis. Although studies of cracked material based on contact mechanics of the crack's lips may lead to a new way for solving practical problems, e.g., [131], use of a homogenization method to obtain a continuum model seems to be lengthy and cumbersome, particularly in the presence of a nonperiodic pattern of microcracks. For methods based on field discontinuity, we use the previously introduced model of a continuum allowing distribution of scalar and vector discontinuity [163]. Indeed, continuum models of damaged material are typically derived within the framework of thermomechanics with internal variables. The present work conforms to this idea.

The goals of this chapter are to compare the micromechanical approach and the continuum approach in modeling the presence of microcrack distribution within a continuum: first, a distribution of microscopic cells including dissipative contacting lips (discontinuity of matter) and second, a distribution of singularity in a continuum (discontinuity of fields). The main result consists in extending the contact thermomechanics of microcrack lips to continuous distribution of discontinuity so as to account for relative translation and relative rotation of the two lips of microcracks.

7.1 Geometry

In this chapter, two models will be considered and compared with respect to their theoretical foundations: the so-called micromechanical model and the continuum model. The micromechanical model is assumed to be an elastic continuum containing a distribution of microcracks: derivative operators on the elastic matrix (bulk) are based on a Euclidean connection [130] and the crack interface is treated as being two contacting lips. Conversely, the continuum model already contains the microcrack distribution based on the model of continuum with discontinuous fields. Therefore, use of an affine connection is more appropriate [148].

For both approaches, consider an initial configuration B_0, and let $\mathbf{u}_0$ be a tangent vector at M_0. Say its motion $\varphi : B_0 \longrightarrow B$. The (vector) tangent space of B (resp. its dual) is denoted $T_M B$ (resp. $T_M B^*$). The linear motion tangent to φ is the application $d\varphi$ from the tangent space $T_{M_0} B_0$ to $T_M B$, which transforms a tangent vector $\mathbf{u}_0$ in M_0 according to, e.g., [159]:

$$\mathbf{u} = d\varphi(\mathbf{u}_0). \tag{7.1}$$

When internal slip surfaces are nucleated during the transformation, we have the basic relation in the presence of microcracks distribution:

$$[\mathbf{u}_a, \mathbf{u}_b] = [d\varphi(\mathbf{u}_{a0}), d\varphi(\mathbf{u}_{b0})] = \aleph^c_{0ab} \mathbf{u}_c \tag{7.2}$$

where $\aleph^c_{0ab}$ are the constants of structure (Cartan), associated to the tangent base $(\mathbf{u}_1, \mathbf{u}_2, \mathbf{u}_3)$ [23].

7.1.1 Micromechanical approach: geometrical variables

The micromechanical approach is based on the discontinuity of matter (micro-slip surface). We consider a "basic cell" $\Sigma = B \cup B'$ containing a microcrack such as that shown in Figure 7.1. The basic characteristics of the cell is that of a cohesion-decohesion phenomenon at the crack level. For each slip surface, the two contacting lips may separate (slipping and debonding). Cohesion-decohesion involves unilateral frictional contact.

Consider two deformable bodies B and B' surrounding the microcrack. The body B is limited by the frontier $\partial B = \partial B_p \cup \partial B_c \cup \partial_c$, where on ∂B_u displacement is imposed, on ∂B_p stress vector is imposed, and ∂B_c is the contact surface between B

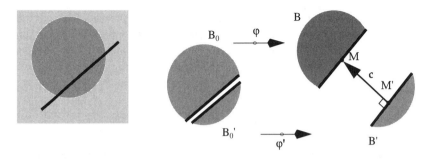

Figure 7.1. The system including the bodies B and B' is disconnected due to the presence of a microcrack. The lips of the crack are represented in dark grey and may separate from each other. The surface in contact ∂B_c is the contact surface and $\partial B'_c$ the target surface. Bodies and boundaries are deformable. Notice that the contact surface and the target surface may vary during the motion of the two bodies. The geometry of two contacting lips surrounding a microcrack modeled with classical contact mechanics is in fact limited to a translation dislocation-like theory.

and B'. The same decomposition holds for B'. B' is taken as a referential body. It is called the target and B the striker. For any material point M located on the boundary $\mathbf{OM} \in \partial B_c$, the proximal point $M^\perp$ the position of which $\mathbf{OM}^\perp = \varphi'(M^\perp, t)$, in the deformed configuration, is the nearest of that to M [38]:

$$\mathbf{OM}^\perp = \arg\min \left[\frac{1}{2} \|\varphi(M, t) - \varphi'(M', t)\|^2 \right]_{M' \in \partial B'_c}. \tag{7.3}$$

Given sufficient conditions of regularity of the contact surfaces, it is possible to establish a bijective mapping between ∂B_c and $\partial B'_c$ at each time t. Proof of existence and uniqueness of this projection may be found in, e.g., [37], [39]. The gap vector $\mathbf{c}$ is defined by, e.g., [37]:

$$\mathbf{c} \equiv \varphi(M, t) - \varphi'(M^\perp, t) = \mathbf{M}^\perp \mathbf{M}$$

where $\mathbf{OM}^\perp(M, t)$. The gap is in some sense the additional variable capturing the jump of the displacement vector field across the interface between ∂B_c and $\partial B'_c$ and, by extension, within the body with microcrack. The normal gap vector c_n is the component of c along the direction $\mathbf{n}'(M^\perp, t)$, normal at $M^\perp$:

$$c_n(M, t) \equiv \mathbf{n}'(M^\perp, t) \cdot [\varphi(M, t) - \varphi'(M^\perp, t)]. \tag{7.4}$$

Three situations may occur: $c_n > 0$ gap, $c_n = 0$ contact, and $c_n < 0$ penetration. The slipping vector is then given by:

$$\mathbf{c}_T(M, t) \equiv \mathbf{c}(M, t) - c_n(M, t)\mathbf{n}'(M^\perp, t). \tag{7.5}$$

It is worthwhile to notice that the definition of the gap vector involves the normal unit vector of the target. It implicitly constitutes a nonlocal formulation.

7.1.2 Microcrack distribution (crack density tensor)

For an isolated crack in an infinite solid, linear and elastic, the three modes of loading (tensile/compression, shear, and anti-plane) produce a crack opening displacement of the same mode only (uncoupled) [87]. In most situations, it is acceptable to neglect the relative influence of any two cracks. In such a case, at distances from the first crack that are must larger than the crack size, the expression for stress simplifies. The far-field asymptotes become applicable at a relatively close distance from the first crack and justify the basic assumption of noninteracting cracks. In reality, interactions of cracks may either increase the stiffness or decrease the stiffness of the material.

The distribution of microcracks is a function of the orientation of the contacting lips. The amount of relative slipping and debonding is assumed to remain small without loss of generality. In this case, the normal unit vectors $\mathbf{n}'(M^\perp, t) \sim \mathbf{n}(M, t)$. The damage at each point M is therefore quantified by a field of doublets $(\mathbf{D}, \mathbf{n})$ where $\mathbf{D}$ is the density of microcrack with a normal $\mathbf{n}$, e.g., [82]. The main problem with these models is to establish a relationship between the density variable $\mathbf{D}$ and the orientation $\mathbf{n}$. Different tensorial measures have been proposed in the literature:

1. **Isotropic damage.** When the microcrack distribution is assumed to be approximately isotropic and the density does not depend on the orientation, the microcrack distribution is fully described by a single scalar field D_0 representing the total microcrack density, averaged over the solid angle around each material point M.

2. **Orthotropic damage.** When the microcrack distribution (discrete or continuous distribution) can be assumed to be orthotropic, it can be supposed to be captured with a second-order symmetric tensor field:

$$\mathbf{D} = \sum_{a,b} \kappa_{ab}\, \mathbf{n}^a \otimes \mathbf{n}^b \quad \mathbf{D} = \int_{-\pi}^{\pi} \mathbf{n}(\alpha) \otimes \mathbf{n}(\alpha)d\alpha.$$

   By extension, microcrack distribution may be captured by a set of tensors:

$$\{\mathbf{D}^{ab}/\mathbf{D}^{ab} \equiv \mathbf{n}^a \otimes \mathbf{n}^b\}.$$

3. **Higher order damage.** Higher order tensors are thought to enhance the precision of the representation of the damage distribution:

$$\mathbf{D} = \kappa_{abcd}\, \mathbf{n}^a \otimes \mathbf{n}^b \otimes \mathbf{n}^c \otimes \mathbf{n}^d.$$

Theories of crack density tensor have been reviewed recently in [87] and [100]. Usually, crack density tensor is limited to second-order tensors since the impact of higher order terms is small and deviations from orthotropy can be neglected.

7.1.3   Continuum approach: geometrical variables

The model of a microcracked continuum is based on the discontinuity of fields (displacement, velocity) rather than on the discontinuity of matter. A continuum with volume distribution of field discontinuity has been modeled by an affinely connected manifold, e.g., [146], [163], [201]. In Chapter 2, use of a path-integral-like method (Cartan) allowed us to obtain the geometrical variables capturing the jump of scalar and vector fields within a continuum [163]: (a) for any discontinuous scalar field, the torsion tensor and (b) for any discontinuous vector field, the torsion and the curvature tensors. The field of microcracks (here scalar and vector discontinuity) is entirely characterized by tensors of torsion and curvature, considered as constitutive primal variables. It follows that the geometry of a continuum with microcracks is defined by, for any vector basis $(\mathbf{u}_1, \mathbf{u}_2, \mathbf{u}_3)$:

1. a metric tensor and a volume-form (usual variables of classical continuum):

$$\mathbf{g} = g_{ab}\mathbf{u}^a \otimes \mathbf{u}^b \qquad \omega_0 = \det(\mathbf{u}_1, \mathbf{u}_2, \mathbf{u}_3)\mathbf{u}^1 \wedge \mathbf{u}^2 \wedge \mathbf{u}^3. \qquad (7.6)$$

2. an affine connection ∇ characterized by the torsion and curvature tensors (additional variables for a continuum with microcrack distribution):

$$\aleph = [(\Gamma^c_{ab} - \Gamma^c_{ba}) - \aleph^c_{0ab}]\mathbf{u}^a \otimes \mathbf{u}^b \otimes \mathbf{u}_c$$
$$\Re = [-\mathbf{u}_b(\Gamma^c_{ad}) + \mathbf{u}_a(\Gamma^c_{bd}) + \Gamma^e_{da}\Gamma^c_{eb} - \Gamma^e_{db}\Gamma^c_{ea} - \aleph^e_{0ab}\Gamma^c_{ed}]$$
$$\mathbf{u}^a \otimes \mathbf{u}^b \otimes \mathbf{u}^d \otimes \mathbf{u}_c \qquad (7.7)$$

where $\Gamma^c_{ab} = \mathbf{u}^c(\nabla_{\mathbf{u}_a}\mathbf{u}_b)$.

Deformation of B includes a transformation of $\mathbf{g}$ and of ω_0 (metric and shape change) and a transformation of ∇ (change of topology). Both "observed internally by B" that is observed with respect to an embedded basis $(\mathbf{u}_1, \mathbf{u}_2, \mathbf{u}_3)$.

7.2   Kinematics

7.2.1   Micromechanical approach: rate of the gap vector

For each basic cell, the rate of the gap vector is given by the derivative with respect to space:

$$\dot{\mathbf{c}} = \frac{d}{dt}\mathbf{c}(M, t) = \frac{d}{dt}\varphi(M, t) - \frac{d}{dt}\varphi'(M^\perp, t). \qquad (7.8)$$

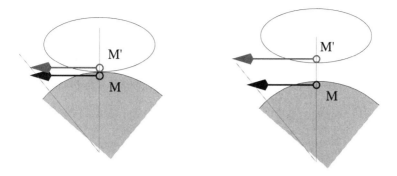

Figure 7.2. The field of relative velocity is not a frame-indifferent variable and thus cannot be considered as a constitutive variable in elaborating constitutive laws for the damaged continuum. When the normal gap vanishes, the usual definitions of objective relative velocity are equivalent.

Projections of $\dot{\mathbf{c}}$ on the normal $\mathbf{n}^{\perp} \equiv \mathbf{n}'(M^{\perp}, t)$ and on the tangent plane perpendicular to $M^{\perp}$ may be written respectively as:

$$\dot{\mathbf{c}}_n = \dot{c}_n(M, t)\mathbf{n}'(M^{\perp}, t) \qquad \dot{\mathbf{c}}_T = c_n(M, t)\dot{\mathbf{n}}'(M^{\perp}, t). \qquad (7.9)$$

The relative velocity of a particle M located in $\varphi(M, t) \in \partial B_c$, with respect to the proximal point $M^{\perp}$, located in $\varphi'(M^{\perp}, t) \in \partial B_c'$, is given by the difference of total derivatives:

$$\mathbf{v}(M, t) - \mathbf{v}'(M^{\perp}, t) = \frac{d}{dt}\varphi(M, t) - \frac{d}{dt}\varphi'(M^{\perp}, t).$$

This relative velocity not being frame indifferent unless the normal gap vanishes (Figure 7.2), an objective contact velocity ζ_c is proposed instead [38]:

$$\zeta_c(M, M^{\perp}, t) = \frac{\partial}{\partial t}\varphi(M, t) - \frac{\partial}{\partial t}\varphi'(M^{\perp}, t) - c_n(M, t)\dot{\mathbf{n}}'(M^{\perp}, t). \qquad (7.10)$$

Normal and tangential projections of ζ_c are respectively:

$$\begin{aligned}
\zeta_{c_n} &= [\mathbf{n}'(M^{\perp}, t) \otimes \mathbf{n}'(M^{\perp}, t)]\,\zeta_c(M, M^{\perp}, t) \\
\zeta_{c_T} &= [I - \mathbf{n}'(M^{\perp}, t) \otimes \mathbf{n}'(M^{\perp}, t)]\,\zeta_c(M, M^{\perp}, t). \qquad (7.11)
\end{aligned}$$

Besides the above definition, numerous measures of objective contact velocity have been introduced. A unified theory of distance and relative velocity measures has been developed recently in [156] in which the contact velocity is defined according to the principles that, first, the contact velocity should be frame indifferent (translation and rotation motions) and, second, that it should coincide with the standard relative velocity $[\mathbf{v}] = \mathbf{v}^+ - \mathbf{v}^-$ (across the contact interface) when contact occurs. To that purpose, the contact relative velocity, decomposed in normal velocity and tangential velocity, was alternatively defined as covariant derivatives of the normal gap and tangential gap between two contacting points. The connection used for the covariant derivative was not specifically named in [156] but it was the Euclidean connection relative to the "target solid."

7.2.2 Continuum approach: objective rate of primal variables

For the continuum approach, we have defined in Chapter 2. a time derivative operator with respect to B such that any tensor embedded in B must have a vanishing derivative [163]. The objective rate of primal variables are the time derivative with respect to the continuum [163]:

$$\zeta_{\omega_0} \equiv \frac{d^B}{dt}\omega_O \quad \zeta_g \equiv \frac{1}{2}\frac{d^B}{dt}\mathbf{g} \quad \zeta_\aleph \equiv \frac{d^B}{dt}\aleph \quad \zeta_\Re \equiv \frac{d^B}{dt}\Re. \qquad (7.12)$$

7.3 Conservation laws

7.3.1 Micromechanical approach: conservation laws

By extending the definition of Coleman and Noll [33], the thermomechanic process at the crack interface may described by means of functions [165]:

1. the position of the material point $\mathbf{OM}(t)$ and the contact gap $\mathbf{c}(M, t)$,

2. the absolute temperature $\theta(M, t)$ and the temperature gap $\delta\theta(M, t)$,

3. the contact stress vectors $\mathbf{p}(M, t)$ and $\mathbf{p}'(M', t)$,

4. the entropy $s(M, t)$ and the contact entropy $s_c(M, t)$,

5. the internal energy $e(M, t)$, or alternatively the Helmholtz free energy $\phi(M, t)$, and the internal contact energy $e_c(M, t)$, or alternatively, the Helmholtz contact free energy $\phi_c(M, t)$,

6. the contact heat fluxes $q(M, t), q'(M', t)$,

7. the body force $\rho\mathbf{b}(M, t)$,

8. the volume heat source $r(M, t)$.

By analogy with the contact gap $\mathbf{c}$, we use the temperature gap $\delta\theta$ instead of the temperature θ':

$$\delta\theta = \delta\theta(M, t) = \theta(M, t) - \theta'(M^\perp(M, t), t). \tag{7.13}$$

Variables $\mathbf{OM}$, $\mathbf{c}$, θ, and $\delta\theta$ are the primary variables while the six functions $\mathbf{p}$, $\mathbf{p}'$, η_c, e_c, $\mathbf{q}$, and $\mathbf{q}'$ are the dual variables. They have to satisfy the axioms of constitutive laws such as the axiom of causality, the axiom determinism, the axiom of equipresence, and the axiom of frame indifference. Consider now a 3-form field $\rho e \omega_0$ defined on $\Sigma = B \cup B'$ and assumed to be an integral invariant in the sense of Poincaré during the motion:

$$\int_\Sigma \rho e \omega_0 = \int_{\varphi(\Sigma)} \rho e \omega_0 \implies \int_\Sigma \frac{d\Sigma}{dt}(\rho e \omega_0) = 0.$$

When expanding the l.h.s. of the second equation, we can write the general balance law:

$$\int_\Sigma \frac{d\Sigma}{dt}(\rho e \omega_0) = \int_\Sigma \frac{\partial}{\partial t}(\rho e)\omega_0 + \int_\Sigma h_0(\rho e \mathbf{v}) = \int_\Sigma r_e(\rho e) + \int_{\partial\Sigma} h_0(\mathbf{J}_e)$$

where $h_0(\rho e \mathbf{v})$ and $h_0(\mathbf{J}_e)$ are 2-form fields on the boundary. These terms may be related to the usual formulation of integral flux by using Cauchy's theorem. In the case in which a singular surface exists within the cell, the boundary may be split into two parts:

$$\partial(B \cup B') = (\partial B - \partial B_c) \cup (\partial B' - \partial B'_c). \tag{7.14}$$

Therefore the above relation gives the general balance of a continuum when there is a singular surface within it:

$$\begin{aligned}
& \int_\Sigma \frac{\partial \rho e}{\partial t}\omega_0 + \int_{\partial B} h_0(\rho e \mathbf{v}) + \int_{\partial B'} h_0(\rho e \mathbf{v}) \\
& - \int_{\partial B_c} h_0(\rho e \mathbf{v}) - \int_{\partial B'_c} h_0(\rho e \mathbf{v}) \\
& = \int_\Sigma r_e \omega_0 + \int_{\partial B} h_0(\mathbf{J}_e) + \int_{\partial B'} h_0(\mathbf{J}_e) \\
& - \int_{\partial B_c} h_0(\mathbf{J}_e) - \int_{\partial B'_c} h_0(\mathbf{J}_e). \tag{7.15}
\end{aligned}$$

When we let the two volumes B and B' shrink down respectively to the target surface ∂B_c and the striker surface $\partial B'_c$, their volume vanishes while the areas of the contact surfaces remain finite. Two situations may occur:

1. first, there is no contact and hence the target surface and the striker surface are empty sets,

2. and second, there is a contact and there is a bijective application between the two contacting surfaces.

In any case, since each body must satisfy the conservation laws, the following holds:

$$\int_{\partial B_c} [h_0(\rho e \mathbf{v}) - h_0(\mathbf{J}_e)] + \int_{\partial B_c'} [h_0(\rho e \mathbf{v}) - h_0(\mathbf{J}_e)] = 0. \qquad (7.16)$$

In local form, for every two points in contact at the contact surface,[2] we obtain the Kotchine theorem. On a singular surface, the general balance equation reduces to (remember we deal with 2-form fields):

$$[h_0(\rho e \mathbf{v}) - h_0(\mathbf{J}_e)] + [h_0(\rho e \mathbf{v}) - h_0(\mathbf{J}_e)]' = 0. \qquad (7.17)$$

7.3.2 Conservation laws at the crack interface

The above laws (7.16) and (7.17) are applied for mass, linear momentum, and angular momentum:

1. **For mass.** By introducing the quantities $e = 1, r_e = 0, \mathbf{J}_e = 0$, in the interface equation (7.16) we obtain:

$$\int_{\partial B_c} [h_0(\rho \mathbf{v})] + \int_{\partial B_c'} [h_0(\rho \mathbf{v})] = 0. \qquad (7.18)$$

We can deduce the local form, related to the so-called first of Cauchy's "shock relation" [179], called also the Stokes–Christoffel condition:

$$[h_0(\rho \mathbf{v})] + [h_0(\rho \mathbf{v})]' = 0. \qquad (7.19)$$

2. **For linear momentum.** By considering the linear momentum per mass unit:

$$e = p_u = \mathbf{g}(\mathbf{v}, \mathbf{u}) = \mathbf{u}^*(\mathbf{v})$$

and $r_e = \mathbf{u}^*(\rho \mathbf{b})$ and $\mathbf{J}_e = \mathbf{p}$, the conservation laws for linear momentum at the interface holds:

$$\int_{\partial B_c} [h_0(\rho \mathbf{u}^*(\mathbf{v})\mathbf{v}) - h_0(\mathbf{p})] + \int_{\partial B_c'} [h_0(\rho \mathbf{u}^*(\mathbf{v})\mathbf{v}) - h_0(\mathbf{p})] = 0$$

---

[2]One must be cautious when handling the 2-form about the orientation of surface.

or in local form:

$$[h_0(\rho \mathbf{u}^*(\mathbf{v})\mathbf{v}) - h_0(\mathbf{p})] + [h_0(\rho \mathbf{u}^*(\mathbf{v})\mathbf{v}) - h_0(\mathbf{p})]' = 0. \qquad (7.20)$$

Again, we recover the second relation in Cauchy's shock relation of linear momentum. This should be linked to the Principle of Action-Reaction.

3. **For angular momentum.** Consider now the angular momentum with respect to any fixed point A and along an axis directed by a vector $\mathbf{u}$ in the ambient space:

$$e = l_\delta = \omega_0(\mathbf{AM}, \mathbf{v}, \mathbf{u}) = (\mathbf{AM} \wedge \mathbf{v})(\mathbf{u})$$

and $r_e = 0$ with $\mathbf{J}_e = \mathbf{AM} \wedge \mathbf{p}$, the conservation laws for angular momentum at the interface holds:

$$\int_{\partial B_c} [h_0(\rho \omega_0(\mathbf{AM}, \mathbf{v}, \mathbf{u})\mathbf{v}) - h_0(\mathbf{AM} \wedge \mathbf{p})] \ +$$

$$\int_{\partial B_c'} [h_0(\rho \omega_0(\mathbf{AM}, \mathbf{v}, \mathbf{u})\mathbf{v}) - h_0(\mathbf{AM} \wedge \mathbf{p})] \ = \ 0$$

and in local form, for any point M at the contact surface:

$$[h_0(\rho \omega_0(\mathbf{AM}, \mathbf{v}, \mathbf{u})\mathbf{v}) - h_0(\mathbf{AM} \wedge \mathbf{p})] \ +$$
$$\left[h_0(\rho \omega_0(\mathbf{AM}, \mathbf{v}, \mathbf{u})\mathbf{v}) - h_0(\mathbf{AM} \wedge \mathbf{p}) \right]' \ = \ 0. \qquad (7.21)$$

Equation (7.21) does not appear in the classical shock relation although expressing the Action-Reaction Principle for the moment and torque at the singular surface. It could be considered as a another formulation of the Principle of Action-Reaction in mechanics [168].

4. **For internal energy.** By applying the same rule as previously, the energy conservation law for the contact surface is reduced to:

$$\int_{\partial B_c} h_0 \left[\rho \left(e + \frac{\mathbf{v}^2}{2} \right) \mathbf{v} \right] + \int_{\partial B_c'} h_0 \left[\rho \left(e + \frac{\mathbf{v}^2}{2} \right) \mathbf{v} \right]$$

$$= \int_{\partial B_c} [h_0(\mathbf{v} \cdot \sigma) - h_0(\mathbf{J}_q)] + \int_{\partial B_c'} [h_0(\mathbf{v} \cdot \sigma) - h_0(\mathbf{J}_q)]. \qquad (7.22)$$

The local form writes:

$$\left[h_0 \left\{ \rho \left(e + \frac{\mathbf{v}^2}{2} \right) \mathbf{v} \right\} - \{h_0(\mathbf{v} \cdot \sigma) - h_0(\mathbf{J}_q)\} \right] \ +$$
$$\left[h_0 \left\{ \rho \left(e + \frac{\mathbf{v}^2}{2} \right) \mathbf{v} \right\} - \{h_0(\mathbf{v} \cdot \sigma) - h_0(\mathbf{J}_q)\} \right]' \ = \ 0. \qquad (7.23)$$

In the above relation, the contact power is first defined as (and in a more classical form):

$$P_c(\partial B_c, t) \equiv \int_{\partial B_c} \mathbf{p}_c \cdot \zeta_c \, dA = \int_{\partial B_c} h_0(\mathbf{v} \cdot \sigma) + \int_{\partial B_c'} h_0(\mathbf{v} \cdot \sigma). \quad (7.24)$$

Second, the derivation of the conservation law of energy allows us to define and assume the existence of interfacial (internal) energy:

$$\dot{E}_c(\partial B_c, t) = \int_{\partial B_c} \dot{e}_c \, dA \equiv \int_{\partial B_c} \left[h_0 \left\{ \rho(e + \frac{\mathbf{v}^2}{2})\mathbf{v} \right\} \right]$$
$$+ \int_{\partial B_c'} \left[h_0 \left\{ \rho(e + \frac{\mathbf{v}^2}{2})\mathbf{v} \right\} \right]. \quad (7.25)$$

Therefore, the usual form of the conservation law of energy at the contact surface is obtained, e.g., [165]:

$$\dot{e}_c = \mathbf{p}_c \cdot \zeta_c - q - q'. \quad (7.26)$$

5. **For entropy.** The global entropy inequality takes the form of:

$$\int_B \frac{d^B}{dt}(\rho s \omega_0) + \int_{B'} \frac{d^{B'}}{dt}(\rho s \omega_0) + \int_{\partial B_c} h_0(\rho s \mathbf{v}) + \int_{\partial B_c'} h_0(\rho s \mathbf{v})$$
$$\geq \int_B \frac{r}{\theta}\omega_0 + \int_{B'} \frac{r}{\theta}\omega_0 - \int_{\partial B} h_0 \left(\frac{\mathbf{J}_q}{\theta} \right) - \int_{\partial B'} h_0 \left(\frac{\mathbf{J}_q}{\theta} \right)$$
$$- \int_{\partial B_c} h_0 \left(\frac{\mathbf{J}_q}{\theta} \right) - \int_{\partial B_c'} h_0 \left(\frac{\mathbf{J}_q}{\theta} \right) \quad (7.27)$$

and in local form:

$$[h_0(\rho s \mathbf{v})] + [h_0(\rho s \mathbf{v})]' \geq \left[h_0 \left(\frac{\mathbf{J}_q}{\theta} \right) \right] + \left[h_0 \left(\frac{\mathbf{J}_q}{\theta} \right) \right]'. \quad (7.28)$$

First, we assume the existence of an interfacial entropy rate which we write:

$$\int_{\partial B_c} \dot{s}_c \, dA \equiv \int_{\partial B_c} h_0(\rho s \mathbf{v}) + \int_{\partial B_c'} h_0(\rho s \mathbf{v}). \quad (7.29)$$

This approach is equivalent to the nonadditivity assumption of the entropy rate of two solids in contact, e.g., [56]. Indeed we assume that the contact surface has an interfacial entropy density projected onto ∂B_c and labeled hereafter $s_c(M, t)$. We can then deduce the entropy inequality at the contact surface in a more usual form [165]:

$$\dot{s}_c \geq - \left(\frac{q}{\theta} + \frac{q'}{\theta'} \right). \quad (7.30)$$

7.3.3   Continuum approach: conservation laws

Following Coleman and Noll [33], the thermomechanic process in a material, with respect to the referential body, is described by eight functions of position of a particle M and a time t [163]:

1. the vector position $\mathbf{OM}(t)$ of M with respect to the referential body, the metric tensor $\mathbf{g}(M, t)$, the volume form $\omega_0(M, t)$, the affine connection characterized by the torsion $\aleph(M, t)$, and the curvature $\Re(M, t)$ as a gauge for irreversible deformation,

2. the absolute temperature $\theta(M, t)$,

3. the stress tensor $\sigma(M, t)$,

4. the entropy $s(M, t)$,

5. the internal energy $e(M, t)$, or alternatively the Helmholtz free energy $\phi(M, t)$,

6. the heat flux vector $\mathbf{J}_q(M, t)$,

7. the body force $\rho\mathbf{b}(M, t)$,

8. the volume heat source $r(M, t)$.

The local conservation laws in the intrinsic form are the following:

1. the conservation of mass:

$$\frac{d^B}{dt}(\rho\omega_0) = 0, \tag{7.31}$$

2. the conservation of linear momentum and moment of momentum (without volume couple stress), $\mathbf{u}^*$ being any 1-form, (Cauchy's equations of motion):

$$\frac{d^B}{dt}[\mathbf{u}^*(\rho\mathbf{v})\omega_0] = \mathbf{u}^*(\mathrm{div}\,\sigma + \rho\mathbf{b}) \qquad \sigma^T = \sigma, \tag{7.32}$$

3. the conservation of total energy:

$$\frac{d^B}{dt}(\rho e\omega_0) + \frac{d^B}{dt}\left(\rho\frac{\mathbf{v}^2}{2}\omega_0\right) = -\mathrm{div}\,\mathbf{J}_q\,\omega_0 + [(\mathrm{div}\,\sigma + \rho\mathbf{b})\cdot\mathbf{v}]\,\omega_0$$
$$+ (\sigma : \nabla\mathbf{v})\,\omega_0 + r\,\omega_0, \tag{7.33}$$

4. the entropy inequality may first be written as follows:

$$\frac{d^B}{dt}(\rho s \omega_0) + \text{div} \left(\frac{\mathbf{J}_q}{\theta}\right) \omega_0 - \frac{r}{\theta} \omega_0 \geq 0. \tag{7.34}$$

Substituting the Helmholtz free energy $\varphi \equiv e - \theta s$, we obtain the inequality of Clausius–Duhem:

$$-\rho s \frac{d^B \theta}{dt} \omega_0 - \frac{d^B}{dt}(\rho \varphi \omega_0) + \sigma : \nabla v \, \omega_0 - \frac{\mathbf{J}_q}{\theta} \cdot \nabla \theta \, \omega_0 \geq 0. \tag{7.35}$$

7.4   Constitutive laws at the crack interface

7.4.1   Micromechanical approach: constitutive laws at the crack

Heat transfer from one lip in contact with another lip is due to heat generation by friction and/or due to the jump of temperature across the interface. Theoretical models of heat generation by friction may be found in, e.g., [85], [211]. Heat conduction and heat generation require to take into consideration the principles of thermodynamics. Bikerman [12] and Klamecki [92], [93] were among the first who proposed a friction description based on thermodynamics. By assuming a fading memory for the thermocontact phenomenon, without going into details, the constitutive functions reduce to [165]:

$$\mathfrak{I}_c = \hat{\mathfrak{I}}_c(c_n, \mathbf{c}_T, \theta, \delta\theta, \zeta_{c_n}, \zeta_{\mathbf{c}_T}, \zeta_\theta, \zeta_{\delta\theta}) \tag{7.36}$$

where:

$$\zeta_\theta \equiv \frac{d^B \theta}{dt} \qquad \zeta_{\delta\theta} \equiv \frac{d^B \delta\theta}{dt}. \tag{7.37}$$

Localization of the two principles of thermodynamics (7.26) and (7.30) give:

$$\dot{e}_c = \mathbf{p}_c \cdot \zeta_c - q - q' \qquad \dot{s}_c \geq -\left(\frac{q}{\theta} + \frac{q'}{\theta'}\right). \tag{7.38}$$

Definition 7.4.1 (*Interfacial free energy*) *The density of an interfacial free energy $\phi_c(M, t)$ distributed on ∂B_c is defined by e_c, s_c, and θ, respectively interfacial internal energy, the entropy of the interface, and the temperature on ∂B_c:*

$$\phi_c \equiv e_c - \theta s_c.$$

By substituting $\dot{s}_c$ from the above definition and $\dot{e}_c$ from (7.38)1, we have:

$$\frac{1}{\theta}(\mathbf{p}_c \cdot \zeta_c - q - q' - \dot{\phi}_c - \dot{\theta}s_c) \geq -\left(\frac{q}{\theta} + \frac{q'}{\theta'}\right). \tag{7.39}$$

By introducing the temperature gap, it reduces to:

$$\mathbf{p}_c \cdot \zeta_c - \dot{\phi}_c - \dot{\theta} s_c + \frac{q'}{\theta'} \delta\theta \geq 0.$$

Axiom 7.4.1 *(Contact thermomechanics) We assume that the interfacial free energy ϕ_c is a function (possibly nondifferentiable) of the normal contact gap c_n and the tangential contact gap $\mathbf{c}_T$ and a function (differentiable) of the interfacial temperature θ on ∂B_c, of the temperature gap $\delta\theta = \theta - \theta'$, and of their rates, respectively:*

$$\phi_c = \hat{\phi}_c(c_n, \mathbf{c}_T, \theta, \delta\theta, \zeta_{c_n}, \zeta_{\mathbf{c}_T}, \zeta_\theta, \zeta_{\delta\theta}). \tag{7.40}$$

Therefore, the objective rate of $\phi_c(M, t)$ may be expressed as:

$$
\begin{aligned}
\dot{\phi}_c &= (\partial_{c_n}\phi_c)\zeta_{c_n} + (\partial_{\mathbf{c}_T}\phi_c)\cdot\zeta_{\mathbf{c}_T} + \frac{\partial\phi_c}{\partial\theta}\zeta_\theta + \frac{\partial\phi_c}{\partial\delta\theta}\zeta_{\delta\theta} \\
&+ (\partial_{\zeta_{c_n}}\phi_c)\frac{d^B}{dt}\zeta_{c_n} + (\partial_{\zeta_{\mathbf{c}_T}}\phi_c)\frac{d^B}{dt}\zeta_{\mathbf{c}_T} \\
&+ \frac{\partial\phi_c}{\partial\zeta_\theta}\frac{d^B}{dt}\zeta_\theta + \frac{\partial\phi_c}{\partial\zeta_{\delta\theta}}\frac{d^B}{dt}\zeta_{\delta\theta}.
\end{aligned}
$$

Splitting the contact power into two parts (normal and tangential), we also obtain the entropy inequality at the interface:

$$
\begin{aligned}
(p_n - \partial_{c_n}\phi_c)\zeta_{c_n} &+ (\mathbf{p}_T - \partial_{\mathbf{c}_T}\phi_c)\cdot\zeta_{\mathbf{c}_T} - \left(s_c + \frac{\partial\phi_c}{\partial\theta}\right)\zeta_\theta + \frac{q'}{\theta'}\delta\theta \\
&- \left(\frac{\partial\phi_c}{\partial\delta\theta}\zeta_{\delta\theta} + (\partial_{\zeta_{c_n}}\phi_c)\frac{d^B}{dt}\zeta_{c_n} + (\partial_{\zeta_{\mathbf{c}_T}}\phi_c)\frac{d^B}{dt}\zeta_{\mathbf{c}_T}\right. \\
&\left. + \frac{\partial\phi_c}{\partial\zeta_\theta}\frac{d^B}{dt}\zeta_\theta + \frac{\partial\phi_c}{\partial\zeta_{\delta\theta}}\frac{d^B}{dt}\zeta_{\delta\theta}\right) \geq 0. \tag{7.41}
\end{aligned}
$$

We extend here the Coleman and Noll method [33] for contact thermomechanics by considering constitutive functions independent of the second-order derivatives of primal variables:

$$\frac{d^B}{dt}\zeta_{c_n} \qquad \frac{d^B}{dt}\zeta_{\mathbf{c}_T} \qquad \frac{d^B}{dt}\zeta_\theta \qquad \frac{d^B}{dt}\zeta_{\delta\theta}.$$

1. Starting with the list of arguments, we can arbitrarily choose second derivatives at the instant t^+ (brusque variation). To satisfy the second principle, it is necessary for second-order derivative coefficients to vanish:

$$\partial_{\zeta_{c_n}}\phi_c = 0 \qquad \partial_{\zeta_{\mathbf{c}_T}}\phi_c = 0 \qquad \frac{\partial\phi_c}{\partial\zeta_\theta} = 0 \qquad \frac{\partial\phi_c}{\partial\zeta_{\delta\theta}} = 0. \tag{7.42}$$

2. Therefore, the free energy ϕ_c, and subsequently the entropy s_c, do not depend on either on ζ_θ or on $\zeta_{\delta\theta}$. We can attribute arbitrary values for ζ_θ and $\zeta_{\delta\theta}$ at a time t^+ without violating the second principle. It implies:

$$s_c = -\frac{\partial \phi_c}{\partial \theta} \qquad \frac{\partial \phi_c}{\partial \delta\theta} = 0. \tag{7.43}$$

Finally, we obtain similar results as with continuum thermomechanics:

1. the interfacial free energy of contact thermomechanics defined by (7.40) necessarily takes the form of:

$$\phi_c = \hat{\phi}_c(c_n, \mathbf{c}_T, \theta), \tag{7.44}$$

2. the entropy inequality reduces to:

$$J_{c_q} \zeta_{c_q} + J_{c_n} \zeta_{c_n} + \mathbf{J}_{\mathbf{c}_T} \cdot \zeta_{\mathbf{c}_T} \geq 0 \tag{7.45}$$

in which:

$$\zeta_{c_q} \equiv \frac{\delta\theta}{\theta'} \qquad J_{c_q} \equiv q' \qquad J_{c_n} \equiv p_n - \partial_{c_n}\phi_c \qquad \mathbf{J}_{\mathbf{c}_T} \equiv \mathbf{p}_T - \partial_{\mathbf{c}_T}\phi_c. \tag{7.46}$$

Theorem 7.4.1 *(Contact irreversible dissipation) The entropy inequality (7.45) expresses the normal contact, the frictional contact, and thermal contact dissipations during the thermomechanic contact of two bodies.*

At this point, it is worthwhile to notice in (7.41) that the normal relative velocity may take any value and that this must not violate the second principle of the thermomechanics theory. We can deduce the following relationships:

$$p_n \in \partial_{c_n}\phi_c \qquad s_c = -\frac{\partial \phi_c}{\partial \theta} \qquad J_{c_q} \zeta_{c_q} + \mathbf{J}_{\mathbf{c}_T} \cdot \zeta_{\mathbf{c}_T} \geq 0. \tag{7.47}$$

The first inclusion (not an equality) is obtained by noticing the inequality $\zeta_{c_n} \geq 0$. Indeed, we can always conclude that $p_n \in \partial_{c_n}\phi_c$, whatever form of ϕ_c is chosen; it is the indicator function of R_+ with respect to the variable c_n. In the particular case in which the free energy does not depend on $\mathbf{c}_T$, we find the classical unilateral frictional laws, e.g., [140]. It has been shown that heat sources may appear in regions where contact occurs. Physically, this is represented by a distribution of heat sources concentrated in the contact surface. Furthermore, the positivity of contact dissipation suggests that the contact heat source is a warm source. This paragraph thus results in a theorem.

Theorem 7.4.2 *(Extension of Coleman and Noll's theorem) Let B and B' be two continua in contact with an interfacial free energy function such as* $\phi_c = \hat{\phi}_c(c_n, \mathbf{c}_T, \theta, \delta\theta, \zeta_{c_n}, \zeta_{\mathbf{c}T}, \zeta_\theta, \zeta_{\delta\theta})$. *Assume that the constitutive functions are not dependent on the second-order time derivatives of the primal variables* $(c_n, \mathbf{c}_T, \theta, \delta\theta)$. *Then, we have:*

$$\phi_c = \hat{\phi}_c(c_n, \mathbf{c}_T, \theta)$$

$$J_{c_q}\, \zeta_{c_q} + \mathbf{J}_{\mathbf{c}T} \cdot \zeta_{\mathbf{c}T} \geq 0$$

$$\zeta_{c_q} \equiv \frac{\delta\theta}{\theta'} \qquad J_{c_q} \equiv q' \qquad \mathbf{J}_{\mathbf{c}T} \equiv \mathbf{p}_T - \partial_{\mathbf{c}_T}\phi_c$$

$$p_n \in \partial_{c_n}\phi_c \qquad s_c = -\frac{\partial\phi_c}{\partial\theta}.$$

Remark. The result obtained in this paragraph constitutes an extension of the Coleman and Noll theorem [33] on continuum thermomechanics to contact thermomechanics [165].

7.4.2 Equations of cracked solids: micromechanical model

For each conservation law, two sets of equations should be considered: bulk equations and interfacial equations. By using the appropriate affine connection (metric connection for the bulk equations), we obtain the following equations, in any base $(\mathbf{u}_1, \mathbf{u}_2, \mathbf{u}_3)$:

1. Mass conservation:

$$\frac{\partial\rho}{\partial t} + \frac{1}{J_u}\left\{ \nabla_{\mathbf{u}_a}[J_u\rho\mathbf{v}(\mathbf{u}^a)] \right\} = 0 \qquad (7.48)$$

$$h_0(\rho\mathbf{v})(\mathbf{u}_a, \mathbf{u}_b) + h_0(\rho'\mathbf{v}')(\mathbf{u}'_a, \mathbf{u}'_b) = 0 \qquad (7.49)$$

2. Linear momentum conservation:

$$\rho\left(\frac{\partial\mathbf{v}}{\partial t} + \nabla_{\mathbf{v}}\mathbf{v} \right) = \frac{1}{J_u}\nabla_{\mathbf{u}_a}[J_u\sigma(\mathbf{u}^a)] + \rho\mathbf{b} \qquad (7.50)$$

$$h_0\left[\rho\mathbf{v} \otimes \mathbf{v}(\mathbf{u}^*) - \mathbf{p} \right](\mathbf{u}_a, \mathbf{u}_b) +$$
$$h_0\left[\rho'\mathbf{v}' \otimes \mathbf{v}'(\mathbf{u}^*) - \mathbf{p}' \right](\mathbf{u}'_a, \mathbf{u}'_b) = 0 \qquad (7.51)$$

3. Angular momentum conservation:

$$\sigma(\mathbf{u}^a, \mathbf{u}^b) = \sigma(\mathbf{u}^b, \mathbf{u}^a) \qquad (7.52)$$

$$h_0\left[\rho(\mathbf{AM} \wedge \mathbf{v})(\mathbf{u})\mathbf{v} - \mathbf{AM} \wedge \mathbf{p} \right](\mathbf{u}_a, \mathbf{u}_b) +$$
$$h_0\left[\rho(\mathbf{AM} \wedge \mathbf{v}')(\mathbf{u}')\mathbf{v}' - \mathbf{AM} \wedge \mathbf{p}' \right](\mathbf{u}'_a, \mathbf{u}'_b) = 0 \qquad (7.53)$$

4. Energy conservation (heat propagation):

$$\rho\left(\frac{\partial \theta}{\partial t} + \nabla_{\mathbf{v}}\theta\right) = -\frac{1}{J_u}\nabla_{\mathbf{u}_a}[J_u\mathbf{J}_q(\mathbf{u}^a)] + \rho\theta\frac{\partial}{\partial\theta}\left(\frac{\sigma}{\rho}\right):\zeta_g$$

$$+ \left[\mathbf{J}_g - \rho\theta\frac{\partial}{\partial\theta}\left(\frac{\mathbf{J}_g}{\rho}\right)\right] + r \qquad (7.54)$$

$$h_0\left[\rho\left(e + \frac{\mathbf{v}^2}{2}\right)\mathbf{v}\right](\mathbf{u}_a, \mathbf{u}_b) + h_0\left[\rho'\left(e' + \frac{\mathbf{v}'^2}{2}\right)\mathbf{v}'\right](\mathbf{u}'_a, \mathbf{u}'_b)$$

$$= [h_0(\mathbf{p})(\mathbf{v})(\mathbf{u}_a, \mathbf{u}_b) + h_0(\mathbf{p}')(\mathbf{v}')(\mathbf{u}'_a, \mathbf{u}'_b)]$$

$$- [h_0(\mathbf{J}_q)(\mathbf{u}_a, \mathbf{u}_b) + h_0(\mathbf{J}'_q)(\mathbf{u}'_a, \mathbf{u}'_b)] \quad (7.55)$$

5. Entropy inequality:

$$\mathbf{J}_q\cdot\zeta_q + \mathbf{J}_g : \zeta_g \geq 0 \qquad (7.56)$$

$$J_{c_q}\zeta_{c_q} + J_{c_n}\zeta_{c_n} + \mathbf{J}_{\mathbf{c}_T}\cdot\zeta_{\mathbf{c}_T} \geq 0. \qquad (7.57)$$

7.4.3   Normal dissipation at the crack interface

Let us focus only on particular classes of normal dissipative mechanisms in contact mechanics.

Axiom 7.4.2 (*Normal dissipation*) *Let* $\zeta = (\zeta_{c_n}, \zeta_{\mathbf{c}_T}, \zeta_{c_q})$ *be the rates that characterize the contact thermomechanics between two continua B and B'. The normal dissipation hypothesis assumes the existence of an internal parametrization by the variables ζ and the existence of a dissipation potential ψ_c, continuous positive (or null) and convex, such that:*

$$\psi_c = \hat{\psi}_c(\zeta_{c_n}, \zeta_{\mathbf{c}_T}, \zeta_{c_q}) \qquad \hat{\psi}_c(0, 0, 0) = 0$$

$$J_{c_q} = \frac{\partial\psi_c}{\partial\zeta_{c_q}} \qquad \mathbf{J}_{\mathbf{c}_T} = \frac{\partial\psi_c}{\partial\zeta_{\mathbf{c}_T}}. \qquad (7.58)$$

The quantities c_n, c_T, and θ may be considered as parameters of the dissipation potential. As previously, we do not mention it explicitly except when necessary. Such a contact model satisfies the second principle of thermodynamics. At first sight, one can observe macroscopically that the behavior of such a contact may change abruptly when the intensity of applied forces overpasses a certain critical value. This brusque variation requires a noncontinuously differentiable model. It is also worthwhile to notice that in Coulomb's classical law, the contact pressure is independent on the friction shear while the friction shear is proportional to the contact pressure. The dissipation potential is accordingly assumed to generally depend on the contact pressure rather than on the normal contact gap.

Definition 7.4.2 *(Conjugate potential of dissipation) The conjugate dissipation potential is defined by the partial Legendre–Fenchel transform, e.g., [170]:*

$$\hat{\psi}_c^*(\zeta_{c_n}, \mathbf{J}_{\mathbf{c}_T}, J_{c_q}) \equiv \mathrm{Sup}_{\zeta_{\mathbf{c}_T}, \zeta_{c_q} \in E_\zeta} [\mathbf{J}_{\mathbf{c}_T} \cdot \zeta_{\mathbf{c}_T} + J_{c_q} \zeta_{c_q}$$
$$- \hat{\psi}_c(\zeta_{c_n}, \zeta_{\mathbf{c}_T}, \zeta_{c_q})]. \tag{7.59}$$

This relation (7.59) is equivalent to Young's inequality:

$$\hat{\psi}_c^*(\zeta_{c_n}, \mathbf{J}_{\mathbf{c}_T}, J_{c_q}) + \hat{\psi}_c(\zeta_{c_n}, \zeta_{\mathbf{c}_T}, \zeta_{c_q}) - \mathbf{J}_{\mathbf{c}_T} \cdot \zeta_{\mathbf{c}_T} + J_{c_q} \zeta_{c_q} \geq 0$$
$$\forall \zeta_{c_n}, \zeta_{\mathbf{c}_T}, \zeta_{c_q}, \mathbf{J}_{\mathbf{c}_T}, J_{c_q}. \tag{7.60}$$

One can in theory find the particular values $(\zeta_{c_n}, \zeta_{\mathbf{c}_T}^*, \zeta_{c_q}^*)$ maximizing the bracketed term of the Legendre–Fenchel transform (7.59) to give:

$$\hat{\psi}_c^*(\zeta_{c_n}, \mathbf{J}_{\mathbf{c}_T}, J_{c_q}) \equiv \mathbf{J}_{\mathbf{c}_T} \cdot \zeta_{\mathbf{c}_T}^* + J_{c_q} \zeta_{c_q}^* - \hat{\psi}_c(\zeta_{c_n}, \zeta_{\mathbf{c}_T}^*, \zeta_{c_q}^*).$$

By substituting this formula in the inequality (7.60), one then obtains $\forall \zeta_{\mathbf{c}_T}, \zeta_{c_q}$:

$$\hat{\psi}_c(\zeta_{c_n}, \zeta_{\mathbf{c}_T}, \zeta_{c_q}) - \hat{\psi}_c(\zeta_{c_n}, \zeta_{\mathbf{c}_T}^*, \zeta_{c_q}^*) - \mathbf{J}_{\mathbf{c}_T} \cdot (\zeta_{\mathbf{c}_T} - \zeta_{\mathbf{c}_T}^*)$$
$$- J_{c_q}(\zeta_{c_q} - \zeta_{c_q}^*) \geq 0. \tag{7.61}$$

In this case, quantities $\mathbf{J}_{\mathbf{c}_T}$ and J_{c_q} are said to be subdifferential elements (partial) of the potential and we write for convenience, e.g., [140]:

$$\mathbf{J}_{\mathbf{c}_T} \in \partial \hat{\psi}_{c\zeta_{\mathbf{c}_T}}(\zeta_{c_n}, \zeta_{\mathbf{c}_T}, \zeta_{c_q}) \qquad J_{c_q} \in \partial \hat{\psi}_{c\zeta_{c_q}}(\zeta_{c_n}, \zeta_{\mathbf{c}_T}, \zeta_{c_q}). \tag{7.62}$$

Inversion of (7.62) and the Fenchel inequality, e.g., [140] directly provides the sliding velocity and the gap temperature, extending the differentiable relations to the noncontinuously differentiable case:

$$\zeta_{\mathbf{c}_T} \in \partial \hat{\psi}_{c\,\mathbf{J}_{\mathbf{c}_T}}^*(\zeta_{c_n}, \mathbf{J}_{\mathbf{c}_T}, J_{c_q}) \qquad \zeta_{c_q} \in \partial \hat{\psi}_{c\,J_{c_q}}^*(\zeta_{c_n}, \mathbf{J}_{\mathbf{c}_T}, J_{c_q}). \tag{7.63}$$

In most situations, the potential dissipation is differentiable with the thermal variables whereas it is not differentiable with respect to the sliding and contact variables:

1. The direct relationships hold:

$$\mathbf{J}_{\mathbf{c}_T} \in \partial \hat{\psi}_{c\zeta_{\mathbf{c}_T}}(\zeta_{c_n}, \zeta_{\mathbf{c}_T}, \zeta_{c_q}) \tag{7.64}$$

$$J_{c_q} = \frac{\partial}{\partial \zeta_{\mathbf{c}_T}} \hat{\psi}_{c\zeta_{c_q}}(\zeta_{c_n}, \zeta_{\mathbf{c}_T}, \zeta_{c_q}). \tag{7.65}$$

2. The inverse relationships can be directly obtained from the derivatives of the Legendre–Fenchel transform (7.59):

$$\zeta_{\mathbf{c}_T} \in \partial \hat{\psi}^*_{c\,\zeta_{\mathbf{c}_T}}(\zeta_{c_n}, \mathbf{J}_{\mathbf{c}_T}, J_{c_q}) \tag{7.66}$$

$$\zeta_{c_q} = \frac{\partial}{\partial \zeta_{\mathbf{c}_T}} \hat{\psi}^*_{c\,\zeta_{c_q}}(\zeta_{c_n}, \mathbf{J}_{\mathbf{c}_T}, J_{c_q}). \tag{7.67}$$

7.4.4 Positively homogeneous potential of dissipation

Consider the case where a sliding of contacting lips is accompanied by a dry friction. The friction force depends only on the direction and orientation of the relative of two contacting points. The friction force does not depend on the intensity of this relative velocity [140]. For this purpose, consider the space of rate of internal variables and the space of dual variables:

$$E_\zeta = (\zeta_{\mathbf{c}_T}, \zeta_{c_q}) \qquad E_J = (\mathbf{J}_{\mathbf{c}_T}, J_{c_q}).$$

Assume now the existence of a contact dissipation potential:

$$\hat{\psi}_c(\zeta_{c_n}, \zeta_{\mathbf{c}_T}, \zeta_{c_q}). \tag{7.68}$$

When a dry friction occurs, the constitutive functions of dissipative mechanisms are homogeneous of degree 0 with respect to internal variables rates. We deduce a dissipation potential of degree 1, $\forall \lambda > 0$:

$$\hat{\psi}_c(\zeta_{c_n}, \lambda\zeta_{\mathbf{c}_T}, \lambda\zeta_{c_q}) \equiv \lambda\hat{\psi}_c(\zeta_{c_n}, \zeta_{\mathbf{c}_T}, \zeta_{c_q}) \qquad \forall \zeta_{\mathbf{c}_T}, \zeta_{c_q} \in E_\zeta.$$

Theorem 7.4.3 *If the dissipation potential ψ_c is positive homogeneous of degree 1, then the conjugate potential satisfies necessarily the relation $\forall \lambda > 0$ and $\forall \mathbf{J}_{\mathbf{c}_T}, J_{c_q} \in E_J$:*

$$\lambda\hat{\psi}^*_c(\zeta_{c_n}, \mathbf{J}_{\mathbf{c}_T}, J_{c_q}) = \hat{\psi}^*_c(\zeta_{c_n}, \mathbf{J}_{\mathbf{c}_T}, J_{c_q}). \tag{7.69}$$

Proof. Using the positive homogeneity of degree 1 of ψ_c, we have:

$$\begin{aligned}
\lambda\psi^*_c &= \lambda\mathrm{Sup}_\zeta[\mathbf{J}_{\mathbf{c}_T} \cdot \zeta_{\mathbf{c}_T} + J_{c_q}\zeta_{c_q} - \hat{\psi}_c(\zeta_{c_n}, \zeta_{\mathbf{c}_T}, \zeta_{c_q})] \\
\lambda\psi^*_c &= \mathrm{Sup}_\zeta[\mathbf{J}_{\mathbf{c}_T} \cdot \lambda\zeta_{\mathbf{c}_T} + J_{c_q}\lambda\zeta_{c_q} - \hat{\psi}_c(\zeta_{c_n}, \lambda\zeta_{\mathbf{c}_T}, \lambda\zeta_{c_q})] \\
\lambda\psi^*_c &= \hat{\psi}^*_c(c_n, \mathbf{c}_T, \theta, \zeta_{c_n}, \mathbf{J}_{\mathbf{c}_T}, J_{c_q}).
\end{aligned}$$

Since the variables $\zeta_{\mathbf{c}_T}, \zeta_{c_q}$ span the set E_ζ, we can introduce a variable transform $(\zeta_{\mathbf{c}_T}, \zeta_{c_q})' = \lambda(\zeta_{\mathbf{c}_T}, \zeta_{c_q})$ and then show the result.

Theorem 7.4.4 *(Homogeneous dissipation potential) The conjugate dissipation potential ψ^*_c is necessarily the indicator function of a convex set C in the space of E_J:*

$$\hat{\psi}^*_c(\zeta_{c_n}, \mathbf{J}_{\mathbf{c}_T}, J_{c_q}) = \begin{cases} 0 & \mathbf{J}_{\mathbf{c}_T}, J_{c_q} \in C \\ \infty & \mathbf{J}_{\mathbf{c}_T}, J_{c_q} \notin C. \end{cases} \tag{7.70}$$

Proof. First, we observe that ψ_c^* can only take the values: $-\infty, 0$, or ∞.

1. If for a particular value of $\mathbf{J_{cT}}, J_{c_q}$, ψ_c^* is equal to $-\infty$, then it remains equal to $-\infty$ for any other value of $\mathbf{J_{cT}}, J_{c_q}$. This amounts to saying thant $\psi_c^* = -\infty$.

2. In the case where ψ_c^* is never equal to $-\infty$, by using the property (7.69), since λ is arbitrary, then ψ_c^* is equal to 0 or to ∞. Being a convex function, such a conjugate potential is necessarily the indicator function of a closed convex set of the space E_J.

Let us now determine the form of the potential. Starting from the conjugation, and by introducing the conjugate (7.59), the indicator of C, we can write:

$$\hat{\psi}_c(\zeta_{c_n}, \zeta_{c_T}, \zeta_{c_q}) \equiv \mathrm{Sup}_{\mathbf{J}_{cT}, J_{c_q} \in C} [\mathbf{J_{cT}} \cdot \zeta_{c_T} + J_{c_q}\zeta_{c_q} - \hat{\psi}_c^*(\zeta_{c_n}, \mathbf{J}_{cT}, J_{c_q})]$$

which gives:

$$\hat{\psi}_c(\zeta_{c_n}, \zeta_{c_T}, \zeta_{c_q}) = \max\{\mathrm{Sup}_{\mathbf{J}_{cT}, J_{c_q} \in C} [\mathbf{J_{cT}} \cdot \zeta_{c_T} + J_{c_q}\zeta_{c_q}], -\infty\}.$$

We deduce the form of the dissipation potential which is positive homogeneous of degree 1 with respect to ζ_{c_T}, ζ_{c_q}:

$$\hat{\psi}_c(\zeta_{c_n}, \zeta_{c_T}, \zeta_{c_q}) = \mathrm{Sup}_{\mathbf{J}_{cT}, J_{c_q} \in C} [\mathbf{J_{cT}} \cdot \zeta_{c_T} + J_{c_q}\zeta_{c_q}]. \tag{7.71}$$

The terms in between brackets are positive and represent the internal dissipation due to thermocontact behavior. Hence, the principle of maximal dissipation holds for a continuum with microcracks distribution, the analogous to the Hill–Mandel Principle, e.g., [39].

Theorem 7.4.5 *(Hill–Mandel Principle) Let there be two states of solicitation* $(\mathbf{J_{cT}}, J_{c_q})$ *(actual) and* $(\mathbf{J_{cT}^*}, J_{c_q}^*)$ *(virtual) belonging to the domain* C *of nonevolution of distribution of microcracks associated to the same rates* $(\zeta_{c_T}, \zeta_{c_q})$. *Then from (4.50), the principle of maximal work of internal microcracks holds:*

$$(\mathbf{J_{cT}} - \mathbf{J_{cT}^*}) \cdot \zeta_{c_T} + (J_{c_q} - J_{c_q}^*)\zeta_{c_q} \geq 0. \tag{7.72}$$

This shows that, given an evolution rate of slipping and contact thermics, the associated dual variables are those that maximize the dissipation due to thermocontact.

Theorem 7.4.6 *("Standard" contact) For normal dissipative mechanisms, constitutive laws of the thermocontacts may be entirely reconstructed from the free energy* ϕ_c *and the dissipation potential* ψ_c, *e.g., [166]:*

$$\phi_c = \hat{\phi}_c(c_n, \mathbf{c}_T, \theta) \qquad \psi_c = \hat{\psi}_c(\zeta_{c_n}, \zeta_{c_T}, \zeta_{c_q}). \tag{7.73}$$

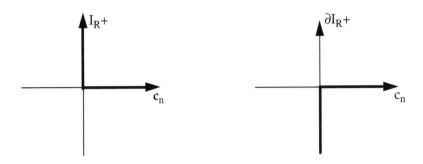

Figure 7.3. Constitutive laws for unilateral perfect contact behavior: the graph on the right-hand side gives the constitutive contact law (multi-valued contact law).

7.4.5 Classical Coulomb dry friction

Various particular laws for unilateral contact mechanics, e.g., [30], [140] may be included in the present framework.

1. **Unilateral contact.** For unilateral contact mechanics, the interfacial free energy is the indicator function of the positive real half line $\phi_c(c_n) = I_{R_+}(c_n)$. Indeed, unilateral contact laws are summarized by the Signorini relationships:

$$c_n \geq 0 \qquad p_n \leq 0 \qquad p_n\, c_n = 0. \qquad (7.74)$$

These are mathematical formulations of the two exclusive states: gap ($c_n > 0$ and $p_n = 0$) or contact ($c_n = 0$ and $p_n \leq 0$). Equivalently, multi-valued contact law $p_n[c_n]$ and its inverse $c_n[p_n]$ may be derived from two pseudo-potentials nondifferentiable in the convex analysis formulation:

$$p_n \in \partial I_{R_+}(c_n) \qquad c_n \in \partial I_{R_-}(p_n) \qquad (7.75)$$

where $I_{R_+}(c_n)$ is the indicator function of real numbers R_+ defined by:

$$I_{R_+}(c_n) = \begin{cases} 0 & c_n \in R_+ \\ \infty & c_n \notin R_+. \end{cases} \qquad (7.76)$$

$\partial I_{R_+}(c_n)$ is the subdifferential of $I_{R_+}(c_n)$ and $I_{R_-}(p_n)$ is its conjugate.

2. **Pure friction.** For pure friction (friction and constant pressure $p_n = c^{ste}$), the dissipation potential is $\psi_c = I^*_{C(p_n)}$. Indeed, pure friction can be considered as friction under a constant pressure. Two cases may occur in the framework of Coulomb friction: adherence $\|\mathbf{p}_T\| + \mu p_n < 0$ and slip $\|\mathbf{p}_T\| + \mu p_n = 0$. These

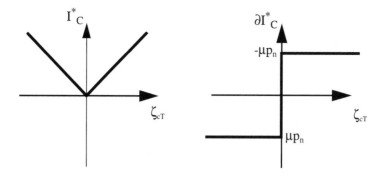

Figure 7.4. Constitutive laws for Coulomb friction behavior: The graph on the right gives the friction contact law (multi-valued frictional law).

two contact situations may be expressed by means of a slip law, a Coulomb criterion, and a complementary condition, μ being the friction coefficient:

$$\begin{aligned}
\zeta_{cT} \|\mathbf{p}_T\| &= \|\zeta_{cT}\|\mathbf{p}_T \\
\|\mathbf{p}_T\| + \mu p_n &\leq 0 \\
\|\zeta_{cT}\|(\|\mathbf{p}_T\| + \mu p_n) &= 0.
\end{aligned} \tag{7.77}$$

By using convex analysis formalism, as for unilateral contact, we can introduce two conjugate potentials and their subgradients as, e.g., [140]:

$$\mathbf{p}_T \in \partial I^*_{C(p_n)}(\zeta_{cT}) \qquad \zeta_{cT} \in \partial I_{C(p_n)}(\mathbf{p}_T) \tag{7.78}$$

where $I^*_{C(p_n)}(\zeta_{cT})$ is the Legendre–Fenchel conjugate of the indicator function $I_{C(p_n)}(\mathbf{p}_T)$ of the convex disk C of radius $-\mu p_n$ centered at the origin:

$$C = \{\mathbf{p}_T \mid \|\mathbf{p}_T\| + \mu p_n \leq 0, \, p_n \leq 0\}. \tag{7.79}$$

3. **Unilateral contact with friction.** In unilateral contact thermomechanics with friction, it is well known that the normal pressure is strongly coupled with tangential friction. First, let us recall the combination of normal contact and pure friction:

$$p_n \in \partial I_{R_+}(c_n) \qquad \mathbf{p}_T \in \partial I^*_{C(p_n)}(\zeta_{cT}) \tag{7.80}$$

where the domain of friction criterion C depends on the contact pressure p_n. Second, heat conduction laws at the interface of two solids were proposed in the works [121] and [122]. The most classical law being, e.g., [60], [135], [206], [207], [208]:

$$J_{c_q} = -\kappa \zeta_{c_q} \qquad \kappa = \hat{\kappa}(p_n)$$

where k is a heat conduction coefficient at the interface also depending on the thermal and mechanical properties of the bodies in contact. In the framework of normal dissipation, we can propose the potential [165]:

$$\psi_c = \frac{1}{2}\hat{\kappa}(p_n)\zeta_{c_q}^2 \qquad J_{c_q} = \hat{\kappa}(p_n)\zeta_{c_q}.$$

4. **Thermocontact with friction.** For a contact thermomechanics law with friction, we combine the above laws to obtain [165]:

$$
\begin{aligned}
\hat{\phi}_c(c_n) &= I_{R_+}(c_n) \\
\hat{\psi}_c(\zeta_{c_T}, \zeta_{c_q}, p_n) &= I^*_{C(p_n)}(\zeta_{c_T}) + \frac{1}{2}\hat{\kappa}(p_n)\zeta_{c_q}^2.
\end{aligned}
\tag{7.81}
$$

7.4.6 Continuum approach: bulk constitutive laws

The histories of dual variables are determined by the histories of the primal variables up to the time t. When the only time derivatives up to nth order of primal variables are involved instead of their entire histories, materials are called of the rate type n, e.g., [194]. The material of the rate type we deal with in this section is assumed to have constitutive laws defined by the tensor function:

$$\Im = \hat{\Im}(\omega_0, \mathbf{g}, \aleph, \Re, \theta, \nabla\theta, \zeta_g, \zeta_\aleph, \zeta_\Re, \zeta_\theta, \zeta_{\nabla\theta}). \tag{7.82}$$

For the models (7.40) and (7.82) to be comparable, we drop the dependence on the strain-rate tensor ζ_g. By applying the classical method of Coleman and Noll [33], we have deduced that:

1. the free energy ϕ of a material defined by (4.1) necessarily takes the form of:

$$\phi = \hat{\phi}(\omega_0, \mathbf{g}, \aleph, \Re, \theta), \tag{7.83}$$

2. the entropy inequality takes the form of, where the last two terms characterize the irreversibility due to microcrack distribution:

$$\mathbf{J}_q \cdot \zeta_q + \mathbf{J}_\aleph : \zeta_\aleph + \mathbf{J}_\Re : \zeta_\Re \geq 0 \tag{7.84}$$

where:

$$\zeta_q \equiv \theta \, \nabla\left(\frac{1}{\theta}\right) \tag{7.85}$$

$$\sigma = \rho\frac{\partial\phi}{\partial\omega_0} : \omega_0\,\mathbf{i} + 2\rho\frac{\partial\phi}{\partial\mathbf{g}} \qquad \mathbf{J}_\aleph \equiv -\rho\frac{\partial\phi}{\partial\aleph} \qquad \mathbf{J}_\Re \equiv -\rho\frac{\partial\phi}{\partial\Re}. \tag{7.86}$$

Theorem 7.4.7 *(Microcrack and dissipation) The entropy inequality (7.84) includes the dissipation due to thermics and to microcrack (scalar and vector discontinuity) distribution. Dissipation due to microcrack distribution is an irreversible dissipation that is induced by the change in the affine structure of the continuum.*

Dual variables, $\mathbf{J}_\aleph$ and $\mathbf{J}_\Re$, are analogous to the Eshelby stress tensor, e.g., [132] in the theory of dislocation elasticity; while primal variables, $\zeta_\aleph$ and $\zeta_\Re$, represent an objective (in the tensorial sense) extension of the classical inhomogeneity velocity gradient, as the evolution rate of the dislocation pattern, e.g., [51].

7.4.7 Basic equations of cracked continuum model

For describing large deformations of solids with microcracks, two possibilities appear for projecting conservation and constitutive laws, e.g., [163]. A usual method consists in choosing the same basis $(\mathbf{u}_1, \mathbf{u}_2, \mathbf{u}_3)$ embedded in B and in adopting the Riemannian connection compatible with the metric $\mathbf{g}$ the coefficients of which are expressed entirely by means of $g_{ab} = \mathbf{g}(\mathbf{u}_a, \mathbf{u}_b)$ and $[\mathbf{u}_a, \mathbf{u}_b] = \aleph^c_{0ab}\mathbf{u}_c$. We can simplify by choosing initially a coordinate basis $(\mathbf{u}_{10}, \mathbf{u}_{20}, \mathbf{u}_{30})$ transformed in $(\mathbf{u}_1, \mathbf{u}_2, \mathbf{u}_3)$ with $\mathbf{u}_a = d\varphi(\mathbf{u}_{a0})$. Actually, since B deforms in holonomic ($\mathbf{g}$ and ω_0) and nonholonomic ($\aleph$ and $\Re$) manners, constants of structure are not necessarily null:

$$[\mathbf{u}_a, \mathbf{u}_b] = \aleph^c_{0ab}\mathbf{u}_c \qquad \mathbf{u}_a \times \mathbf{u}_b = \omega_0(\mathbf{u}_a, \mathbf{u}_b, \mathbf{u}_c)\,\mathbf{u}^c.$$

We then project the conservation laws onto the basis $(\mathbf{u}_1, \mathbf{u}_2, \mathbf{u}_3)$ keeping in mind that the torsion and curvature tensors do not necessarily vanish. The mass conservation (3.36) reads:

$$\frac{\partial \rho}{\partial t} + \frac{1}{J_u}\nabla_{\mathbf{u}_a}[J_u\rho\mathbf{v}(\mathbf{u}^a)] + \rho\tilde{\aleph}_{0a}v^a = 0 \tag{7.87}$$

in which a source-like mass appears as in an open system (flux of microcrack distribution). The conservation of linear momentum (3.37) yields:

$$\rho\left(\frac{\partial \mathbf{v}}{\partial t} + \nabla_{\mathbf{v}}\mathbf{v}\right) = \frac{1}{J_u}\nabla_{\mathbf{u}_a}[J_u\sigma(\mathbf{u}^a)] + \sigma(\tilde{\aleph}_0) + \rho\mathbf{b}. \tag{7.88}$$

We notice the occurrence of internal forces due to nonhomogeneity (crack distribution), e.g., [132], [148], [194]. It is not always possible to eliminate such a supplementary force field by merely using an "appropriate" affine connection and an "appropriate" basis, e.g., [3]. From the internal energy $\hat{e}(\omega_0, \mathbf{g}, \aleph, \Re, \theta)$ and by decomposing (3.38) on $(\mathbf{u}_1, \mathbf{u}_2, \mathbf{u}_3)$, we obtain [163]:

$$\rho\left(\frac{\partial \theta}{\partial t} + \nabla_{\mathbf{v}}\theta\right) = -\frac{1}{J_u}\nabla_{\mathbf{u}_a}[J_u\mathbf{J}_q(u^a)] - \tilde{\aleph}_0(\mathbf{J}_q) + \tilde{\mathbf{J}}_\aleph : \zeta_\aleph + \tilde{\mathbf{J}}_\Re : \zeta_\Re \tag{7.89}$$

in which are defined the Eshelbian pseudo-"dual" variables:

$$\tilde{\mathbf{J}}_{\aleph} \equiv \mathbf{J}_{\aleph} - \rho\theta\frac{\partial}{\partial\theta}\left(\frac{\mathbf{J}_{\aleph}}{\rho}\right) \qquad \tilde{\mathbf{J}}_{\Re} \equiv \mathbf{J}_{\Re} - \rho\theta\frac{\partial}{\partial\theta}\left(\frac{\mathbf{J}_{\Re}}{\rho}\right). \tag{7.90}$$

The classical theory of thermoelasticity induces a heat propagation equation:

$$\rho\left(\frac{\partial\theta}{\partial t} + \nabla_{\mathbf{v}}\theta\right) = -\overline{\mathrm{div}}\,\mathbf{J}_q + r.$$

Then, we may conclude that the heat volume production rate, r, is identified to the term:

$$r = -\tilde{\aleph}_0(\mathbf{J}_q) + \tilde{\mathbf{J}}_{\aleph} : \zeta_{\aleph} + \tilde{\mathbf{J}}_{\Re} : \zeta_{\Re}.$$

Again, equation (7.89) shows that heat sources may appear in regions where the density of microcrack varies. Physically, this represents a distribution of heat sources concentrated in cracked regions. Furthermore, the positivity of internal dissipation suggests that the heat source of the microcrack distribution is a warm source. For each conservation law, equations may be divided into two sets: bulk equations, which can be related to the material without cracks and interfacial equations, here pointed out by the presence of the constants of structure.

7.4.8 Continuum with microcracks (normal dissipation)

The constitutive laws are defined by the free energy potential and a dissipation potential. To extend the theory of the thermoelastic continua to damaged continuum, we focus on a particular class of dissipative materials:

$$\begin{aligned} \Im &= \hat{\Im}(\omega_0, \mathbf{g}, \aleph, \Re, \theta, \zeta_{\aleph}, \zeta_{\Re}, \zeta_q) \\ \Im &= \rho, \sigma, s, \phi, \mathbf{J}_{\aleph}, \mathbf{J}_{\Re}, \mathbf{J}_q. \end{aligned} \tag{7.91}$$

For a material characterized by the existence of a stress threshold, the behavior may change abruptly when the intensity of applied forces goes beyond a certain critical value. This brusque variation requires a noncontinuously differentiable model [61]. In this case, the dual variables $\mathbf{J}_{\aleph}$ and $\mathbf{J}_{\Re}$, subdifferential (partial) elements of the dissipation potential, may be written as, e.g., [140]:

$$\mathbf{J}_{\aleph} \in \partial\hat{\psi}_{\aleph}(\theta, \zeta_{\aleph}, \zeta_{\Re}, \zeta_q) \qquad \mathbf{J}_{\Re} \in \partial\hat{\psi}_{\Re}(\zeta_{\aleph}, \zeta_{\Re}, \zeta_q). \tag{7.92}$$

Or inversely:

$$\zeta_{\aleph} \in \partial\hat{\psi}_{\aleph}^*(\mathbf{J}_{\aleph}, \mathbf{J}_{\Re}, \zeta_q) \qquad \zeta_{\Re} \in \partial\hat{\psi}_{\Re}^*(\mathbf{J}_{\aleph}, \mathbf{J}_{\Re}, \zeta_q). \tag{7.93}$$

These laws connect the evolution of the dislocations and disclinations pattern with the variables similar to the Eshelby stress-like tensors. In fact, these laws extend similar laws developed in, e.g., [51] which were restricted to the dislocation pattern evolution.

7.4.9 Positively homogeneous potential of dissipation

Consider the particular case where the occurrence of singularity is only accompanied by dry friction, the friction force depends only on the direction and orientation of the relative velocity of contacting points and not on the intensity of this relative velocity [140]. We introduced the dissipation potential ψ:

$$\hat{\psi}(\zeta_\aleph, \zeta_\Re, \zeta_q).\tag{7.94}$$

For the sake of simplicity, let us focus on the dissipation due to microcrack. Assuming the friction is a dry friction, the constitutive functions of dissipative mechanisms are homogeneous of degree 0 with respect to the internal variables rates. We then deduce results analogous to previous developments.

Theorem 7.4.8 *(Homogeneous dissipation potential) If the dissipation potential ψ is positive homogeneous of degree 1, then the conjugate potential dissipation potential ψ^* is necessarily the indicator function of a convex set in the space of E_J:*

$$\hat{\psi}^*(\mathbf{J}_\aleph, \mathbf{J}_\Re, \zeta_q) = \begin{cases} 0 & \mathbf{J}_\aleph, \mathbf{J}_\Re \in C \\ \infty & \mathbf{J}_\aleph, \mathbf{J}_\Re \notin C. \end{cases}\tag{7.95}$$

We also deduce, by explicitly writing the internal variables, the dissipation potential, due only to the nonhomogeneity, positive homogeneous of degree 1 with respect to $(\zeta_\aleph, \zeta_\Re)$:

$$\hat{\psi}(\zeta_\aleph, \zeta_\Re) = \mathrm{Sup}_{\mathbf{J}_{c_T}, \mathbf{J}_{c_q} \in C}(\mathbf{J}_\aleph : \zeta_\aleph + \mathbf{J}_\Re : \zeta_\Re).\tag{7.96}$$

The terms in between brackets are positive and represent the internal dissipation due to singularity distribution. For normal dissipative materials, constitutive laws of the microcracked continuum (4.1) may be entirely reconstructed from the free energy ϕ and the dissipation potential ψ, e.g., [64], [113], [209]:

$$\phi = \hat{\phi}(\omega_0, \mathbf{g}, \aleph, \Re, \theta) \qquad \psi = \hat{\psi}(\zeta_\aleph, \zeta_\Re, \zeta_q).\tag{7.97}$$

The generalized form of constitutive equations of damaged continuum in the framework of normal dissipative process are then summarized by:

$$\sigma = \rho \frac{\partial \phi}{\partial \omega_0} : \omega_0 \, \mathbf{i} + 2\rho \frac{\partial \phi}{\partial \mathbf{g}} \qquad s = \frac{\partial \phi}{\partial \theta}\tag{7.98}$$

$$-\rho \frac{\partial \phi}{\partial \aleph} \in \partial \hat{\psi}_\aleph \qquad -\rho \frac{\partial \phi}{\partial \Re} \in \partial \hat{\psi}_\Re \qquad \mathbf{J}_q = \frac{\partial \psi}{\partial \zeta_q}.\tag{7.99}$$

Example. As a simple example, one can propose the following free energy function and the dissipation potential. The dissipation is defined by a convex region C defining a zone of noncreation and nonevolution of microdiscontinuity of scalar and vector fields:

$$\phi = \frac{1}{2}\hat{\lambda}(\omega_0, \aleph, \Re, \theta)\mathrm{tr}^2\left(\frac{\mathbf{g}-\mathbf{I}}{2}\right) + \frac{1}{2}\hat{\mu}(\omega_0, \aleph, \Re, \theta)\mathrm{tr}\left(\frac{\mathbf{g}-\mathbf{I}}{2}\right)^2 \quad (7.100)$$

$$\psi = \frac{1}{2}\hat{\kappa}(\omega_0, \mathbf{g}, \theta)\|\zeta_q\|^2 + \mathrm{Sup}_{\mathbf{J}_{c_T}, J_{c_q} \in C}(\mathbf{J}_\aleph : \zeta_\aleph + \mathbf{J}_\Re : \zeta_\Re). \quad (7.101)$$

This model defines a Kelvin–Voigt material with an elastic behavior of Kirchhoff–St-Venant, a heat conduction law of Fourier and finally a rate-independent "rigid-plastic" behavior of the damage evolution.

7.5   Concluding remarks

In short, we can give a table of correspondences between the contact mechanics based model and the weakly continuous based model by comparing the expression of the sources of dissipation within the continuum.

The present section constitutes an attempt to give a theoretical basis to elaboration of the material thermomechanics in the presence of microcrack fields. Two models are presented. The main difference between these two methods resides in the choice of internal variables capturing the interstitial dissipation: torsion and curvature of the affine connection in the continuum approach and jump of velocity and jump of temperature across the cracks for the contact mechanics approach. In a sense, all crack

	Continuum	Micromechanics
Entropy inequality	$\mathbf{J}_q \cdot \zeta_q + \mathbf{J}_\aleph : \zeta_\aleph + \mathbf{J}_\Re : \zeta_\Re \geq 0$	$J_{c_q}\zeta_{c_q} + J_{c_n}\zeta_{c_n} + \mathbf{J}_{\mathbf{c}_T} \cdot \zeta_{\mathbf{c}_T} \geq 0$
Intrinsic dissipation	$\mathbf{J}_\aleph : \zeta_\aleph + \mathbf{J}_\Re : \zeta_\Re$	$J_{c_n}\zeta_{c_n} + \mathbf{J}_{\mathbf{c}_T} \cdot \zeta_{\mathbf{c}_T}$
Thermal dissipation	$\mathbf{J}_q \cdot \zeta_q$	$J_{c_q}\zeta_{c_q}$

Table 7.1. Comparison of dissipation terms between the micromechanical model and the continuum model.

types are spanned by the continuum model. The continuum approach involves more details than the micromechanical approach due to the fact that it accounts for the three translations and the three rotations of the contacting lips of the microcracks. Indeed, the occurrence of rotations allows us to capture a "nonlocal" behavior of microcracks: both relative translation and relative rotations, e.g., [166].

Previous studies have already shown the influence of the curvature of the internal interface, being considered as a potential location of crack surface within a continuum undergoing plane deformation, e.g., [129]. With a different approach brought by this book, the displacement continuity conditions could be replaced by equivalent equations involving the change in principal curvatures, in the mean geodesic torsion of internal curved crack interface [204]. Independently of this approach, it is worthwhile to consider the torsion and curvature tensors to capture the internal distortion of an interface embedded in a continuum undergoing large strains.

8
Conclusion

Any rational theory of continuum thermomechanics presupposes, for the formulation of both conservation laws and constitutive laws, some basic geometry. The choice of this geometical structure predetermines the effectiveness of the model to describe "abnormal phenomena" due to defects, which can be attributed to the change of local topology in the continuum. The present work is an attempt at giving some theoretical basis to the elaboration of the thermomechanics of a material in the presence of field singularity (distribution of scalar and vector discontinuity). Defects in various mechanical systems, such as fluids or solids, have the common property that, within the continuum limit, certain closed contour integrals over field either scalar or vector variables do not vanish because of the presence of singularity (in other words, there are jumps of fields). This process comes under the range of the Cartan path. Use of this path-dependent integral method leads us to obtain results, summarized in few points:

1. Deformation of a continuum with singularity has been shown to be characterized by: the metric tensor $\mathbf{g}$, the volume form ω_0, and an affine (not necessarily metric) connection ∇. This latter, by means of its torsion $\aleph$ and curvature tensors $\mathfrak{R}$, captures singularity as translational dislocation (scalar discontinuity) and rotational dislocation (vector discontinuity) referred to as disclinations in solids as well as the vorticity source in fluid flow. Torsion and curvature measure irreversible deformation (i.e., destruction of the affine equivalence between a

referential body and the deformed continuum). This could be related to the dissipation phenomenon within a material bulk.

2. Following Euler's original idea, we define an objective time derivative with respect to the continuum B. Two transposition theorems are proposed, which lead to the obtainment of a compatible time derivative in the theory of deformable solids undergoing large strains. The main applications are the time derivative of primal variables (metric, volume form, torsion and curvature tensors) and the localization of conservation laws.

3. The action of the external environment on the body through its boundary is assumed to be modeled by 2-form fields (scalar-valued for thermics and vector-valued for mechanics). The present work has shown that this assumption allows us to predict the existence of a vector heat flux and a stress vector and consequently of a stress tensor. The Cauchy theorem has been demonstrated within this framework.

4. In the presence of field singularity, a balance law of linear momentum may be found in classical works [148]. In the present study, we first define an intrinsic divergence and rotational operators by means of an affine connection and the Lie–Jacobi bracket. Then, it is possible to obtain compatible forms of conservation laws when singularity occurs, namely for the mass conservation equation and for the heat propagation equation.

5. In this book, we develop a generalized standard material belonging to the class of continuum with singularity. Such an approach could be used in modeling internal change of affine structure such as in damaged continuum or elastic-plastic continuum.

6. Comparison of two approaches, the micromechanics method based on contact thermomechanics and the continuum method in the presence of singularity fields, has been sketched in the last chapter of this book. Conservation laws and constitutive laws in their local forms have been explicitly formulated. Nevertheless, a more effective comparison should be conducted within the framework of the homogenization theory which is beyond the scope of the present work.

In short, we extend here the classical Cauchy equations of balance of linear momentum for nonhomogeneous elastic materials developed by Noll to a class of dissipating continua with singularity distribution including mass, linear momentum, energy, and entropy balance (in)-equations. The basic guideline is to relate the irreversible part of deformation to the change in the affine structure of the continuum. Indeed, such an approach has been proposed initially by Cartan as early as 1923 to extend the concept of spacetime of classical mechanics (Galilean group) to a spacetime appropriated to nonrelativistic gravitational mechanics (Cartan's group). In a sense, affine connection in continuum theory, besides vector and tensor fields, are an attractive and alternative way suitable to representing a dissipating continua with continuous distribution of singularity.

The guiding feature of the present work is that the evolution of continuum topology, quantified by the torsion tensor and the curvature tensor (Cartan), should be intensively used to model the discontinuous transformations. The fundamental conclusion is that the creation of singularity distribution, and its evolution, involves a topological change (connectivity) rather than the classical geometric change (shape and size). Finally, the topological change induces the irreversibility of the continuum evolution. The irreversibility of continuum thermomechanics is indeed linked to the study of topological properties of matter, either solid or fluid.

Appendix A
Mathematical Preliminaries

A.1 Vectors and tensors

A.1.1 Vector, space, basis

A vector space V over the real R is a set in which two operations, addition and multiplication by an element of R (called scalar), are defined. The elements of V (called vectors) satisfy the following axioms, for any vectors $\mathbf{u}$, $\mathbf{v}$, and $\mathbf{w}$ and any scalars λ and μ.

1. $\mathbf{u} + \mathbf{v} = \mathbf{v} + \mathbf{u}$

2. $(\mathbf{u} + \mathbf{v}) + \mathbf{w} = \mathbf{u} + (\mathbf{v} + \mathbf{w})$

3. There exists a zero vector 0 such that $\mathbf{u} + 0 = \mathbf{u}$

4. $\forall \mathbf{u} \in V$, there exists $-\mathbf{u}$, such that $\mathbf{u} + (-\mathbf{u}) = 0$

5. $\lambda(\mathbf{u} + \mathbf{v}) = \lambda\mathbf{u} + \lambda\mathbf{v}$

6. $(\lambda + \mu)\mathbf{u} = \lambda\mathbf{u} + \mu\mathbf{u}$

7. $(\lambda\mu)\mathbf{u} = \lambda(\mu\mathbf{u})$

8. $1\mathbf{u} = \mathbf{u}$.

Let there be $(\mathbf{u}_1, \ldots, \mathbf{u}_n)$, a set of n vectors of V. This set is linearly independent if:

$$\lambda^1 \mathbf{u}_1 + \cdots + \lambda^n \mathbf{u}_n = 0 \quad \Longrightarrow \quad \lambda^1 = \cdots = \lambda^n = 0. \tag{A.1}$$

Definition A.1.1 *(Vector basis, components) A set of n linearly independent vectors $(\mathbf{u}_1, \ldots, \mathbf{u}_n)$ is called a vector basis of V if any element $\mathbf{v}$ of V is written uniquely as a linear combination:*

$$\mathbf{v} = \lambda^1 \mathbf{u}_1 + \cdots + \lambda^n \mathbf{u}_n. \tag{A.2}$$

The numbers $\lambda^1, \ldots, \lambda^n$ are the components of $\mathbf{v}$ with respect to the basis.

This set is unique for a given basis, and $n = \dim V$ (dimension of V). Although V may be of infinite dimensional, we are mainly concerned with finite dimensional spaces in this book.

A.1.2   Linear maps and dual vector spaces

Definition A.1.2 *Given the vector space V, a map L is called a linear map if it satisfies (L is a real-valued linear function):*

$$L(\lambda^1 \mathbf{u}_1 + \lambda^2 \mathbf{u}_2) = \lambda^1 L(\mathbf{u}_1) + \lambda^2 L(\mathbf{u}_2)$$
$$\forall \lambda^1, \lambda^2 \in R \quad \forall \mathbf{u}_1, \mathbf{u}_2 \in V. \tag{A.3}$$

From the linearity of L, we can write for any basis $(\mathbf{u}_1, \ldots, \mathbf{u}_n)$:

$$L(\lambda^1 \mathbf{u}_1 + \cdots + \lambda^n \mathbf{u}_n) = \sum_{i=1}^{n} \lambda^i L(\mathbf{u}_i). \tag{A.4}$$

By introducing $L_i \equiv L(\mathbf{u}_i)$, we remark that $L(\mathbf{u})$ is a linear function of the components of $\mathbf{u}$:

$$L(\mathbf{u}) = \lambda^i L_i.$$

The set of all linear functions on the vector space V constitutes a vector space. Namely a linear combination of two linear functions is also a linear function:

$$(\lambda^1 L_1 + \lambda^2 L_2)(\mathbf{v}) = \lambda^1 L_1(\mathbf{v}) + \lambda^2 L_2(\mathbf{v})$$
$$\forall \lambda^1, \lambda^2 \in R \quad \forall \mathbf{v} \in V.$$

Definition A.1.3 *(Dual vector space) The set of all linear functions on the vector space V is called the dual vector space of V and is labeled V^*. An element of the dual vector space V^* is called a 1-form, a covariant vector, or a covector.*

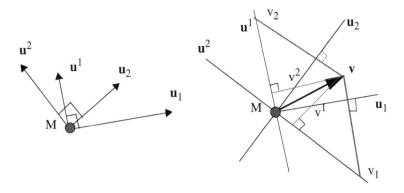

Figure A.1. Vector base and its dual. Vector decomposition on contravariant and covariant bases.

Remarks.

1. If the dimension of V is finite, dim $V^* = $ dim V. The dual basis (also named reciprocal basis) is completely defined by:

$$\mathbf{u}_i\,(\mathbf{u}^j) \equiv \delta_i^j \qquad i, j = 1, \ldots, n. \tag{A.5}$$

2. Geometrically, each one of the vectors $\mathbf{u}^j$ is perpendicular to the plane built from all other vectors $\mathbf{u}_{i \neq j}$. The magnitude of $\mathbf{u}^j$ is equal to the area of the plane constructed on the vectors $\mathbf{u}_{i \neq j}$, divided by the volume of parallelotope constructed on $\mathbf{u}_j$. This interpretation is important when dealing with the components of stress tensor σ.

A.1.3 Tensors, tensor product

Consider now the vector space product $E = V \times \cdots \times V$ on R of dimension n.

Definition A.1.4 (*p-covariant tensor*) *Consider any application* $\mathbf{A}$ *from* V^p *to* R *such that* $\mathbf{A}$ *is linear with respect to each argument.* $\mathbf{A}$ *is said to be a p-covariant tensor on* V:

$$\mathbf{A} : (\mathbf{u}_1, \ldots, \mathbf{u}_p) \in V^p \longrightarrow \mathbf{A}(\mathbf{u}_1, \ldots, \mathbf{u}_p) \in R.$$

The values of the function $\mathbf{A}(\mathbf{u}_1, \ldots, \mathbf{u}_n)$ must be independent of the vector basis onto which every vector is decomposed. In the same way, a q-contraviant tensor is a multi-linear real-valued function on V^{*q}:

$$\mathbf{A} : (\mathbf{u}^1, \ldots, \mathbf{u}^q) \in V^{*q} \longrightarrow \mathbf{A}(\mathbf{u}^1, \ldots, \mathbf{u}^q) \in R.$$

Any $(p+q)$-linear form on $V^p \times V^{*q}$ is labeled a q-contravariant, p-covariant tensor on V:

$$\mathbf{A} : (\mathbf{u}_1, \ldots, \mathbf{u}_p, \mathbf{u}^1, \ldots, \mathbf{u}^q) \in V^p \times V^{*q} \longrightarrow \mathbf{A}(\mathbf{u}_1, \ldots, \mathbf{u}_p, \mathbf{u}^1, \ldots, \mathbf{u}^q) \in R.$$

Definition A.1.5 *(Tensor product) For any two tensors* $\mathbf{A}$ *and* $\mathbf{B}$, *respectively* p *and* p' *covariant on* V, *we define an application from* $V^{p+p'}$ *into* R, *called the tensor product of* $\mathbf{A}$ *and* $\mathbf{B}$, *with:*

$$\mathbf{A} \otimes \mathbf{B}(\mathbf{u}_1, \ldots, \mathbf{u}_p, \mathbf{u}_{p+1}, \ldots, \mathbf{u}_{p+p'}) \equiv \mathbf{A}(\mathbf{u}_1, \ldots, \mathbf{u}_p) \cdot \mathbf{B}(\mathbf{u}_{p+1}, \ldots, \mathbf{u}_{p+p'}).$$

This application is $(p + p')$-linear on V and consequently is a $(p + p')$-covariant tensor. The tensor product is not commutative. Any p-covariant tensor $\mathbf{A}$ may be written in the form:

$$\mathbf{A} = A_{i_1 \ldots i_p} \mathbf{u}^{i_1} \otimes \cdots \otimes \mathbf{u}^{i_p}. \tag{A.6}$$

The set of tensors $\left\{ \mathbf{u}^{i_1} \otimes \cdots \otimes \mathbf{u}^{i_p} \right\}$ constitutes a tensor base for the vector space of p-covariant tensors. The set of numbers $A_{i_1 \ldots i_p}$ are said to be the covariant components of the tensor $\mathbf{A}$ on this tensor base. The extension of the tensor product is straightforward for any contravariant and mixed tensors. Then, any q-contravariant, p-covariant tensor may be decomposed along the tensor base:

$$\mathbf{u}^{i_1} \otimes \cdots \otimes \mathbf{u}^{i_p} \otimes \mathbf{u}_{j_1} \otimes \cdots \otimes \mathbf{u}_{j_q}. \tag{A.7}$$

Example. If $\mathbf{e}$ is a unit vector in a 3-dimensional space, the tensor product $\mathbf{e} \otimes \mathbf{e}$ is the projection onto the space of $\mathbf{e}$, whereas $\mathbf{I} - \mathbf{e} \otimes \mathbf{e}$ is the projection onto the plane normal to $\mathbf{e}$. The tensor $\mathbf{I} - 2\mathbf{e} \otimes \mathbf{e}$ is the reflection across the plane which is orthogonal to the unit vector.

A.1.4   p-forms, wedge product

Definition A.1.6 *(Symmetry) Consider* $\mathbf{A}$, *a* p-*covariant tensor.* $\mathbf{A}$ *is symmetric if, for any permutation* $(i_1, \ldots, i_p)$ *of* $(1, \ldots, p)$ *and for any* p-*plet of vectors* $(\mathbf{u}_1, \ldots, \mathbf{u}_p)$ *of* V:

$$\mathbf{A}(\mathbf{u}_{i_1}, \ldots, \mathbf{u}_{i_p}) = \mathbf{A}(\mathbf{u}_1, \ldots, \mathbf{u}_p).$$

For usual tensors in continuum mechanics, we can write symmetry for the "tensor"(order 2) and the "tangent tensor"(order 4) on a vector space V:

$$\mathbf{A}(\mathbf{u}_1, \mathbf{u}_2) = \mathbf{A}(\mathbf{u}_2, \mathbf{u}_1) \qquad \forall \mathbf{u}_1, \mathbf{u}_2 \in V \tag{A.8}$$

$$\mathbf{A}(\mathbf{u}_1, \mathbf{u}_2, \mathbf{u}_3, \mathbf{u}_4) = \mathbf{A}(\mathbf{u}_2, \mathbf{u}_1, \mathbf{u}_3, \mathbf{u}_4) = \mathbf{A}(\mathbf{u}_1, \mathbf{u}_2, \mathbf{u}_4, \mathbf{u}_3)$$

$$\mathbf{A}(\mathbf{u}_1, \mathbf{u}_2, \mathbf{u}_3, \mathbf{u}_4) = \mathbf{A}(\mathbf{u}_3, \mathbf{u}_4, \mathbf{u}_1, \mathbf{u}_2) \qquad \forall \mathbf{u}_1, \mathbf{u}_2, \mathbf{u}_3, \mathbf{u}_4 \in V. \tag{A.9}$$

The first equalities of the second relation constitute a minor symmetry, while the second relations the major symmetry.

Definition A.1.7 *(Skew symmetry) Consider a p-covariant tensor* **A**. **A** *is skew symmetric if, whenever there exists i $\neq$ j with* $\mathbf{u}_i = \mathbf{u}_j$, *we necessarily have:*

$$\mathbf{A}(\mathbf{u}_1, \ldots, \mathbf{u}_p) = 0.$$

A similar definition is available for q-contravariant tensor fields (but not for mixed tensors).

Definition A.1.8 *Any skew-symmetric p-covariant tensor is called a p-form on V. 0-forms are scalar.*

Definition A.1.9 *(Skew-symmetrization) Consider* **A** *a p-covariant tensor. The tensor* **B** *is called the skew symmetry of tensor* **A** *defined by, ε being the permutation tensor of a p-plet* $(1, 2, \ldots, p)$:

$$\mathbf{B}(\mathbf{u}_1, \ldots, \mathbf{u}_p) = \frac{1}{p!} \sum \varepsilon^{i_1 \ldots i_p}_{1 \ldots p} \, \mathbf{A}(\mathbf{u}_{i_1}, \ldots, \mathbf{u}_{i_p}).$$

This skew-symmetric application is p-linear and is therefore a p-form. This definition is intrinsic since it does not depend on any chosen base. We notice $\mathbf{B} = \mathcal{A}(\mathbf{A})$. Skew-symmetrization is linear and involutive.

Definition A.1.10 *(Wedge product) Let there be two tensors, a p-form* **A** *and a p'-form* **B** *on the vector space V. The field defined hereafter is a* $(p + p')$*-form and is called the wedge product of* **A** *and* **B**:

$$\mathbf{A} \wedge \mathbf{B} \equiv \frac{(p + p')!}{p! \, p'!} \mathcal{A}(\mathbf{A} \otimes \mathbf{B}). \tag{A.10}$$

For example, for two 1-forms ω and ω', we have:

$$\omega \wedge \omega'(\mathbf{u}, \mathbf{v}) = \omega(\mathbf{u})\omega'(\mathbf{v}) - \omega(\mathbf{v})\omega'(\mathbf{u}).$$

The exterior (or wedge) product is associative and anti-commutative:

$$(\mathbf{A} \wedge \mathbf{B}) \wedge \mathbf{C} = \mathbf{A} \wedge (\mathbf{B} \wedge \mathbf{C}) \qquad \mathbf{A} \wedge \mathbf{B} = (-1)^{pp'} (\mathbf{B} \wedge \mathbf{A}).$$

Moreover, the exterior product is right and left distributive for addition. The multiplication with any real number may be done at any moment.

Definition A.1.11 *(Base of p-forms) The following set of p-form fields constitutes a base of the vector space of p-forms:*

$$\mathbf{u}^{i_1} \wedge \cdots \wedge \mathbf{u}^{i_p} \qquad i_1 < \cdots < i_p. \tag{A.11}$$

For a space of dimension $m \leq 3$, the only existing bases are:

$$\{\mathbf{u}^1, \mathbf{u}^2, \mathbf{u}^3\} \qquad \{\mathbf{u}^1 \wedge \mathbf{u}^2, \mathbf{u}^2 \wedge \mathbf{u}^3, \mathbf{u}^3 \wedge \mathbf{u}^1\} \qquad \{\mathbf{u}^1 \wedge \mathbf{u}^2 \wedge \mathbf{u}^3\}. \tag{A.12}$$

A.2   Topological spaces

The main purpose of topology is to classify objects as spaces, in our case as continuum models. During its deformation, a continuum B is transformed from its initial configuration B_0 to its actual state $B = \varphi(B_0)$.[1] Therefore it is quite natural to borrow the concept of equivalence relation used in topology to analyze continuum deformation.

A.2.1   Topological spaces

To begin with, an open ball in R^n, of radius r, centered at $\mathbf{u} \in R^n$ is the set $B_v(r) = \{\mathbf{v} \in R^n \mid \|\mathbf{v} - \mathbf{u}\| < r\}$. A closed ball is defined by $\overline{B}_v(r) = \{\mathbf{v} \in R^n \mid \|\mathbf{v} - \mathbf{u}\| \leq r\}$. In other words, a closed ball is the union of an open ball and its edge or its boundary. A set $B_i \subset R^n$ is declared open if for any given element $\mathbf{u} \in B_i$ there is an open ball of a radius such that $r > 0$, centered at $\mathbf{u}$, that lies entirely in B_i. Both the real line R^n and the empty set $\emptyset$ are open sets. The union of any collection of open sets of R^n is open and the intersection of any finite number of open sets of R^n is also open. Extended material on topology and differential geometry may be found in, e.g., [14] for junior level and in, e.g., [144] for senior level. Let B be any set and $\Im = \{B_i, i \in I\}$ indicate a certain collection of subsets of B, called open sets. The pair $(B, \Im)$ is a topological space if $\Im$ satisfies the three following requirements:

1. $(\emptyset, B) \in \Im$.

2. If J is any (maybe infinite) subcollection of I, the family $\{B_j, j \in J\}$ satisfies $\bigcup_{j \in J} B_j \in \Im$.

3. If K is any finite subcollection of I, the family $\{B_k, k \in K\}$ satisfies $\bigcap_{k \in K} B_k \in \Im$.

B alone is often called a topological space. The B_i are called open sets and $\Im$ is said to give B a topology. A metric $\mathbf{g} : B \times B \longrightarrow R$ is a function that satisfies the axioms, for any elements $\mathbf{u}$, $\mathbf{v}$ and $\mathbf{w}$ of B:

1. $\mathbf{g}(\mathbf{u}, \mathbf{v}) = \mathbf{g}(\mathbf{v}, \mathbf{u})$.

2. $\mathbf{g}(\mathbf{u}, \mathbf{v}) \geq 0$ where the equality holds if and only if $\mathbf{u} = \mathbf{v}$.

3. $\mathbf{g}(\mathbf{u}, \mathbf{v}) + \mathbf{g}(\mathbf{v}, \mathbf{w}) \geq \mathbf{g}(\mathbf{u}, \mathbf{w})$.

---

[1] An equivalence relation in topology is a relation under which geometrical objects, such as a continuum, are classified according to whether it is possible to deform the continuum from an initial configuration to the actual configuration by continuous mapping.

Then, if B is endowed with a metric it is possible to define a collection of open discs and therefore a topology on B:

$$B_\varepsilon(\mathbf{v}) = \{\mathbf{u} \in B \mid \mathbf{g}(\mathbf{u}, \mathbf{v}) < \varepsilon\}.$$

The topology $\mathfrak{I}$ thus defined is called the metric topology determined by $\mathbf{g}$. The topological space is called the metric space in this particular case. In this book, we always assume a metric space, even when it is if not specially mentioned.

A.2.2   Continuous maps

Let B and B' be topological spaces. Basically, a map is a rule by which one assigns $\mathbf{v} \in B'$ for each element of $\mathbf{u} \in B$. We write $\mathbf{f} : B \longrightarrow B'$.

A map $\mathbf{f} : B \longrightarrow B'$ is continuous if the inverse image of an open set in B' is an open set in B. Suppose $\mathfrak{I}$ gives a topology to B. Let $\mathbf{u}$ be an element of B. $N(\mathbf{u})$ is a neighborhood of $\mathbf{u}$ if $N(\mathbf{u})$ is a subset of B and $N(\mathbf{u})$ contains at least one open set B_i to which $\mathbf{u}$ belongs. A topological space $(B, \mathfrak{I})$ is a Hausdorff space if, for any pair of distinct points $\mathbf{u}$ and $\mathbf{u}'$ of B, there always exist neighborhoods $N(\mathbf{u})$ and $N(\mathbf{u}')$ such that $N(\mathbf{u}) \cap N(\mathbf{u}') = \varnothing$. Continuous maps are to topological spaces what linear maps are to vector spaces, namely those that preserve the fundamental structures under consideration. To classify the various deformations $\varphi : B \longrightarrow B'$ of a continuum, let us recall the particular names of the maps:

1. φ is called injective (or one-to-one) map if $\mathbf{u} \neq \mathbf{v}$ implies $\varphi(\mathbf{u}) \neq \varphi(\mathbf{v})$,

2. φ is called surjective (or onto) map if for each $\mathbf{v} \in B'$ there exists at least one element of $\mathbf{u} \in B$ such that $\mathbf{u} = \varphi(\mathbf{u})$,

3. φ is bijective map if it is both injective and surjective.

A.2.3   Compactness

Let $(B, \mathfrak{I})$ be a topological space. A family $\{B_i\}$ of subsets of B is called a covering of B, if:

$$\bigcup_{i \in I} B_i = B. \tag{A.13}$$

If all B_i are open sets of the topology, the covering is called an open covering. Consider now all possible coverings of B. The set is compact if, for every open coverings of $\{B_i, i \in I\}$, there exists a finite subset J of I such that $\{B_j, j \in J\}$ is also a covering of B.

For example in R, the open interval $]a, b[$, considered as subspace of R, is not compact since it is not possible to extract a finite subcovering from the open covering given by the sets $B_n = \{u \in R \mid \frac{b-a}{n} < u < b\}$, $n = 1, \ldots, N$. Conversely, the closed interval $[a, b]$ is a compact subspace of R. It is worthwhile to cite two basic theorems on compactness.

Theorem A.2.1 *(Borel covering theorem) Every bounded closed subset $B_i \subset R^n$ is compact.*

Theorem A.2.2 *(Tychonov's theorem) The product of any nonempty class of compact spaces is itself compact.*

Without going into details, these two basic theorems on compact sets allows us to show that for any subset $B_i \subset R^n$ is compact if and only if:

1. B_i is a closed subset of R^n, and,

2. B_i is a bounded subset, that is, $\exists c > 0$, $\|\mathbf{u}\| < c, \forall \mathbf{u} \in B_i$.

Intuitively, the compactness is a kind of generalization of the notion of finiteness to topological setting. The sets under consideration generally have infinitely many points, but we can nonetheless define a notion of finiteness. An important property of compactness is that it is preserved under continuous mappings. In the case where we deal with a noncompact domain (for instance an initial configuration of a continuum), it is convenient to embed the domain in a layer space which is compact (compactification).

A.2.4 Connectedness

Roughly speaking, B is said to be connected if any two points in B may be connected by a continuous path lying in B. A topological space B is connected if it cannot be written as $B = B_1 \bigcup B_2$ where B_1 and B_2 are both open and $B_1 \bigcap B_2 = \varnothing$. This may be rephrased as follows:

Theorem A.2.3 *A topological space B is not connected if and only if there are nonempty open sets B_1 and B_2 such that $B_1 \bigcap B_2 = \varnothing$ and $B_1 \bigcup B_2 = B$.*

A loop in a topological space B is a continuous map $f : [0, 1] \longrightarrow B$ such that $\mathbf{f}(0) = \mathbf{f}(1)$. If any loop in B can be continuously shrunk to a point, B is called simply connected. Intuitively, a subset of Euclidean space is connected if it is made of "one piece." Otherwise B is called disconnected.

Remark. A space B is said to be totally disconnected if its only nonempty connected subsets consist of simple points.

A.2.5 Homeomorphisms and topological invariance

Consider two topological spaces (B_1, B_2). A map $\mathbf{f} : B_1 \longrightarrow B_2$ is a homeomorphism if it is continuous and has an inverse $\mathbf{f}^{-1} : B_2 \longrightarrow B_1$ that is also continuous. If there is a homeomorphism between (B_1, B_2), B_1 is said to be homeomorphic to B_2 and vice versa. Homeomorphisms are to topological spaces what isomorphisms are to groups (a linear isomorphism of vector spaces is linear map that is bijective). Topological invariants are those quantities that are conserved under homeomorphisms.

A basic problem in continuum mechanics modeling is to characterize the equivalent classes of homeomorphism by observing topological invariants. We remember the statement that if two spaces are not homeomorphic to each other, they have different topological invariants. Topological invariance constitutes the basic concept of this work. For example, the Hausdorff property and compactness are invariant under homeomorphisms. Historically, Noll's basic idea was to build the mathematical structure of continuum, evolving with respect to a Euclidean referential body Σ, on a set of mappings (T), called hereafter transformations, satisfying the following properties:

1. (T) is a set of invertible maps the domain and the image of which are subsets of the Euclidean space E^3;

2. Each element of (T) is C^1 (case of strongly continuous media);

3. For each element of (T), its inverse is also an element of (T);

4. When the image of a mapping of (T) is the domain of another map of (T), then their composition is also an element of (T).

It is straightforward to verify that the set of isometric transformations is a subset of (T). Starting from (T), Noll defined the mathematical structure of continua as follows. A material system B is a continuum of type (T) if it is endowed with a structure defined by a class of mappings (Π), called the placement of B in the space Σ which satisfies the following conditions:

1. (Π) is a set of invertible maps from B onto open subsets of Σ;

2. For any pair $(\mathbf{f}, \mathbf{g})$ of (Π), $\mathbf{f} \circ \mathbf{g}^{-1}$ is also an element of (T);

3. If $\mathbf{f}$ is an element of (Π) and $\mathbf{g}$ an element of (T), and if the image of $\mathbf{f}$ is equal to the domain of $\mathbf{g}$ then the composition $\mathbf{f} \circ \mathbf{g}$ is also a member of (Π).

The mathematical structure of continuum deformations, as developed by Noll, is basically built on the following theorem, e.g., [159], [205].

Theorem A.2.4 *(Topology invariance) Let B_1, B_2, and B_3 be topological spaces and let $f_1 : B_1 \longrightarrow B_2$ and $f_2 : B_2 \longrightarrow B_3$ be continuous mappings. If B is any subset of B_1, then we the following properties:*

1. *If B is connected, then $f_1 (B)$ is connected;*

2. *If B is compact and B_2 a Hausdorff space, then $f_1 (B)$ is compact;*

3. *The mapping $f_2 \circ f_1 : B_1 \longrightarrow B_3$ is continuous.*

Remarks. The connectedness of space is a topological invariant, meaning that if B_1 is homeomorphic to B_2, then B_2 is connected if and only if B_1 is connected. Conversely, if transformations of (T) are only piecewise C^1 (we can define them as weakly continuous media), their inverse does not necessarily exist and some of the Noll structure should be revisited.

A.3   Manifolds

The existence of a biunivoque map between the material points and the geometrical points in the space allows us to transfer the geometrical properties of the continuum onto the vector space underlying the Euclidean referential body. Theory of manifolds is essentially based on the transfer onto an abstract object, or onto the continuum, the elementary structure defined by the open subsets of R^n isomorphic to these objects.

A.3.1   Definition of manifold

Definition A.3.1 *(Differentiable manifold) B is an m-differentiable manifold if:*

1. *B is a topological space;*

2. *B is provided by a family of pairs $\{(B_i, \varphi_i)\}$;*

3. *$\{B_i\}$ is a family of open sets that covers B, that is, $\bigcup_i B_i = B$. φ_i is a homeomorphism from B_i onto an open subset B_i^r of R^m;*

4. *Given B_i and B_j such that $B_i \bigcap B_j = \varnothing$, the map $\varphi_{ij} \equiv \varphi_i \varphi_j^{-1}$ from $\varphi_j(B_i \bigcap B_j)$ to $\varphi_i(B_i \bigcap B_j)$ is infinitely differentiable.*

Some remarks hold:

1. Intuitively, a manifold is an extension of usual curves and surfaces to arbitrary dimensional objects (3-dimensional in this monograph);

2. Basically, a curve (resp. surface) in the 3-dimensional space is locally homeomorphic to R (resp. R^2). A manifold is a topological space locally homeomorphic to R^n (use of n parameters as local coordinates);

3. Globally, it is necessary to introduce several sets of coordinates at each point, the transition between these different sets of coordinates being smooth;

4. (B_i, φ_i) is called a chart and the set (B_i, φ_i) is called an atlas.

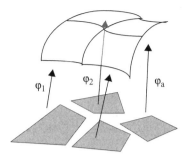

Figure A.2. The pair (B_i, φ_i) is called a chart while the whole family $\{(B_i, \varphi_i)\}$ is called an atlas. The subset B_i is called the coordinate neighborhood while φ_i is the coordinate function, or simply the coordinate.

A.3.2   Tangent vector

Motivation for a definition of tangent vectors on manifolds is to generalize the notion of directional derivative in R^n of a differential function and to consider tangent vectors as linear derivations of functions. A function ψ on B is a smooth (differentiable as many times as necessary) map from B to R. ψ is a real scalar-valued map. The set of functions on B is denoted $F(B)$. Let $\mathbf{u}$ be a vector with $M \in B$ for origin. The usual directional derivative of function ψ with respect to the vector $\mathbf{u}$ at the point M is defined as:

$$D_u \psi \equiv \frac{d}{d\varepsilon}[\psi(M + \varepsilon\mathbf{u})]\,|_{\varepsilon=0}\;.$$

Projected onto any base $(\mathbf{e}_1, \mathbf{e}_2, \ldots, \mathbf{e}_n)$ of the space, this derivative writes:

$$D_u \psi \equiv \mathbf{u}(\psi) = \sum_{i=1}^{n} \frac{\partial\psi}{\partial x^a}(M)u^a.$$

This defines an operator that is independent of the system of coordinates $(x^1, \ldots, x^n)$ which is used.

Definition A.3.2 *(Tangent vector) Consider a manifold B. A vector (tangent to B) at the point M is an application, denoted $\mathbf{u}$, from the set $F(B)$ defined on B into the set of real numbers R such that:*

$$\forall\psi_1, \psi_2 \in F(B) \qquad \mathbf{u}(\lambda_1\psi_1 + \lambda_2\psi_2) = \lambda_1\mathbf{u}(\psi_1) + \lambda_2\mathbf{u}(\psi_2)$$
$$\forall\lambda_1, \lambda_2 \in R \qquad \mathbf{u}(\psi_1\psi_2) = \lambda_1\mathbf{u}(\psi_1)\psi_2 + \psi_1\mathbf{u}(\psi_2). \qquad (A.14)$$

A vector field on the manifold B is the differentiable assignment of a vector $\mathbf{u}$ to each point $M \in B$. The tangent space may be endowed with two operators, addition and external multiplication by a real number, and therefore endowed with a structure of vector space:

$$\forall \psi \in F(B) \qquad \forall \lambda \in R \qquad \forall \mathbf{u}_1, \mathbf{u}_2 \in T_M B$$
$$(\mathbf{u}_1 + \mathbf{u}_2)(\psi) = \mathbf{u}_1(\psi) + \mathbf{u}_2(\psi) \qquad (\lambda \mathbf{u}_1)(\psi) = \lambda \mathbf{u}_1(\psi). \tag{A.15}$$

Definition A.3.3 *(Tangent vector space) The set of tangent vectors at M to B is denoted $T_M B$ and is called tangent space at M to the manifold B. It has the structure of a vector space.*

A.3.3 Tangent dual vector

Let us introduce the set of 1-forms $\omega : T_M B \longrightarrow R$ such that the image of any vector $\mathbf{u}$ is transcribed $\omega(\mathbf{u})$. Let us write:

$$\forall \mathbf{u} \in T_M B \qquad \forall \lambda \in R \qquad \forall \omega_1, \omega_2 \in T_M B^*$$
$$(\omega_1 + \omega_2)(\mathbf{u}) = \omega_1(\mathbf{u}) + \omega_2(\mathbf{u}) \qquad (\lambda \omega_1)(\mathbf{u}) = \lambda \omega_1(\mathbf{u}). \tag{A.16}$$

Definition A.3.4 *(Tangent dual vector space) The set of 1-forms on $T_M B$ defines the tangent dual vector space (space of linear maps), denoted $T_M B^*$, which is also endowed with the structure of a vector space.*

For a finite dimension manifold, the dimension of $T_M B^*$, such as m, is the same as the dimension of $T_M B$. $T_M B^*$ is also called cotangent space. Elements of $T_M B$ are contravariant vectors and those of $T_M B^*$ are covariant vectors. The dual basis is completely defined by the relation:

$$\mathbf{u}_a(\mathbf{u}^b) = \delta_a^b \qquad a, b = 1, \dots, m. \tag{A.17}$$

Consider now the vector space product $E = T_M B \times \cdots \times T_M B$ on R of dimension n.

Definition A.3.5 *(Tensor on B) Consider a manifold B and any application $\mathbf{A}_M$ from E to R such that $\mathbf{A}$ is linear with respect to each argument. $\mathbf{A}_M$ is said to be a p-linear form on $T_M B$ or a p-covariant tensor field on B.*

In the same way, we can define a q-contravariant, p-covariant tensor on $T_M B$ such as any $(p + q)$-linear form on:

$$T_M B^p \times (T_M B^*)^q.$$

Definition A.3.6 *(Tensor field) A tensor field on a manifold B is defined by the attribution of a tangent tensor on any point M of the manifold.*

A differentiable tensor field on a manifold B has components that vary differentiably on B. In continuum mechanics, we usually have to deal with order 0 (scalar), order 1 (vector and 1-form), order 2 (tensor), order 3 (volume form), and order 4 (tangent tensor).

Definition A.3.7 *(Tensor product) For any two tensor fields* **A** *and* **B**, *respectively p and p' covariant on $T_M B$, we define an application from $T_M B^{(p+p')}$ into R, called the tensor product of* **A** *and* **B**, *with:*

$$\mathbf{A} \otimes \mathbf{B}(\mathbf{u}_1, \ldots, \mathbf{u}_p, \mathbf{u}_{p+1}, \ldots, \mathbf{u}_{p+p'}) \equiv \mathbf{A}(\mathbf{u}_1, \ldots, \mathbf{u}_p) \cdot \mathbf{B}(\mathbf{u}_{p+1}, \ldots, \mathbf{u}_{p+p'}).$$

This application is $(p + p')$-linear on $T_M B$ and is therefore a $(p + p')$-covariant tensor. Any p-covariant tensor **A** may then be projected onto the local tensor base, at any point M,:

$$\mathbf{A} = A_{i_1 \ldots i_p} \, \mathbf{u}^{i_1} \otimes \cdots \otimes \mathbf{u}^{i_p}. \tag{A.18}$$

More familiar notation takes the form $\mathbf{A} = A_{i_1} \ldots i_p \, d\mathbf{x}^{i_1} \otimes \ldots \otimes d\mathbf{x}^{i_p}$. Tensors $\mathbf{u}^{i_1} \otimes \ldots \otimes \mathbf{u}^{i_p}$ (or $d\mathbf{x}^{i_1} \otimes \cdots \otimes d\mathbf{x}^{i_p}$) constitute a base of the p-covariant tensor space at any point M. The extension of the tensor product is straightforward for any contravariant and mixed tensors. Then, any q-contravariant, p-covariant tensor may be decomposed along the tensorial base:

$$\{\mathbf{u}^{i_1} \otimes \cdots \otimes \mathbf{u}^{i_p} \otimes \mathbf{u}_{j_1} \otimes \cdots \otimes \mathbf{u}_{j_q}\}. \tag{A.19}$$

An alternative notation for a coordinate base stands:

$$\left\{ d\mathbf{x}^{i_1} \otimes \cdots \otimes d\mathbf{x}^{i_p} \otimes \frac{\partial}{\partial \mathbf{x}^{j_1}} \otimes \cdots \otimes \frac{\partial}{\partial \mathbf{x}^{j_q}} \right\}.$$

A.3.4 Field of p-form

Definition A.3.8 *(Symmetry) Consider a manifold B and* **A** *a field of p-covariant tensor.* **A** *is symmetrical if, for any permutation $(i_1, \ldots, i_p)$ of $(1, \ldots, p)$ and for any p-plet of vector fields of $T_M B$, at any point $M \in B$:*

$$\mathbf{A}(\mathbf{u}^{i_1}, \ldots, \mathbf{u}^{i_p}) = \mathbf{A}(\mathbf{u}_1, \ldots, \mathbf{u}_p)$$

Again, for usual tensors in continuum mechanics, we can find symmetry for the "tensor"(order 2) and the "tangent tensor"(order 4):

$$\mathbf{A}(\mathbf{u}_1, \mathbf{u}_2) = \mathbf{A}(\mathbf{u}_2, \mathbf{u}_1) \qquad \forall \mathbf{u}_1, \mathbf{u}_2 \in T_M B \tag{A.20}$$

$$\mathbf{A}(\mathbf{u}_1, \mathbf{u}_2, \mathbf{u}_3, \mathbf{u}_4) = \mathbf{A}(\mathbf{u}_2, \mathbf{u}_1, \mathbf{u}_3, \mathbf{u}_4) = \mathbf{A}(\mathbf{u}_1, \mathbf{u}_2, \mathbf{u}_4, \mathbf{u}_3)$$

$$\mathbf{A}(\mathbf{u}_1, \mathbf{u}_2, \mathbf{u}_3, \mathbf{u}_4) = \mathbf{A}(\mathbf{u}_3, \mathbf{u}_4, \mathbf{u}_1, \mathbf{u}_2) \qquad \forall \mathbf{u}_1, \mathbf{u}_2, \mathbf{u}_3, \mathbf{u}_4 \in T_M B. \tag{A.21}$$

Definition A.3.9 *(Skew symmetry) Consider a p-covariant tensor field* **A** *on B. A is skew symmetric if, whenever there exists i $\neq j$ with* $\mathbf{u}_i = \mathbf{u}_j$, *we necessarily have at any point* $M \in B$:

$$\mathbf{A}(\mathbf{u}_1, \ldots, \mathbf{u}_p) = 0.$$

A similar definition is available for q-contravariant tensor fields.

Definition A.3.10 *(Skew-symmetrization) Consider* **A**, *a p-covariant tensor field on B. We call the skew-symmetrization of* **A** *the tensor field* **B** *defined by,* ε *being the permutation tensor of p-plet* $(1, 2, \ldots, p)$:

$$\mathbf{B}(\mathbf{u}_1, \ldots, \mathbf{u}_p) = \frac{1}{p!} \sum \varepsilon_{1,\ldots,p}^{i_1,\ldots,i_p} \mathbf{A}(\mathbf{u}^{i_1}, \ldots, \mathbf{u}^{i_p}).$$

Definition A.3.11 *(p-form) A p-form field on B is a p-covariant tensor field that is skew symmetric on B at any point* $M \in B$.

This definition is intrinsic since it does not depend on any base. We notice $\mathbf{B} = \mathcal{A}(\mathbf{A})$. Skew-symmetrization is linear and involutive.

Definition A.3.12 *(Wedge product) Let there be two tensor fields, p-form* **A** *and p'-form* **B**, *on* $T_M B$. *The field hereafter defined is a* $(p + p')$-*form field on the manifold B and is called the wedge product of* **A** *and* **B**:

$$\mathbf{A} \wedge \mathbf{B} = \frac{(p + p')!}{p!\, p'!} \mathcal{A}(\mathbf{A} \otimes \mathbf{B}). \tag{A.22}$$

Example. For any two 1-forms $\omega = d\mathbf{x}^i$ and $\omega' = d\mathbf{x}^j$, we have:

$$d\mathbf{x}^i \wedge d\mathbf{x}^j (\mathbf{u}, \mathbf{v}) = d\mathbf{x}^i(\mathbf{u}) d\mathbf{x}^j(\mathbf{v}) - d\mathbf{x}^i(\mathbf{v}) d\mathbf{x}^j(\mathbf{u}) = \det \begin{pmatrix} \mathbf{u}^i & \mathbf{v}^j \\ \mathbf{u}^j & \mathbf{v}^j \end{pmatrix}.$$

This is equal to the $\pm$ area of the parallelogram spanned by the projections of the vectors **u** and **v** onto the $x^i x^j$ plane.

As for with tensor algebra, the exterior product is associative and not commutative:

$$(\mathbf{A} \wedge \mathbf{B}) \wedge \mathbf{C} = \mathbf{A} \wedge (\mathbf{B} \wedge \mathbf{C}) \qquad \mathbf{A} \wedge \mathbf{B} = (-1)^{pp'} (\mathbf{B} \wedge \mathbf{A}).$$

Moreover, the exterior product is right and left distributive for addition and the multiplication with a real number may be done at any moment.

Theorem A.3.1 *(Base of p-forms) The following set of p-form fields constitutes a base of the p-forms space on B:*

$$\mathbf{u}^{i_1} \wedge \cdots \wedge \mathbf{u}^{i_p} \qquad i_1 < \cdots < i_p. \tag{A.23}$$

For a manifold of dimension $m \leq 3$, the only existing bases are:

$$\left\{\mathbf{u}^1, \mathbf{u}^2, \mathbf{u}^3\right\} \quad \left\{\mathbf{u}^1 \wedge \mathbf{u}^2, \mathbf{u}^2 \wedge \mathbf{u}^3, \mathbf{u}^3 \wedge \mathbf{u}^1\right\} \quad \left\{\mathbf{u}^1 \wedge \mathbf{u}^2 \wedge \mathbf{u}^3\right\}. \qquad \text{(A.24)}$$

Examples. Let $U \in R^3$ be an open set and (x^1, x^2, x^3) Cartesian coordinates of any point M of U (manifold). The canonical bases of 1-forms, 2-forms, and 3-forms are respectively:

$$\left\{d\mathbf{x}^1, d\mathbf{x}^2, d\mathbf{x}^3\right\} \quad \left\{d\mathbf{x}^1 \wedge d\mathbf{x}^2, d\mathbf{x}^2 \wedge d\mathbf{x}^3, d\mathbf{x}^3 \wedge d\mathbf{x}^1\right\} \quad \left\{d\mathbf{x}^1 \wedge d\mathbf{x}^2 \wedge d\mathbf{x}^3\right\}.$$

At any point, not on the z-axis, spherical coordinates (r, θ, φ) are admissible coordinates and the basis of 1-forms is:

$$\{dr, r\, d\theta, r\, \sin\theta\, d\varphi\}.$$

A.3.5 Mappings between manifolds

Let $\varphi : B \longrightarrow E$ be a map from a m-dimensional manifold B to a n-dimensional manifold E, $m \leq n$. A point $M \in B$ is mapped to a point $\varphi(M) \in E$. To a smooth map corresponds a linear map, called the differential map and written $d\varphi : T_M B \longrightarrow T_{\varphi(M)} E$. The differential map is one of the most important concepts in the continuum modeling. We thus proceed step by step. To start with, let $\psi : B \longrightarrow E$ be an element of $F(B)$. The differential of ψ at a point M, denoted $d\psi$, is the linear map $d\psi : T_M B \longrightarrow R$ defined as $d\psi(\mathbf{u}) \equiv \mathbf{u}(\psi)$. This definition can be brought down to the usual concept of differential:

$$d\psi(\mathbf{u}) \equiv \mathbf{u}(\psi) = \sum_{a=1}^{n} \frac{\partial \psi}{\partial x^a} u^a \qquad u^a = dx^a.$$

For the sake of clarity, we now consider a map from R^m to R^n. Of course, the tangent space to R^m in $\mathbf{x}$ is by definition the vector space of all vectors in R^m based in $\mathbf{x}$ (that is, a copy of R^m with origin shifted to $\mathbf{x}$). Let:

$$\varphi : (x^1, \ldots, x^m) \in R^m \longrightarrow (y^1, \ldots, y^n) \in R^n$$

be a smooth map, $\mathbf{y} = \mathbf{y}(\mathbf{x})$. Let $\mathbf{u}$ be a tangent vector to R^m at $\mathbf{x}_0$. Consider now any smooth curve $\mathbf{x}(t)$ such that $\mathbf{x}_0 = \mathbf{x}(0)$ and that $\dot{\mathbf{x}}(0) \equiv \frac{d}{dt}\mathbf{x}(t)\,|_{t=0} = \mathbf{u}_0$. The image of such a curve is transcribed:

$$\mathbf{y}(t) = \varphi[\mathbf{x}(t)].$$

It has a tangent vector $\mathbf{u}$ at $\mathbf{y}_0$ calculated by the chain rule:

$$\mathbf{u} = \frac{d}{dt}\mathbf{y}(t) \mid_{t=0} = \sum_{i=1}^{m} \frac{\partial y^j}{\partial x^i}(\mathbf{x}_0)\dot{x}^i(0) = \sum_{i=1}^{m} \frac{\partial y^j}{\partial x^i} u_0^i.$$

$\mathbf{u}_0 \longrightarrow \mathbf{u} = d\varphi(\mathbf{u}_0)$ is, in this equation, independent of the curve $\mathbf{x}(t)$, and defines a linear transformation called the differential of φ at x_0:

$$d\varphi : \mathbf{u}_0 \longrightarrow \mathbf{u} = d\varphi(\mathbf{u}_0).$$

Definition A.3.13 *(Differential map) Let φ be a map from B to E. The differential $d\varphi$ of φ has the same meaning as in the case of $R^m \longrightarrow R^n$. Let $\mathbf{u}_0 \in T_M B$, $\mathbf{x}(t)$ be a curve on B with $\mathbf{x}(0) = 0$ with the velocity vector $\dot{\mathbf{x}}(0) = \mathbf{u}_0$. Then $\mathbf{u} = d\varphi(\mathbf{u}_0)$ is the velocity vector $\frac{d}{dt}\varphi[\mathbf{x}(t)] \mid_{t=0}$ of the image curve at $\mathbf{y} = \varphi(\mathbf{x})$ on E. This vector $\mathbf{u}$ is independent of the chosen curve $\mathbf{x}(t)$, as long as $\dot{\mathbf{x}}(0) = \mathbf{u}_0$.*

The matrix of this differential map is given by:

$$d\varphi_a^i = \frac{\partial y^i}{\partial x^a}.$$

An important application of the differential map comes from the inverse function theorem (recalled without proof), which is useful for a change of coordinates on a manifold.

Definition A.3.14 *(Diffeomorphism) Let $\varphi : B \longrightarrow E$, (dim B = dim E), be a map. φ is a diffeomorphism if φ is one-to-one map, onto map, and if, in addition, φ^{-1} is also differentiable.*

In other words, a diffeomorphism φ is a differentiable homeomorphism with a differentiable inverse φ^{-1}. When a diffeomorphism exists between two differential manifolds B_0 and B with dim B_0 = dim B, they are said to be diffeomorphic. Such is the case for the various configurations of a continuum undergoing nonlinear elastic deformation.

Theorem A.3.2 *(Inverse function theorem) If the map $\varphi : B \longrightarrow E$ (dim B = dim E) is a differentiable map between manifolds of the same dimension, and if at $\mathbf{x}_0 \in B$, $d\varphi$ is an isomorphism (one-to-one and onto), then φ is a local diffeomorphism near $\mathbf{x}_0$.*

Practically, this means that there is a neighborhood $N(\mathbf{x})$ of $\mathbf{x}$ such that $\varphi[N(\mathbf{x})]$ is open in E and $\varphi : N(\mathbf{x}) \longrightarrow \varphi[N(\mathbf{x})]$ is a diffeomorphism. Therefore, it is possible to define new coordinates in a neighborhood of $\mathbf{x}$.

Definition A.3.15 *(Immersion) Let $\varphi : B \longrightarrow E$ be a smooth map and let* dim $B \leq$ dim E. *The map φ is called an immersion of B into E if its differential $d\varphi : T_M B \longrightarrow T_{\varphi M} E$ is a one-to-one (injection) map at every point M of B.*

For every point $\mathbf{x} \in B$, the differential $d\varphi$ has a rank equal to the dimension of B (n). It means that the matrix representing this differential has n linearly independent rows (or equivalently, n linearly independent columns):

$$[d\varphi] = \begin{pmatrix} \frac{\partial \varphi^1}{\partial x^1} & \cdots & \frac{\partial \varphi^1}{\partial x^n} \\ \cdots & \cdots & \cdots \\ \frac{\partial \varphi^n}{\partial x^1} & \cdots & \frac{\partial \varphi^n}{\partial x^n} \\ \cdots & \cdots & \cdots \\ \frac{\partial \varphi^m}{\partial x^1} & \cdots & \frac{\partial \varphi^m}{\partial x^n} \end{pmatrix}.$$

Definition A.3.16 *(Embedding) Let $\varphi : B \longrightarrow E$ be a smooth map and let* dim $B \leq$ dim E. *The map φ is called an embedding of B into E if $\varphi : B \longrightarrow E$ is an injection and an immersion.*

Now we introduce the dual notion of submersion by considering the following diagram and starting from the immersion $\varphi : B \longrightarrow E$ and $\psi : E \longrightarrow B$.

Definition A.3.17 *(Submersion) Let $\psi : E \longrightarrow B$ be a smooth map and let* dim $B \leq$ dim E. *The map ψ is called a submersion if its differential $d\psi : T_M E \longrightarrow T_{\psi M} B$ is an onto map.*

For every point $y \in E$, dim $E = m$, and dim $B = n$, the differential $d\psi$ has equal rank to the dim $B = n$. This means that the matrix representing $d\psi$ has n independent columns (or equivalently n independent rows):

$$[d\varphi] = \begin{pmatrix} \frac{\partial \psi^1}{\partial y^1} & \cdots & \frac{\partial \psi^1}{\partial y^n} & \cdots & \frac{\partial \psi^1}{\partial y^m} \\ \cdots & \cdots & \cdots & \cdots & \cdots \\ \frac{\partial \psi^n}{\partial y^1} & \cdots & \frac{\partial \psi^n}{\partial y^n} & \cdots & \frac{\partial \psi^n}{\partial y^m} \end{pmatrix}.$$

Appendix B
Invariance Group and Physical Laws

B.1 Conservation laws and invariance group

Conservation laws are essentially derived from the existence of an equivalence class of referential bodies (ambient space) [48]. This ambient space Σ is directed by a Euclidean vector space endowed with a metric (associated to an orthogonal group) and a volume form (orientation). Moreover, chronology is an operation of temporal classification of events. Classical mechanics admits a unique chronology of the events independently of the position of the point within the ambient space. The clock measuring time is then introduced accordingly. A spacetime structure is the couple formed by the referential body and the clock.

B.1.1 Newton spacetime

Newton's original point of view assumes the existence of at least one fixed referential body, that is the existence of an absolute ambient space and an absolute chronology of events.

Axiom B.1.1 *(Newton) There exists at least one fixed referential body (absolute) endowed with a Euclidean structure (metric, orthogonal group). With respect to two referential bodies belonging to the equivalence class, the vector position of any material point M is related by:*

$$\overline{\mathbf{OM}} = \alpha \mathbf{OM} + \beta. \tag{B.1}$$

The transformations α (rotation) and β (translation) do not depend on the time t and express the indifference of the origin position O (homogeneity of space) and of the referential body orientation (isotropy of space). The group of transformations that keeps the Newtonian spacetime structure invariant is defined by: dilatation (3 parameters), rotation (3 parameters), and translation (3 parameters) of space and the shift of the time origin (1 parameter). This group of transformations is called the elementary group of Weyl.

B.1.2   Leibniz spacetime

The existence of an absolute space was questioned among others by Leibniz, who proposed a spacetime structure where neither translation nor rotation is preferred. The associated group is called the kinematics group. It contains 3 parameters of dilatations and 6 functions depending on the time: 3 for the rotation velocity and 3 for the translation velocity. The shift of the time origin is characterized by 1 parameter.

Axiom B.1.2 *(Leibniz) There exists at least one equivalence class of referential bodies (ambient space). The referential rigid bodies may have any rigid motion, rotation or/and translation, with respect to each other. The vector position of a material point expressed in any two referential bodies belonging to this equivalence class is transformed as follows:*

$$\overline{\mathbf{OM}} = \alpha(t)\mathbf{OM} + \beta(t). \tag{B.2}$$

The Leibniz group constitutes a homogeneous invariance group defined by the Euclidean displacement: an isometric transformation that keeps constant the distance between any two material points of the continuum. Any Euclidean displacement can be brought down to a composition of translation and orthogonal transformation.

B.1.3   Galilean spacetime

Euler and later Lange recognized the importance of affinity property in the elaboration of spacetime structure of dimension 4. The basic idea was that the affine lines of ambient space that are not contained in the hyperplanes t = constant should be able to represent free motions of material points. This axiom expresses the inertial law without any reference to any coordinate system.

Axiom B.1.3 *(Classical Mechanics-Galilee) There exists at least one equivalence class of referential bodies endowed with a Euclidean structure (metric, orthogonal group). The referential bodies may have relative uniform translation motion with respect to each other. The representations of any vector position in any two referential bodies of this equivalence class are related by:*

$$\overline{\mathbf{OM}} = \alpha\mathbf{OM} + \beta t + \mathbf{u}_0. \tag{B.3}$$

The spacetime of classical mechanics has been based and elaborated on this inertial law. The associated invariance group is called the Galilean group. On the one hand, it includes the elementary group of Weyl and on the other hand, it is included in the group of Leibniz. This Galilean group allows us to define entirely the equivalence class of inertial referential bodies.

B.1.4 Physical roots of conservation laws

Bessel and Hagen have extensively resorted to the use of the Galilean group of transformations to derive conservation laws when dealing with material discrete points without dissipation. In the same way, relations between an invariance group and conservation laws are based essentially on the Noether theorem in analytical mechanics. Mathematically, the invariance subgroups are the following:

1. **Energy.** The time origin shift, equivalent to a uniform flow of time $\overline{\mathbf{OM}}(\bar{t}) = \mathbf{OM}(t)$, $\bar{t} = t - a$, induces the energy conservation. Implicitly, it assumes the existence of a referential body endowed with a clock so that the time is uniform. This particular transformation has two specific properties: the invariance with respect to the choice of time origin and the limitation to the chronology orientation. In fact, two conservation laws are associated to it: the conservation of energy and the production of internal entropy. An extended invariance group including the change of the time orientation $\overline{\mathbf{OM}}(\bar{t}) = \mathbf{OM}(t)$ and $\bar{t} = \pm t - a$, includes some features of the second law of thermodynamics.

2. **Linear momentum.** The indifference of the space origin $\overline{\mathbf{OM}}(\bar{t}) = \mathbf{OM}(t) + \mathbf{u}_0$, $\bar{t} = t$ or of the space homogeneity implies the conservation of linear momentum.

3. **Angular momentum.** The indifference of the referential body orientation $\overline{OM}(\bar{t}) = \alpha \mathbf{OM}(t)$, $\bar{t} = t$ or of the space isotropy gives the conservation of angular momentum.

4. **Center of mass.** The indifference with respect to the uniform translation of any two referential bodies $\overline{\mathbf{OM}}(\bar{t}) = \mathbf{OM}(t) + \beta t$, $\bar{t} = t$ induces the uniformity (constant velocity) of the center of mass of the system.

B.2 Constitutive laws and invariance group

Material properties, idealized by constitutive functions, should not depend on the referential body and on the ambient space definition. To derive the objectivity, we retain hereafter the motion and the temperature histories as primal variables of constitutive functions. Implicitly, authors have utilized an extended invariance group for this purpose.

B.2.1   Spacetime of Cartan

The spacetime of Cartan is inspired from the nonhomogeneous invariance group. Due to strain effects, continuum deformations are usually not isometric. It is therefore impossible to compensate for the body deformation with a Euclidean displacement of the referential body to recover the initial configuration. An extension of the homogeneous group may be defined by cutting the continuum in infinitesimal small pieces (microcosms) such that the motion of each piece could be assumed as a Euclidean displacement. Therefore, we obtain a group of transformations at each point:

$$\overline{\mathbf{OM}}(t) = \alpha(M, t)\mathbf{OM}(t) + \beta(M, t).$$

Each microcosm is endowed with a metric tensor and a volume form. Moreover, a local group of Euclidean transformations is defined. The continuity asks a basic problem: How does one connect any two microcosms surrounding any two points M and M'? The connection of these two microcosms can be assumed to be realized with a transformation map that differs from the homogeneous group as follows:

$$\alpha(M', t) = \alpha(M, t) + \delta\alpha(M, t) \qquad \delta\alpha(M, t) = \mathbf{A}(M, t)[u]$$
$$\beta(M', t) = \beta(M, t) + \delta\beta(M, t) \qquad \delta\beta(M, t) = \mathbf{B}(M, t)[u]$$

in which $\mathbf{u} = \mathbf{MM'} \in T_M B$. The problem focuses on defining the two linear maps $\mathbf{A}(M, t)$ and $\mathbf{B}(M, t)$ at each point M at a given time t. Resorting of two transformations enables us to transport various tensor fields from one microcosm to another and to define entirely the connection of the continuum.

Axiom B.2.1 *(Cartan) The Cartan group is defined as the nonhomogeneous invariance group extended from the homogeneous group of Galilee. The relationship between two representations of the vector position with respect to any two referential bodies belonging to the equivalence class is given by:*

$$\overline{\mathbf{OM}} = \alpha(M)\mathbf{OM} + \beta(M)t + \mathbf{u}_0(M). \tag{B.4}$$

At each point, the invariance group has 10 parameters, all being in turn functions of the point M. The global scheme is transferred to the local level "microcosm." For practical purposes, the affine connection is more or less familiar with the above notion.

B.2.2   Objectivity (frame indifference) of constitutive laws

The invariance group used for constitutive laws is closer to the Cartan group than to the Galilean group. The Galilean group permits us to define the physical variables as forces while the group of Cartan should be used to warrant the frame indifference of the other variables:

$$\overline{\mathbf{OM}}(\bar{t}) = \alpha(M, t)\mathbf{OM}(t) + \beta(M, t)$$
$$\bar{t} = t - a(M) \qquad \alpha\alpha^T = i \qquad \det \alpha = 1.$$

$\overline{\mathbf{OM}}(\overline{t})$ and $\mathbf{OM}(t)$ are respectively the positions of M with respect to the two referential bodies. $\alpha(M, t)$ and $\beta(M, t)$ are an orthogonal transformation and a translation. $a(M)$ is the shift of the time origin.

Axiom B.2.2 (*Objectivity*) *Constitutive functions must have an invariant formulation during a local translation $\beta(M, t)$, a local rotation $\alpha(M, t)$ of the referential body, and a local time origin shift $a(M)$.*

During such transformations, constitutive variables, the components of which satisfy the tensor rule although α and β are functions of time, are said to be objective variables. The objectivity axiom imposes that the form of constitutive functions should be invariant when the set of these transformations is applied:

$$\hat{\Im}\left[\mathbf{OM}'(t'), \theta(M', t'), M, t\right] = \hat{\overline{\Im}}\left[\overline{\mathbf{OM}'}(\overline{t}'), \overline{\theta}(M', t'), M, \overline{t}\right]$$

$$\overline{\mathbf{OM}'}(\overline{t}') = \alpha(M', t')\mathbf{OM}'(t') + \beta(M', t')$$

$$\overline{\theta}(M', t') = \theta(M', t')$$

$$\hat{\overline{\Im}}(M', \overline{t}') = \alpha(M', t')\left[\hat{\Im}(M', t')\right]$$

$$\forall \alpha \ (\alpha\alpha^T = \alpha^T\alpha = i) \qquad \forall \beta$$

$$\forall a \quad \overline{t}' = t' - a(M') \qquad \overline{t} = t - a(M).$$

For the convenience, the transformation of a tensor T by the orthogonal transformation α will be written $\alpha(T)$.

1. **Local translation.** Consider the particular translation defined by:

$$\alpha = i \qquad a = 0 \qquad \beta(t') = -\mathbf{OM}(t')$$

in which the space origin at the time t' is defined by the position $\mathbf{OM}(t')$ thus depending on M and on t. We deduce:

$$\overline{t}' = t' \qquad \overline{\theta}(M', t') = \theta(M', t')$$

$$\overline{\mathbf{OM}'}(\overline{t}') = \mathbf{OM}'(t') - \mathbf{OM}(t') = \mathbf{MM}'(t').$$

This leads to the more restrictive form of intrinsic constitutive functions such as:

$$\Im(M, t) = \hat{\Im}[\mathbf{MM}'(t'), \theta(M', t'), M, t]. \tag{B.5}$$

Theorem B.2.1 *The relative motion of the point M' with respect to the point M is more relevant than the global motion for the constitutive laws formulation. Constitutive functions are independent of the continuum's global motions.*

2. **Time origin shift.** This transformation has already been used for the conservation laws. For constitutive laws, consider a particular shift of the time origin:

$$\alpha = i \qquad a = t \qquad \beta(t') = 0.$$

The time origin is the actual time. The frame indifference implies the following
condition:

$$\bar{t}' = t' - t \qquad \bar{t} = 0 \qquad -\infty < \bar{t} < 0$$
$$\bar{\theta}(M', \bar{t}') = \theta(M', t') \qquad \overline{\mathbf{OM}'}(\bar{t}') = \mathbf{OM}'(t').$$

Therefore, by introducing the time delay $\tau' \equiv t - t' > 0$:

$$\Im(M, t) = \hat{\Im}[\mathbf{OM}'(t - \tau'), \theta(M', t - \tau'), M, 0].$$

Theorem B.2.2 *The constitutive functions should not depend explicitly on the
time t. The constitutive functions can be expressed as follows, where M' is a
current point of the continuum and t' varies from 0 to ∞:*

$$\Im(M, t) = \hat{\Im}[\mathbf{MM}'(t - \tau'), \theta(M', t - \tau'), M]. \tag{B.6}$$

3. **Local rotation.** Consider now the effects of the local rotation on the constitutive
functions:

$$\alpha(t') \qquad a = 0 \qquad \beta(t') = 0.$$

It is a particular case of time-dependent and point-dependent rotation of the
group of Leibniz:

$$\bar{\theta}(M', \bar{t}') = \theta(M', t') \qquad \overline{\mathbf{OM}'}(\bar{t}') = \alpha(t')\mathbf{OM}'(t').$$

Theorem B.2.3 *The objectivity imposes that the rotation must not have any
influence on the form of constitutive functions. The form of constitutive functions
must then satisfy the following relationships:*

$$\hat{\Im}\left[\mathbf{MM}'(t - \tau'), \theta(M', t - \tau'), M\right] =$$
$$\alpha(t - \tau') \left(\hat{\Im}[\alpha(t - \tau')\,\mathbf{MM}'(t - \tau'), \theta(M', t - \tau'), M]\right)$$
$$\forall \alpha(t - \tau') \in O \qquad 0 \le \tau' < \infty. \tag{B.7}$$

Appendix C
Affinely Connected Manifolds

C.1 Riemannian manifolds

C.1.1 Metric tensor

Definition C.1.1 *(Metric tensor) Let B be a differentiable manifold. A Riemann metric* **g** *on B is a 2-covariant tensor field on B that satisfies the following properties at each point M of B, where the tangent space at each point M is denoted $T_M B$:*

1. $\mathbf{g}(\mathbf{u}, \mathbf{v}) = \mathbf{g}(\mathbf{v}, \mathbf{u})$, $\forall \mathbf{u}, \mathbf{v} \in T_M B$,

2. $\mathbf{g}(\mathbf{u}, \mathbf{u}) \geq 0$, $\forall \mathbf{u} \in T_M B$ *in which the equality holds only when* $\mathbf{u} = 0$.

If a smooth manifold B admits a Riemannian metric $\mathbf{g}$, then the pair $(B, \mathbf{g})$ is called a Riemannian manifold. For a submanifold B of a manifold E, the metric induced by the space E may be defined as a metric for the submanifold. At any point $M \in B$ the metric tensor $\mathbf{g}$ may be decomposed on the tangent vector base $(\mathbf{u}_1, \mathbf{u}_2, \mathbf{u}_3)$:

$$g_{ab} = \mathbf{g}(\mathbf{u}_a, \mathbf{u}_b). \tag{C.1}$$

The matrix g_{ab} is invertible and its elements are not constant but depend on the coordinates of the point M and the time t, for which they have been defined.

C.2   Affine connection

A tangent vector on a manifold is a directional derivative on the set of functions defined on B. For a tensor of higher order, an extra structure called the connection is needed, this structure specifies how the tensor fields are transported along a path within the manifold. In other words, a connection is the geometric structure used to answer the basic question: What is the constant field in the neighborhood of a given point M on B?

C.2.1   Metric connection, Levi-Civita connection

We start with an intuitive introduction to the Christoffel symbols by considering a manifold B, endowed with a metric tensor $\mathbf{g}$. In curvilinear coordinates, differentiation of a vector field $\mathbf{v}$ on a manifold B gives:

$$\frac{\partial \mathbf{v}}{\partial x^a} = \frac{\partial}{\partial x^a}(v^b \mathbf{u}_b) = \frac{\partial v^b}{\partial x^a}\mathbf{u}_b + v^b \frac{\partial \mathbf{u}_b}{\partial x^a}$$

in which (x^1, x^2, x^3) are the curvilinear coordinates defined on B and $(\mathbf{u}_1, \mathbf{u}_2, \mathbf{u}_3)$ the associated tangent basis. These derivatives may be resolved in components with respect to the same basis $(\mathbf{u}_1, \mathbf{u}_2, \mathbf{u}_3)$. We write:

$$\frac{\partial \mathbf{u}_b}{\partial x^a} = \Gamma^c_{ba}\mathbf{u}_c$$

where Γ^c_{ba} are the Christoffel and are not a third-order tensor ($\Gamma^c_{ba} \equiv 0$ in a Cartesian system of axes). It is thus easy to write the derivative:

$$\frac{\partial \mathbf{v}}{\partial x^a} = \left(\frac{\partial v^c}{\partial x^a} + v^b \Gamma^c_{ba}\right)\mathbf{u}_c$$

in which $\nabla_a v^c \equiv \frac{\partial v^c}{\partial x^a} + v^b \Gamma^c_{ba}$ is called the covariant derivative of v^c with respect to x^a. A similar expression leads to the covariant derivative of a "covector" v_c as a follows $\nabla_a v_c \equiv \frac{\partial v_c}{\partial x^a} - v_b \Gamma^b_{ca}$. $\nabla_a v^c$ and $\nabla_a v_c$ are components of second-order tensors. Roughly speaking, the definition of the Christoffel symbols (number n^3), and more generally of the affine connection in the sense of Levi-Civita, gives a structure that enables us to differentiate vector (and tensor) fields. A classic calculus gives the Christoffel symbols, e.g., [144], [159]:

$$\Gamma^c_{ab} = \frac{1}{2} g^{cd}\left(\frac{\partial g_{db}}{\partial x^a} + \frac{\partial g_{ad}}{\partial x^b} - \frac{\partial g_{ab}}{\partial x^d}\right).$$

Remark. The affine connection in the sense of Levi-Civita is not a sufficiently powerful tool to work with space with singularity.

C.2.2 Affine connections

The affine connection we seek is a map on a differentiable manifold B, possibly with singularity distribution, defining a structure that enables us to characterize a law of covariant differentiation.

Definition C.2.1 *An affine connection ∇ on a manifold B is a map $\nabla : T_M B \times T_M B \longrightarrow T_M B$ that associates to any couple of vectors $\mathbf{u}$ and $\mathbf{v}$ on B a vector $\nabla_{\mathbf{u}}\mathbf{v}$ such that, λ and μ being any two real numbers and ψ any scalar field on B, e.g., [29]:*

1. $\nabla_{\lambda\mathbf{u}_1 + \mu\mathbf{u}_2}\mathbf{v} = \lambda\nabla_{\mathbf{u}_1}\mathbf{v} + \mu\nabla_{\mathbf{u}_2}\mathbf{v}$

2. $\nabla_{\mathbf{u}}(\lambda\mathbf{v}_1 + \mu\mathbf{u}_2) = \lambda\nabla_u\mathbf{v}_1 + \mu\nabla_{\mathbf{u}}\mathbf{v}_2$

3. $\nabla_{\psi\mathbf{u}}\mathbf{v} = \psi\nabla_{\mathbf{u}}\mathbf{v}$

4. $\nabla_{\mathbf{u}}(\psi\mathbf{v}) = \psi\nabla_{\mathbf{u}}\mathbf{v} + \mathbf{u}(\psi)\,\mathbf{v}$.

An infinite number of connections may exist for a manifold B, such as affine connection, geodesic connection, and crystal connection. For practical purposes, let $(\mathbf{u}_1, \mathbf{u}_2, \mathbf{u}_3)$ be any vector basis of the tangent space $T_M B$. The nine functions Γ_{ab}^c defined by $\nabla_{\mathbf{u}_a}\mathbf{u}_b \equiv \Gamma_{ab}^c\mathbf{u}_c$ are called connection coefficients. The connection coefficients depend on the point $\mathbf{x}$ and specify how the vector base $(\mathbf{u}_1, \mathbf{u}_2, \mathbf{u}_3)$ changes from point to point.

C.2.3 Covariant derivative of tensor fields

The quantity $\nabla_{\mathbf{u}}$ is a derivative along the direction $\mathbf{u}$. It is natural to define the covariant derivative of a scalar field on B:

$$\nabla_{\mathbf{u}}\phi \equiv \mathbf{u}(\phi). \tag{C.2}$$

By extension, we define the covariant derivative for all types of tensor fields, such as vectors and 1-forms, as follows:

$$\nabla\mathbf{v}(\mathbf{u}) \equiv \nabla_{\mathbf{u}}\mathbf{v} \qquad (\nabla_{\mathbf{u}}\omega)(\mathbf{v}) = \mathbf{u}[\omega(\mathbf{v})] - \omega(\nabla_{\mathbf{u}}\mathbf{v}) \qquad \forall\mathbf{u} \in T_M B. \tag{C.3}$$

The covariant derivative satisfies the following properties:

1. $\nabla_{\mathbf{u}}\phi \equiv \mathbf{u}(\phi), \phi \in C^1(B), \forall\mathbf{u} \in T_M B$

2. $\nabla\mathbf{v}(\mathbf{u} + \mathbf{w}) = \nabla\mathbf{v}(\mathbf{u}) + \nabla\mathbf{v}(\mathbf{w}), \forall\mathbf{u}, \mathbf{v}, \mathbf{w} \in T_M B$

3. $\nabla_{\mathbf{v}}(\mathbf{u} \otimes \mathbf{w}) = \nabla_{\mathbf{v}}\mathbf{u} \otimes \mathbf{w} + \mathbf{u} \otimes \nabla_{\mathbf{v}}\mathbf{w}, \forall\mathbf{u}, \mathbf{v}, \mathbf{w} \in T_M B$

4. The covariant derivative commutes with the operation of contracted multiplicator.

Definition C.2.2 *(Covariant derivative) Let* **A** *be a q-covariant and p-contravariant tensor field on a manifold B. The covariant derivative* $\nabla\mathbf{A}$ *is a q + 1-covariant and p-contravariant tensor field defined by:*

$$(\nabla\mathbf{A})(\mathbf{u}, \mathbf{v}_1, \ldots, \mathbf{v}_q, \omega^1, \ldots, \omega^p) \equiv \nabla_{\mathbf{u}}\mathbf{A}(\mathbf{v}_1, \ldots, \mathbf{v}_q, \omega^1, \ldots, \omega^p)$$

$$\forall \mathbf{u}, \mathbf{v}_1, \ldots, \mathbf{v}_q \in T_M B \qquad \forall \omega^1, \ldots, \omega^p \in T_M B^*. \tag{C.4}$$

The covariant derivative is a proper extension of the directional derivative of function to tensor fields of all ranks.

Examples. The covariant derivatives of various tensor fields along the vector field **v** may be expressed in terms of their components on the base $(\mathbf{u}_1, \mathbf{u}_2, \mathbf{u}_3)$:

1. scalar field $\psi(\mathbf{x})$

$$\nabla_{\mathbf{v}}\psi = \frac{\partial\psi}{\partial x^a}v^a$$

2. vector field $\mathbf{w}(\mathbf{x})$

$$\nabla_{\mathbf{v}}\mathbf{w} = \left(\frac{\partial w^a}{\partial x^b} + \Gamma^a_{bc}w^c\right)v^b\mathbf{u}_a$$

3. 1-form $\omega(\mathbf{x})$

$$\nabla_{\mathbf{v}}\omega = \left(\frac{\partial\omega_a}{\partial x^b} - \Gamma^c_{ab}\omega_c\right)v^b\mathbf{u}^a$$

4. tensor of second rank $\sigma(\mathbf{x})$

$$\nabla_{\mathbf{v}}\sigma = \left(\frac{\partial\sigma^{ab}}{\partial x^c} + \Gamma^a_{cd}\sigma^{db} + \Gamma^b_{cd}\sigma^{ad}\right)v^c\,\mathbf{u}_a \otimes \mathbf{u}_b.$$

Remark on metric connection. Let $(\mathbf{u}_1, \mathbf{u}_2, \mathbf{u}_3)$ be a vector basis on B. For a Riemannian manifold, it is possible to put restrictions on the possible form of connections. We demand that the metric $\mathbf{g}(\mathbf{u}_a, \mathbf{u}_b) = g_{ab}$ be covariantly constant:

$$\nabla_w[\mathbf{g}(\mathbf{u}, \mathbf{v})] = w^a u^b v^c (\nabla_a\mathbf{g})_{bc} \equiv 0. \tag{C.5}$$

Since this is true for all curves and vectors, we thus have:

$$(\nabla_a\mathbf{g})_{bc} = 0. \tag{C.6}$$

The affine connection is said to be metric compatible or a metric connection.

Theorem C.2.1 *(Fundamental theorem of Riemannian geometry) On a Riemannian manifold, there exists a unique torsion-free connection* $\overline{\nabla}$ *that is compatible with the metric. This connection is called the Levi-Civita connection.*

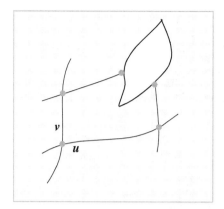

Figure C.1. In continuum mechanics theory, the Lie–Jacobi bracket [**u**, **v**] measures the failure of closure of a parallelogram (initially closed). This happens when the continuum deformation includes the nucleation of microcracks or the occurrence of dislocations within crystalline solids.

C.3   Curvature and torsion

The connection is not a tensor and thus cannot have an intrinsic geometrical meaning such as the measure of distortion and curving of manifolds. Torsion and curvature are the proper variables for a geometric interpretation.

C.3.1   Lie–Jacobi bracket of two vector fields

Definition C.3.1 *(Lie–Jacobi bracket) The bracket of two vector fields* **u** *and* **v** *on B is a vector field labeled* [**u**, **v**] *and defined by,* ψ *being any scalar field:*

$$[\mathbf{u}, \mathbf{v}](\psi) \equiv \mathbf{u}\mathbf{v}(\psi) - \mathbf{v}\mathbf{u}(\psi). \tag{C.7}$$

Example. Let $(\mathbf{e}_1, \mathbf{e}_2, \mathbf{e}_3)$ be a Cartesian base of the space and consider two vector fields $\mathbf{u}(\mathbf{y})$ and $\mathbf{v}(\mathbf{y})$ of class C^1. For any material point M, the vector position is decomposed as:

$$\mathbf{OM} = y^i\,\mathbf{e}_i \qquad \mathbf{u} = u^i\,\mathbf{e}_i \qquad \mathbf{v} = v^j\,\mathbf{e}_j.$$

The decomposition of the Lie–Jacobi bracket of the two vector fields on the Cartesian base gives (summation over the two indices):

$$[\mathbf{u}, \mathbf{v}] = \left(v^j\,\frac{\partial u^i}{\partial y^j} - u^j\,\frac{\partial v^i}{\partial y^j} \right)\,\mathbf{e}_i.$$

The definition (C.7) yields the Jacobi identity:

$$[\mathbf{u}, [\mathbf{v}, \mathbf{w}]] + [\mathbf{v}, [\mathbf{w}, \mathbf{u}]] + [\mathbf{w}, [\mathbf{u}, \mathbf{v}]] = 0, \tag{C.8}$$

and moreover, the bracket is bilinear. At each material point M of B, the constants of structure (Cartan), associated to a basis $(\mathbf{u}_1, \mathbf{u}_2, \mathbf{u}_3)$ and labeled $\aleph^c_{0ab}$, are defined by [22], [23]:

$$[\mathbf{u}_a, \mathbf{u}_b] \equiv \aleph^c_{0ab}\, \mathbf{u}_c. \tag{C.9}$$

If $\aleph^c_{0ab} \neq 0$, the base $(\mathbf{u}_1, \mathbf{u}_2, \mathbf{u}_3)$ is called a noncoordinate basis.

Lie–Jacobi brackets are closely related to integrability conditions. This raises the question whether the condition $[\mathbf{u}_a, \mathbf{u}_b] \equiv 0$ for vector fields $\mathbf{u}_1, \mathbf{u}_2, \mathbf{u}_3$ implies that the latter are coordinate vector fields associated with a system of coordinates. Let there be a set of C^∞ vector fields $(\mathbf{u}_1, \mathbf{u}_2, \mathbf{u}_3)$ linearly independent at the point $M \in B$ that satisfy the conditions:

$$[\mathbf{u}_a, \mathbf{u}_b] \equiv 0 \qquad a, b = 1, 2, 3 \tag{C.10}$$

in the neighborhood of M. By virtue of the theorem of Frobenius, the following theorem holds, e.g., [29].

Theorem C.3.1 *If $(\mathbf{u}_1, \mathbf{u}_2, \mathbf{u}_3)$ are linearly independent and satisfy the relation (C.10), there exist coordinates (y^1, y^2, y^3) in the neighborhood of M, relative to which $(\mathbf{u}_1, \mathbf{u}_2, \mathbf{u}_3)$ constitutes a field of vector basis.*

C.3.2 Exterior derivative

Definition C.3.2 *(Exterior derivative) Let B be a manifold. The exterior derivative of a p-form ω may then be defined (intrinsically) by the relationship, for any vector fields $(\mathbf{u}_1, \ldots, \mathbf{u}_{p+1}) \in T_M B$ on the manifold B (the vector with a hat is omitted), e.g., [120], [123]:*

$$d\omega(\mathbf{u}_1, \ldots, \mathbf{u}_{p+1}) = \sum_{i=1}^{p+1}(-1)^{i+1}\mathbf{u}_i[\omega_0(\mathbf{v}, \mathbf{u}_1, \ldots, \hat{\mathbf{u}}_i, \ldots, \mathbf{u}_{p+1})]$$

$$+ \sum_{i<j=1}^{p+1}(-1)^{i+j}\omega_0(\mathbf{v}, [\mathbf{u}_i, \mathbf{u}_j], \mathbf{u}_1, \ldots, \hat{\mathbf{u}}_i, \ldots, \hat{\mathbf{u}}_j, \ldots, \mathbf{u}_{p+1}). \tag{C.11}$$

The exterior derivative of a p-form is defined without any reference. The operator d exists and is unique as a map from the set of p-forms on the continuum B to the set of $(p + 1)$-forms on B. This operator satisfies the following properties (which may be considered as definition):

1. For a scalar field ψ (a 0-form):

$$d\psi(\mathbf{u}) \equiv \mathbf{u}(\psi) \qquad \forall \mathbf{u} \in T_M B \qquad \text{(C.12)}$$

2. For a p-form ω_p and a q-form ω_q:

$$d(\omega_p \wedge \omega_q) = d\omega_p \wedge \omega_q + (-1)^{pq}\omega_p \wedge d\omega_q \qquad \text{(C.13)}$$

3. For any p-form ω:

$$d(d\omega) = 0 \qquad \text{(C.14)}$$

4. For any pair of p-forms, d is linear:

$$d(\omega + \omega') = d\omega + d\omega'. \qquad \text{(C.15)}$$

This operator is one of the most useful in the mathematical formulation of physics equations (gradient, divergence, curl, Laplacian operators).

Example. The exterior derivatives of fields on a 3-dimensional continuum may be rewritten in a more familiar form in an Euclidean manifold R^3:

1. 0-form (scalar, $p = 0$)

$$d\omega = \frac{\partial \psi}{\partial y^1}dy^1 + \frac{\partial \psi}{\partial y^2}dy^2 + \frac{\partial \psi}{\partial y^3}dy^3$$

2. 1-form ($p = 1$)

$$
\begin{aligned}
d\omega \;=\; & \left(\frac{\partial \omega_2}{\partial y^1} - \frac{\partial \omega_1}{\partial y^2}\right)dy^1 \wedge dy^2 + \left(\frac{\partial \omega_3}{\partial y^2} - \frac{\partial \omega_2}{\partial y^3}\right)dy^2 \wedge dy^3 \\
+ & \left(\frac{\partial \omega_1}{\partial y^3} - \frac{\partial \omega_3}{\partial y^1}\right)dy^3 \wedge dy^1
\end{aligned}
$$

3. 2-form ($p = 2$)

$$d\omega = \left(\frac{\partial \omega_{23}}{\partial y^1} + \frac{\partial \omega_{31}}{\partial y^2} + \frac{\partial \omega_{12}}{\partial y^3}\right)dy^1 \wedge dy^2 \wedge dy^3$$

4. 3-form $(p = 3)$

$$dw = 0.$$

Remark. For deriving these relations, notice that the basis $(\mathbf{u}_1, \mathbf{u}_2, \mathbf{u}_3)$ is associated to a system of coordinates (as is the case in R^3) if and only if for any two vectors $[\mathbf{u}_a, \mathbf{u}_b] = 0$, e.g., [23], [29].

Let w_p and w_q be any 0-forms or 1-forms on R^3. Then the relation on exterior derivative $d(w_p \wedge w_q) = dw_p \wedge w_q + (-1)^{pq} w_p \wedge dw_q$ contains the following formulae of classical vector analysis, e.g., [205]:

$$
\begin{aligned}
\nabla\,(\varphi\psi) &= \nabla\varphi\,(\psi) + \varphi\,\nabla\psi \\
\mathrm{rot}\,(\varphi\,\mathbf{u}) &= \nabla\varphi \times \mathbf{u} + \varphi\,\mathrm{rot}\,\mathbf{u} \\
\mathrm{div}\,(\varphi\,\mathbf{u}) &= \nabla\varphi\,(\mathbf{u}) + \varphi\,\mathrm{div}\,\mathbf{u} \\
\mathrm{div}\,(\mathbf{u} \times \mathbf{v}) &= \mathbf{v}\cdot\mathrm{rot}\,\mathbf{u} - \mathbf{u}\cdot\mathrm{rot}\,\mathbf{v}.
\end{aligned}
$$

C.3.3 Poincaré Lemma.

From basic vector analysis, there are some accepted results in $U \in R^3$:

1. If $\nabla\varphi = 0$, then φ is a uniform scalar field.

2. If $\mathrm{rot}\,\mathbf{v} = 0$, then there is a scalar field φ such that $\mathbf{v} = \nabla\varphi$.

3. If $\mathrm{div}\,\mathbf{v} = 0$, then there exists a vector field $\mathbf{u}$ such that $\mathrm{rot}\,\mathbf{u} = \mathbf{v}$.

4. For every scalar field φ, there is a vector field $\mathbf{v}$ such that $\mathrm{div}\,\mathbf{v} = \varphi$.

These vector relations not only require the differentiability assumption but also may fail if their domain U does not fullfill some topological conditions (namely connectedness).

Lemma C.3.1 *(Poincaré lemma) Let B be a manifold and w a p-form on B. Assume that there exists a $(p − 1)$-form α on B such that $d\alpha = w$. Then $d(dw) = 0$.*

In the Euclidean space R^3, the Poincaré lemma $d(dw) = 0$ has the well-known interpretation in electromagnetism and fluid mechanics (these results may be deduced from above exterior derivative formulae on 1-forms and 2-forms):

$$\mathrm{rot}\,(\nabla\varphi) = 0 \qquad \mathrm{div}\,(\mathrm{rot}\,\mathbf{v}) = 0.$$

Definition C.3.3 *(Closed form, exact form) A form w is closed if $dw = 0$. A p-form w is exact if there is a $(p − 1)$-form α such that $w = d\alpha$.*

Theorem C.3.2 *(Converse of Poincaré lemma) If ω is a p-form on an open set $U \in B$ (which is contractible to a point, star-shaped) such that $d\omega = 0$, then there exists a $(p-1)$-form α such that $\omega = d\alpha$.*

There are some important observations:

1. Every exact form is closed (the converse is not always true).

2. The product of two closed forms is closed.

3. The product of a closed form and an exact form is exact.

Some of these results are questionable when we deal with continuum with singularity, which are not simply connected manifolds.

C.3.4 Torsion and curvature

Consider on B an affine connection ∇, a 1-form field ω, and vector fields $\mathbf{u}_1$, $\mathbf{u}_2$, and $\mathbf{w}$.

Definition C.3.4 *(Torsion) Let ∇ be an affine connection on a manifold B. The field of torsion tensor $\aleph$, a 2-covariant 1-contravariant, associated to ∇, is defined by:*

$$\aleph(\mathbf{u}, \mathbf{v}, \omega) \equiv \omega(\nabla_{\mathbf{u}}\mathbf{v} - \nabla_{\mathbf{v}}\mathbf{u} - [\mathbf{u}, \mathbf{v}]) \qquad \forall \mathbf{u}, \mathbf{v} \in T_M B \qquad \forall \omega \in T_M B^*. \quad (C.16)$$

It is also usual to define the vector field $\aleph(\mathbf{u}, \mathbf{v}) \equiv \nabla_{\mathbf{u}}\mathbf{v} - \nabla_{\mathbf{v}}\mathbf{u} - [\mathbf{u}, \mathbf{v}]$. In any basis $(\mathbf{u}_1, \mathbf{u}_2, \mathbf{u}_3)$, the torsion components are, e.g., [29]:

$$\aleph = \aleph_{ab}^c \mathbf{u}^a \otimes \mathbf{u}^b \otimes \mathbf{u}_c \qquad \aleph_{ab}^c = (\Gamma_{ab}^c - \Gamma_{ba}^c) - \aleph_{0ab}^c. \quad (C.17)$$

Let B be any differentiable manifold endowed with a metric and a metric connection. If the torsion tensor vanishes on the manifold B, the metric connection ∇ is called the Levi-Civita connection.

Definition C.3.5 *(Curvature) Let ω be any 1-form field on B and $\mathbf{u}_1$, $\mathbf{u}_2$, and $\mathbf{w}$ any vector fields on B. The curvature tensor, labeled $\Re$ and associated to the affine connection, is defined by:*

$$\Re(\mathbf{u}_1, \mathbf{u}_2, \mathbf{w}, \omega) \equiv \{\nabla_{\mathbf{u}_1}\nabla_{\mathbf{u}_2}\mathbf{w} - \nabla_{\mathbf{u}_2}\nabla_{\mathbf{u}_1}\mathbf{w} - \nabla_{[\mathbf{u}_1, \mathbf{u}_2]}\mathbf{w}\}(\omega). \quad (C.18)$$

The components of the curvature tensor in any vector base $(\mathbf{u}_1, \mathbf{u}_2, \mathbf{u}_3)$ are:

$$\Re_{abd}^c = [-\mathbf{u}_b(\Gamma_{ad}^c) + \mathbf{u}_a(\Gamma_{bd}^c) + \Gamma_{da}^e\Gamma_{eb}^c - \Gamma_{db}^e\Gamma_{ea}^c - \aleph_{0ab}^e\Gamma_{ed}^c].$$

Under a more familiar form, we have, in a coordinate base:

$$\Re_{abd}^c = -\frac{\partial \Gamma_{ad}^c}{\partial x^b} + \frac{\partial \Gamma_{bd}^c}{\partial x^a} + \Gamma_{da}^e\Gamma_{eb}^c - \Gamma_{db}^e\Gamma_{ea}^c.$$

The number of independent components of the curvature tensor $\mathfrak{R}$ in an n-dimensional space is $\frac{1}{12}n^2(n^2-1)$. For a 2-dimensional and 3-dimensional continuum, we respectively have $N=1$ and $N=6$ independent curvature components.

Remark. For any affine connection ∇ we can define two tensor fields of torsion $\aleph$ and curvature $\mathfrak{R}$ (and more generally higher order of covariant derivatives). The converse problem, which is very important, is to determine an affine connection (at least locally) if the torsion and curvature are given. First, an affine connection ∇ does not define the local geometric properties of the manifold B. Intrinsic geometric properties are entirely, at least locally, defined by the fields of torsion $\aleph$ and curvature $\mathfrak{R}$ associated to ∇. A fundamental theorem of differential geometry states that an analytical connection ∇ is locally defined thanks the knowledge of $\aleph$ and $\mathfrak{R}$ and their high order derivatives, e.g., [55].

C.3.5 Holonomy group

The manifold B endowed with the affine connection ∇ is called an affinely connected manifold. For any affinely connected manifold, the connection defines a transformation group at each tangent space as follows. Let M be a point of the Riemannian manifold $(B, \mathbf{g})$ and consider the set of closed loops at M:

$$\{\Gamma(t) \mid 0 \le t \le 1, \Gamma(0) = \Gamma(1) = M\}. \tag{C.19}$$

Take a vector $\mathbf{u} \in T_M B$ and parallel transport $\mathbf{u}$ along a curve $\Gamma(t)$. After a trip along this curve, we end up with a new vector $\mathbf{u}_\Gamma \in T_M B$. Thus the loop $\Gamma(t)$ and the affine connection ∇ induces a linear transformation. The set of these linear transformations is called the holonomy group at the point M.

C.4 Integration of p-form

Conceptually, one integrates forms but not vectors on manifold. However, for metric manifolds, it is possible to rephrase integration of 1-forms and 2-forms in terms of vector integration (arc lengths, surface areas).

C.4.1 Orientation on connected manifolds

Let B be a connected 3-dimensional differentiable manifold. At each point M of B, the tangent space is spanned by the basis $(\mathbf{u}_1, \mathbf{u}_2, \mathbf{u}_3)$. For any 3-dimensional orientable manifold B, there exists a 3-form ω_0 that never vanishes in any case. This form is called the volume element or the volume form. It should be stressed that the idea of orientability does not generally depend on size and shape. Consequently, volume form and metric tensor need not be related each other.

Definition C.4.1 *(Volume form) A volume form on a manifold B (of dimension n) is a n-form field ω_0 such that $\omega_0(M) \neq 0$ for any point M of B. B is called orientable if there exists some volume form on B.*

In the same way, $\omega_0(\mathbf{u}_1, \ldots, \mathbf{u}_n)$ is $\pm$ the n-dimensional volume of the parallelepiped spanned by the projections of vectors $(\mathbf{u}_1, \ldots, \mathbf{u}_n)$ onto the $x^1, \ldots, x^n$ coordinates.

Let B be a connected 3-dimensional differentiable manifold, endowed with the volume form ω_0. The field of vector product of two ordered vector fields $(\mathbf{u}_1, \mathbf{u}_2)$ is the 1-form field defined on the tangent space as follows:

$$(\mathbf{u}_1 \times \mathbf{u}_2)(\mathbf{v}) \equiv \omega_0(\mathbf{u}_1, \mathbf{u}_2, \mathbf{v}) \qquad \forall \mathbf{v} \in T_M B. \tag{C.20}$$

Examples.

1. **Volume form.** Given any basis on a tangent space endowed with a metric tensor, there exists a "natural"volume form which is invariant under coordinate transformation, e.g., [159]:

$$\omega_0 = \sqrt{\det \begin{pmatrix} \mathbf{g}(\mathbf{u}_1, \mathbf{u}_1) & \mathbf{g}(\mathbf{u}_1, \mathbf{u}_2) & \mathbf{g}(\mathbf{u}_1, \mathbf{u}_3) \\ \mathbf{g}(\mathbf{u}_2, \mathbf{u}_1) & \mathbf{g}(\mathbf{u}_2, \mathbf{u}_2) & \mathbf{g}(\mathbf{u}_2, \mathbf{u}_3) \\ \mathbf{g}(\mathbf{u}_3, \mathbf{u}_1) & \mathbf{g}(\mathbf{u}_2, \mathbf{u}_2) & \mathbf{g}(\mathbf{u}_3, \mathbf{u}_3) \end{pmatrix}} \; \mathbf{u}^1 \wedge \mathbf{u}^2 \wedge \mathbf{u}^3. \tag{C.21}$$

This is a very particular case of volume form.

2. **Vector product.** Let $(\mathbf{u}_1, \mathbf{u}_2, \mathbf{u}_3)$ be a vector base (possibly nonorthogonal) on the manifold B. Application of the vector product allows us to write:

$$\mathbf{u}_1 \times \mathbf{u}_2 = J_u \mathbf{u}^3 \qquad \mathbf{u}_2 \times \mathbf{u}_3 = J_u \mathbf{u}^1 \qquad \mathbf{u}_3 \times \mathbf{u}_2 = J_u \mathbf{u}^2$$

where $J_u = \omega_0(\mathbf{u}_1, \mathbf{u}_2, \mathbf{u}_3)$ and $(\mathbf{u}^1, \mathbf{u}^2, \mathbf{u}^3)$ the dual vector base.

Remarks. From the above definition we essentially observe that:

1. A connected manifold B has two orientations $\pm \omega_0$.

2. A nonconnected manifold is said to be orientable if each connected (component) part of B is orientable. The choice of the orientation on each component defines the global orientation of the manifold (two possibilities $\pm \omega_0$ on each component).

3. Assume that B is covered by the set $\{B_i\}$. B is orientable iff, for any overlapping charts $\{B_i\}$ and $\{B_j\}$, there exist local coordinates $\{x^a\}$ for $\{B_i\}$ and $\{y^b\}$ for $\{B_j\}$ such that $J = \det \frac{\partial x^a}{\partial y^b} > 0$. In other words, we can write for dim $B = n$:

$$d\mathbf{x}^1 \wedge \cdots \wedge d\mathbf{x}^n = \det \left(\frac{\partial x^a}{\partial y^b} \right) d\mathbf{y}^1 \wedge \cdots \wedge d\mathbf{y}^n$$

This remark applies for both connected and nonconnected manifolds.

C.4.2   Integration of p-form in R^n

Consider a change of variables $\varphi : U \in R^p \longrightarrow U' \in R^p$, which is a diffeomorphism of U. The tangent linear mapping $d\varphi$ has the Jacobian:

$$J \equiv \det \left(\frac{\partial \varphi^i}{\partial x^i} \right) > 0$$

where $(x^1, \ldots, x^p)$ are the coordinates of the point M transformed to $(\varphi^1, \ldots, \varphi^p)$. Let $\omega_0 \equiv d\mathbf{x}^1 \wedge \cdots \wedge d\mathbf{x}^p$ be the canonical volume form on R^p and $\omega = \rho(\mathbf{x}) \, d\mathbf{x}^1 \wedge \cdots \wedge d\mathbf{x}^p$ for any p-form defined on $U \in R^p$. First, the integral of the p-form ω on U is defined by the relation:

$$\int_U \omega \equiv \int_U \rho(\mathbf{x}) \, d\mathbf{x} = \int_U \rho(x^1, \ldots, x^p) \, dx^1 \cdots dx^p.$$

Second we have the theorem of change of variables.

Theorem C.4.1 *Let φ be a diffeomorphism (which preserves orientation) of an open set $U \in R^p$ to $U' \in R^p$ and ω a p-form on the set U'. Then:*

$$\int_{\varphi(U)} \omega = \int_U \varphi^* \omega.$$

Proof. See in, e.g., [159], [205].

C.4.3   Integral of p-form on manifold

Definition C.4.2 *(Partition of unity) Let $\{B_i\}$ be an open covering of B such that each point M of B is covered by a finite number of B_i. If a family of differentiable functions $\varepsilon_i(M)$ satisfies:*

   1. $0 \leq \varepsilon_i(M) \leq 1$

   2. $\varepsilon_i(M) = 0$ if $M \notin B_i$

   3. $\sum_i \varepsilon_i(M) = 1 \; \forall M \in B$.

The family $\{\varepsilon_i(M)\}$ is called a partition of unity subordinate to $\{B_i\}$.

The integration of a p-form over a manifold B is defined only when B admits a volume form, labeled ω_0. The integral of any p-form $\rho\omega_0$ on a manifold B is given by:

$$\int_B \rho\omega_0 = \sum_i \int_{B_i} \rho_i \omega_0$$

which is independent of the choice of the covering. We now focus on the integration on a connected component B_i of B. In a coordinate neighborhood B_i with the coordinates $\{\mathbf{y}\}$, we can bring out the integration of a p-form $\rho\omega_0$ by:

$$\int_{B_i} \rho\omega_0 \equiv \int_{\varphi_i(B_i)} \rho[\varphi_i^{-1}(\mathbf{y})]h[\varphi_i^{-1}(\mathbf{y})]dy^1\cdots dy^p \tag{C.22}$$

where the right-hand side is an ordinary multiple integration of a function of a p-variables. Once the integration of ρ over B_i is defined, the integral over the entire B is given with the help of the partition of unity. Let ω be a p-form on $\varphi(B)$ and $(\mathbf{u}_1,\ldots,\mathbf{u}_p)$ a vector base of the tangent space $T_M B$. The integral of the p-form holds:

$$\int_{\varphi(B)} \omega \equiv \int_B \omega(\mathbf{u}_1,\ldots,\mathbf{u}_p)dy^1\cdots dy^p$$

Example. Line integral. Let $C = \{\mathbf{y} = \varphi(s)\}$ be a curve for $a \leq s \leq b$ in the space R^3, oriented so that the derivative $\frac{d}{ds}$ defines a positive orientation on the curve. Let $\omega = \omega_a dy^a$ be a 1-form on the curve. The line integral of ω over the curve C is calculated as follows:

$$\int_C \omega = \int_C \omega_a dy^a = \int_a^b \omega\left(\frac{d\mathbf{y}}{ds}\right)ds.$$

This is the usual formula for evaluating a line integral over an oriented parametrized curve. Rigorously, line integration is defined on a 1-form and not on a vector field. However, in a Riemannian metric space such as R^3, a 1-form field is associated to a vector field by means of the metric tensor $\mathbf{v}^* \equiv \mathbf{g}(\mathbf{v}, .)$. Then we may write the following:

$$\int_C \mathbf{v}^* = \int_C \mathbf{g}(\mathbf{v}, .).$$

Or in a more practical form, as the circulation of the vector field $\mathbf{v}$:

$$\int_C \mathbf{v}^*\left(\frac{d\mathbf{y}}{ds}\right)ds = \int_C \mathbf{g}\left(\mathbf{v}, \frac{d\mathbf{y}}{ds}\right)ds = \int_C \mathbf{v}\cdot d\mathbf{y}$$

If we want to insist on integrating a vector field (circulation calculus) on a curve, we need to do so on a Riemanian manifold so as to convert the contravariant vector into a covariant vector one.

Example. Surface integral. Consider an oriented parametrized surface $S = \{\mathbf{y} = \varphi(u^1, u^2)\} = \varphi(U)$ in R^3, with $\mathbf{y}$ being any coordinate system. Let ω be a 2-form on this surface, $\omega = \omega_{23}dy^2 \wedge dy^3 + \omega_{31}dy^3 \wedge dy^1 + \omega_{12}dy^1 \wedge dy^2$. The integral of ω on the surface holds:

$$\int_S \omega = \int_S \omega_{23}dy^2 \wedge dy^3 + \omega_{31}dy^3 \wedge dy^1 + \omega_{12}dy^1 \wedge dy^2.$$

It may be also written as:

$$\int_{S=\varphi(U)} \omega = \int_U \omega \left(\frac{\partial \mathbf{y}}{\partial u^1}, \frac{\partial \mathbf{y}}{\partial u^2} \right) du^1 du^2.$$

To draw a comparison with the classical formula of vector flux, we can introduce the vector field $\mathbf{w}$ associated to the 2-form ω written as:

$$\omega_0 \left(\mathbf{w}, \frac{\partial \mathbf{y}}{\partial u^1}, \frac{\partial \mathbf{y}}{\partial u^2} \right) \equiv \omega \left(\frac{\partial \mathbf{y}}{\partial u^1}, \frac{\partial \mathbf{y}}{\partial u^2} \right).$$

By observing the usual Euclidean volume form, we find:

$$\omega_0 \left(\mathbf{w}, \frac{\partial \mathbf{y}}{\partial u^1}, \frac{\partial \mathbf{y}}{\partial u^2} \right) = \mathbf{w} \cdot \left(\frac{\partial \mathbf{y}}{\partial u^1} \times \frac{\partial \mathbf{y}}{\partial u^2} \right).$$

By introducing the area element, we recover the usual formula of vector flux:

$$d\mathbf{S} \equiv \mathbf{n} dA \equiv \frac{\partial \mathbf{y}}{\partial u^1} \times \frac{\partial \mathbf{y}}{\partial u^2} du^1 du^2 \qquad \int_{\varphi(U)} \omega = \int_U \mathbf{w} \cdot d\mathbf{S} = \int_U \mathbf{w} \cdot \mathbf{n} dA.$$

The vector product of the two tangent vectors weighted by $du^1 du^2$ is the basic element of surface. The use of the "flux vector" concept requires both the orientation (volume form) and the metric.

Then, the integral of $\rho \omega_0$ on the manifold B is given by:

$$\int_B \rho \omega_0 = \sum_i \int_{B_i} \rho_i \omega_0 \qquad\qquad (C.23)$$

which is independent of the choice of the covering.

C.4.4 Stokes' theorem

One of the most powerful integral calculus of p-forms is Stokes' theorem. The divergence theorem is a special case of Stokes' theorem. Whereas the formulation of the divergence theorem depends on the Riemannian metric of the manifold, the theorem of Stokes is completely independent of any metric tensor concept. This point is crucial in this book since the domain of integration is weakly continuous and may be difficult to capture with the metric only.

Theorem C.4.2 *(Stokes' theorem) Let B be a $(p+1)$-dimensional region, star-shaped with respect to an interior point O. Its boundary ∂B is a closed p-dimensional subspace admitting a parametric representation of class C^1. Then, for any p-form ω defined on B with continuously differentiable coefficients, we have:*

$$\int_B d\omega = \int_{\partial B} \omega. \qquad\qquad (C.24)$$

To understand Stokes' theorem, it helps to remember the three particular cases which are most known.

1. **Fundamental theorem.** When $p = 0$, we have the fundamental theorem of integral calculus. It expresses that, for every smooth function $\omega = \psi$ on an open interval of the real line containing a closed interval $B = [a, b]$ with $\partial B = \{a, b\}$, we have the fundamental theorem on integral calculus:

$$\int_a^b d\psi = \psi(b) - \psi(a). \tag{C.25}$$

2. **Green's theorem.** When $p = 1$, we obtain Green's theorem. For every compact oriented 2-dimensional submanifold B of the plane and for every 1-form field:

$$\omega = \omega_1 dx^1 + \omega_2 dx^2 + \omega_3 dx^3$$

identified with a "vector field" $\vec{\omega} = (\omega_1, \omega_2, \omega_3)$ by means of the metric tensor, on a neighborhood of B, we may write:

$$\int_B d\omega = \int_{\partial B} \omega \qquad \Longleftrightarrow \qquad \int_B \operatorname{rot} \vec{\omega} \cdot d\mathbf{S} = \int_{\partial B} \vec{\omega} \cdot d\mathbf{s} \tag{C.26}$$

where the elementary surface dS on the body B and the elementary line ds on the board ∂B are introduced.

3. **Gauss' theorem.** When $p = 2$, we have Gauss' theorem, also named divergence theorem. For every compact oriented 3-dimensional submanifold B of R^3 and for every 2-form field:

$$\omega = \omega_{12} dx^1 \wedge dx^2 + \omega_{23} dx^2 \wedge dx^3 + \omega_{31} dx^3 \wedge dx^1$$

identified with a "vector field" $\vec{\omega} = (\omega_{12}, \omega_{23}, \omega_{31})$ by means of the volume form on the neighborhood of B, we have:

$$\int_B d\omega = \int_{\partial B} \omega \qquad \Longleftrightarrow \qquad \int_B \operatorname{div} \vec{\omega} \, dv = \int_{\partial B} \vec{\omega} \cdot d\mathbf{s} \tag{C.27}$$

in which the elementary volume dv has been introduced. Of course, it appears that these particular cases are only valid if the manifold is endowed with a metric tensor and a volume form.

C.5   Brief history of connection

Implicitly, the notion of connection first emerged in the work of Newton (1687) when he introduced the transport (parallel) of instantaneous velocity of a material point before calculating the instantaneous acceleration. Any two instantaneous velocities of a given material point at different times do not have the same origin (spatially) and have two different spatial points. Thus no operation can be done before the parallel transport from one point to the other.

A second step in the connection theory development can be attributed to the work of Gauss (1827) when he studied the geodetic curves on a surface and particularly the influence of the curvature on the properties of triangles embedded in the surface. However, he did not explicitly propose the method that consists in comparing the tangent planes at different locations (edges of the triangles). A few years later, Riemann extended to a multi-dimensional space the work of Gauss on a surface (1854). He introduced namely the notion of a metric field on a manifold. The change of the metric along an embedded curve on the manifold can be expressed in terms of curvature. Considering the formulations of the metric fields at any two neighboring points on the manifold, Christoffel (1869) found that the mathematical condition of the two metrics should verify to be transformed into each other (smoothly). He then suggested the celebrated Christoffel symbols. This was the starting point of the modern connection theory. Necessary conditions for the equivalence of the two metrics were used to extract the definition of the curvature tensor.

Ricci and Levi-Civita laid down the basic concept of connection by considering the invariance of the Laplace–Beltrami operator during a change of frame. They formally derived the concept of connection (1888). Indeed, they developed the concept of covariant equations and of the covariant derivative. Levi-Civita in 1917 and a few years later related the connection to the affine connection which was formally derived by Weyl (1918) and which was used to formulate the basic equations of gravitation and electromagnetism.

Elie Cartan was among the first to emphasize the need of separation between the metric and the connection concepts. A fact that was neglected by his predecessors is that the connection exists independently of the metric, (1923, 1924). Cartan proposed the notion of a holonomy group. The affine connection plays a role in the fiber bundle concept as it was proposed by Esherman (1950) and in the physics gauge theory of Yang and Mills (1954). However, the precise relation between the affine connection and the physics theory proposed by these authors was only pointed out recently (1980).

Another application of the affine connection in field theory in physics was proposed by Weinberg and Salam (1980) and now constitutes the base of modern development of the concept. The starting point of the application of affine connection in continuum mechanics can be attributed to the work of Noll (1958), particularly on the theory of nonhomogeneous bodies, a class of continuum with dislocations. Since then, numerous authors have contributed to develop the continuum theory with various types of singularity.

Appendix D
Bianchi Identities

In this appendix, we derive the Bianchi identities with respect to any basis $(\mathbf{u}_1, \mathbf{u}_2, \mathbf{u}_3)$ embedded in the continuum B, $\forall \omega \in T_M B^*$, $\forall \mathbf{v} \in T_M B$.

D.1   Skew symmetry

For any two vector fields $(\mathbf{u}_a, \mathbf{u}_b)$ on B, we have:

$$\Re(\mathbf{u}_a, \mathbf{u}_b, \mathbf{v}, \omega) = -\Re(\mathbf{u}_b, \mathbf{u}_a, \mathbf{v}, \omega)$$

$$\forall \omega \in T_M B^* \qquad \forall \mathbf{v} \in T_M B \qquad \forall \mathbf{u}_a, \mathbf{u}_b \in T_M B. \tag{D.1}$$

D.2   First identities of Bianchi

The curvature tensor is defined by:

$$\Re(\mathbf{u}_a, \mathbf{u}_b, \mathbf{u}_c, \omega) = \omega(\nabla_{\mathbf{u}_a} \nabla_{\mathbf{u}_b} \mathbf{u}_c - \nabla_{\mathbf{u}_b} \nabla_{\mathbf{u}_a} \mathbf{u}_c - \nabla_{[\mathbf{u}_a, \mathbf{u}_b] \mathbf{u}_c}).$$

By introducing the torsion definition and by adding up terms after a circular permutation:

$$\sum_{(abc)} \Re(\mathbf{u}_a, \mathbf{u}_b, \mathbf{u}_c, \omega) = \omega\{\nabla_{\mathbf{u}_1}[\aleph(\mathbf{u}_2, \mathbf{u}_3)] + \nabla_{\mathbf{u}_2}[\aleph(\mathbf{u}_3, \mathbf{u}_1)] + \nabla_{\mathbf{u}_3}[\aleph(\mathbf{u}_1, \mathbf{u}_2)]$$

$$+ \quad \aleph(\mathbf{u}_1, [\mathbf{u}_2, \mathbf{u}_3]) + \aleph(\mathbf{u}_2, [\mathbf{u}_3, \mathbf{u}_1]) + \aleph(\mathbf{u}_3, [\mathbf{u}_1, \mathbf{u}_2])$$

$$+ \quad [\mathbf{u}_1, [\mathbf{u}_2, \mathbf{u}_3]] + [\mathbf{u}_2, [\mathbf{u}_3, \mathbf{u}_1]] + [\mathbf{u}_3, [\mathbf{u}_1, \mathbf{u}_2]]\}.$$

Then:

$$\sum_{(abc)} \Re(\mathbf{u}_a, \mathbf{u}_b, \mathbf{u}_c, \omega) = \omega\{\sum_{(abc)} \nabla_{\mathbf{u}_a}[\aleph(\mathbf{u}_b, \mathbf{u}_c)] + \aleph(\mathbf{u}_a, [\mathbf{u}_b, \mathbf{u}_c])\}.$$

The first term of the right-hand side can be written as follows:

$$\sum_{(abc)} \nabla_{\mathbf{u}_a}[\aleph(\mathbf{u}_b, \mathbf{u}_c)] = \sum_{(abc)} (\nabla_{\mathbf{u}_a}\aleph)(\mathbf{u}_b, \mathbf{u}_c) + \aleph(\nabla_{\mathbf{u}_a}\mathbf{u}_b, \mathbf{u}_c) + \aleph(\mathbf{u}_b, \nabla_{\mathbf{u}_a}\mathbf{u}_c)$$

$$= \sum_{(abc)} (\nabla_{\mathbf{u}_a}\aleph)(\mathbf{u}_b, \mathbf{u}_c) + \aleph(\aleph(\mathbf{u}_a, \mathbf{u}_b), \mathbf{u}_c) + \aleph([\mathbf{u}_a, \mathbf{u}_b], \mathbf{u}_c).$$

Finally, we obtain the first identities of Bianchi as follows, $\forall \omega \in T_M B^*$:

$$\sum_{(abc)} \Re(\mathbf{u}_a, \mathbf{u}_b, \mathbf{u}_c, \omega) = \sum_{(abc)} (\nabla_{\mathbf{u}_a}\aleph)(\mathbf{u}_b, \mathbf{u}_c, \omega) + \aleph(\aleph(\mathbf{u}_a, \mathbf{u}_b), \mathbf{u}_c, \omega). \quad \text{(D.2)}$$

D.3   Second identities of Bianchi

On the one hand, the directional derivative of $\Re(\mathbf{u}_a, \mathbf{u}_b, \mathbf{v}, \omega)$ holds:

$$\mathbf{u}_c[\Re(\mathbf{u}_a, \mathbf{u}_b, \mathbf{v}, \omega)] \equiv \nabla_{\mathbf{u}_c}[\Re(\mathbf{u}_a, \mathbf{u}_b, \mathbf{v}, \omega)].$$

Distribution of the covariant derivative and adding up the terms after a circular permutation give:

$$\sum_{(abc)} = (\nabla_{\mathbf{u}_3}\Re)(\mathbf{u}_1, \mathbf{u}_2, \mathbf{v}, \omega) + (\nabla_{\mathbf{u}_1}\Re)(\mathbf{u}_2, \mathbf{u}_3, \mathbf{v}, \omega) + (\nabla_{\mathbf{u}_2}\Re)(\mathbf{u}_3, \mathbf{u}_1, \mathbf{v}, \omega)$$

$$+ \quad \Re(\nabla_{\mathbf{u}_3}\mathbf{u}_1, \mathbf{u}_2, \mathbf{v}, \omega) + \Re(\nabla_{\mathbf{u}_1}\mathbf{u}_2, \mathbf{u}_3, \mathbf{v}, \omega) + \Re(\nabla_{\mathbf{u}_2}\mathbf{u}_3, \mathbf{u}_1, \mathbf{v}, \omega)$$

$$+ \quad \Re(\mathbf{u}_1, \nabla_{\mathbf{u}_3}\mathbf{u}_2, \mathbf{v}, \omega) + \Re(\mathbf{u}_2, \nabla_{\mathbf{u}_3}\mathbf{u}_1, \mathbf{v}, \omega) + \Re(\mathbf{u}_3, \nabla_{\mathbf{u}_1}\mathbf{u}_2, \mathbf{v}, \omega)$$

$$+ \quad \Re(\mathbf{u}_1, \mathbf{u}_2, \nabla_{\mathbf{u}_3}\mathbf{v}, \omega) + \Re(\mathbf{u}_2, \mathbf{u}_3, \nabla_{\mathbf{u}_1}\mathbf{v}, \omega) + \Re(\mathbf{u}_3, \mathbf{u}_1, \nabla_{\mathbf{u}_2}\mathbf{v}, \omega)$$

$$+ \quad \Re(\mathbf{u}_1, \mathbf{u}_2, \mathbf{v}, \nabla_{\mathbf{u}_3}\omega) + \Re(\mathbf{u}_2, \mathbf{u}_3, \mathbf{v}, \nabla_{\mathbf{u}_1}\omega) + \Re(\mathbf{u}_3, \mathbf{u}_1, \mathbf{v}, \nabla_{\mathbf{u}_2}\omega).$$

On the other hand, consider the curvature operator which holds:

$$\Re(\mathbf{u}_a, \mathbf{u}_b) \equiv \nabla_{\mathbf{u}_a}\nabla_{\mathbf{u}_b} - \nabla_{\mathbf{u}_b}\nabla_{\mathbf{u}_a} - \nabla_{[\mathbf{u}_a, \mathbf{u}_b]}$$

$$\mathbf{u}_c[\Re(\mathbf{u}_a, \mathbf{u}_b, \mathbf{v}, \omega)] = \mathbf{u}_c[\omega\{\Re(\mathbf{u}_a, \mathbf{u}_b)\,\mathbf{v}\}].$$

Provided the relation on the vector derivative, we have:

$$
\begin{aligned}
\mathbf{u}_c[\omega\{\Re(\mathbf{u}_a, \mathbf{u}_b)\mathbf{v}]] &= (\nabla_{\mathbf{u}_c}\omega)\{\Re(\mathbf{u}_a, \mathbf{u}_b)\mathbf{v}\} + \omega\{\nabla_{\mathbf{u}_c}[\Re(\mathbf{u}_a, \mathbf{u}_b)\mathbf{v}]\} \\
&= \Re(\mathbf{u}_a, \mathbf{u}_b, \mathbf{v}, \nabla_{\mathbf{u}_c}\omega) + \omega\{\nabla_{\mathbf{u}_c}[\Re(\mathbf{u}_a, \mathbf{u}_b)\mathbf{v}]\}.
\end{aligned}
$$

By introducing the definition of the curvature operator and by adding up the terms after a circular permutation, the following equation holds:

$$
\sum_{(abc)} \omega\{\nabla_{\mathbf{u}_c}[\Re(\mathbf{u}_a, \mathbf{u}_b)\mathbf{v}]\} = \sum_{(abc)} \Re(\mathbf{u}_c, \mathbf{u}_a, \nabla_{\mathbf{u}_b}\mathbf{v}, \omega)
$$

$$
+ \sum_{(abc)} \Re([\mathbf{u}_c, \mathbf{u}_a], \mathbf{u}_b, \mathbf{v}, \omega) + \omega \left\{ \sum_{(abc)} \nabla_{[[\mathbf{u}_c, \mathbf{u}_a], \mathbf{u}_b]}\mathbf{v} \right\}
$$

in which the last term vanishes thanks to the Jacobi identity. Finally, we obtain the second identities of Bianchi:

$$
\sum_{(abc)} (\nabla_{u_c}\Re)(\mathbf{u}_a, \mathbf{u}_b, \mathbf{v}, \omega) + \sum_{(abc)} \Re(\aleph(\mathbf{u}_a, \mathbf{u}_b), \mathbf{u}_c, \mathbf{v}, \omega) = 0. \qquad \text{(D.3)}
$$

Identities (D.2) and (D.3) extend the Bianchi formulae of Riemannian geometry in the presence of field singularity.

First, these two sets of differential identities are not differential equations for the determination of $\aleph(M, t)$ and $\Re(M, t)$ since the covariant derivatives associated to the affine connection, and the torsion and curvature tensors are not independent [194].

Second, the existence of these identities was used to show that the classical compatibility equations of deformations of nonlinear elasticity were not independent equations, e.g., [124]. In the particular case where the torsion tensor vanishes, relations (D.1), (D.2), and (D.3) can be rewritten in the more usual form:

$$
\begin{aligned}
\Re(\mathbf{u}_a, \mathbf{u}_b, \mathbf{v}, \omega) + \Re(\mathbf{u}_b, \mathbf{u}_a, \mathbf{v}, \omega) &= 0 \\
\sum_{(abc)} \Re(\mathbf{u}_a, \mathbf{u}_b, \mathbf{u}_c, \omega) &= 0 \\
\sum_{(abc)} (\nabla_{u_c}\Re)(\mathbf{u}_a, \mathbf{u}_b, \mathbf{v}, \omega) &= 0. \qquad \text{(D.4)}
\end{aligned}
$$

Historically, use of the Bianchi identities (D.4)3 and the definition of Ricci's tensor make way for the introduction of Einstein's tensor and the establishment of the so-called conservation laws in the general relativistic theory, e.g., [116]. In the field of solid defects, Bianchi identities have been developed by Kröner [102].

Appendix E
Theorem of Cauchy–Weyl

In this section, we consider the isotropic property of a scalar function defined on a manifold B the arguments of which are vector fields. The theorem of Cauchy–Weyl allows us to define the representation of any scalar function. We will recall the main features of this theorem, for a 3-dimensional referential body without loss of generality, and we will split into two parts the development of the proof. Consider m vector fields $(\mathbf{v}_1, \ldots, \mathbf{v}_m)$ defined on B that span a subspace of dimension p. For the following theorems, we will consider a scalar function Φ from $T_M B \times \cdots \times T_M B$ onto R.

E.1 Theorem of Cauchy (1850)

Theorem E.1.1 *(Cauchy 1850) Consider a continuum B endowed with a metric tensor $\mathbf{g}$ and a volume form ω_0, induced from the referential body. For any scalar function Φ the arguments of which are elements of $T_M B$, the following equivalence holds, O being the orthogonal group of the referential body:*

$$\Phi = \hat{\Phi}(\mathbf{v}_1, \ldots, \mathbf{v}_m) = \hat{\Phi}(q\mathbf{v}_1, \ldots, q\mathbf{v}_m) \qquad \forall \mathbf{Q} \in O$$

$$\Longleftrightarrow$$

$$\exists \tilde{\Phi} \qquad \Phi = \tilde{\Phi}[\mathbf{g}(\mathbf{v}_k, \mathbf{v}_l), k, l = 1, \ldots, m]. \tag{E.1}$$

Proof. We first give a proof of the following lemma.

Theorem E.1.2 *(Lemma) Consider two sets of ordered vector fields of* $T_M B$ *{$\mathbf{v}_1, \ldots, \mathbf{v}_m$} and {$\mathbf{u}_1, \ldots, \mathbf{u}_m$}. Then, we can write:*

$$\mathbf{g}(\mathbf{v}_k, \mathbf{v}_l) = \mathbf{g}(\mathbf{u}_k, \mathbf{u}_l) \qquad k, l = 1, \ldots, m$$

$$\Longleftrightarrow$$

$$\exists \mathbf{Q} \in O \mid q(\mathbf{v}_k) = \mathbf{u}_k \qquad k = 1, \ldots, m. \tag{E.2}$$

Indeed, let V be the vector subspace with dimension $p \le m$ generated by vectors {$\mathbf{v}_1, \ldots, \mathbf{v}_m$} and U by {$\mathbf{u}_1, \ldots, \mathbf{u}_m$}. Suppose the vectors are ordered such that the first ones are linearly independent. Using the induction of the volume form of the referential body ω_{0V} on V, and ω_{0U} on U, we obtain:

$$[\omega_{0V}(\mathbf{v}_1, \ldots, \mathbf{v}_p)]^2 = \det [\mathbf{g}(\mathbf{v}_k, \mathbf{v}_l)] \ne 0$$
$$\Longrightarrow [\omega_{0V}(\mathbf{u}_1, \ldots, \mathbf{u}_p)]^2 = [\omega_{0V}(\mathbf{v}_1, \ldots, \mathbf{v}_p)]^2 \ne 0$$

which shows that the two subspaces U and V have the same dimension p and thus that there exists a unique linear transformation (endomorphism) $\mathbf{Q}$ from U onto V such that:

$$\mathbf{Q}(\mathbf{v}_k) = \mathbf{u}_k \qquad k = 1, \ldots, m.$$

1. This transformation is orthogonal since:

$$\mathbf{g}(\mathbf{Q}(\mathbf{v}_k), \mathbf{Q}(\mathbf{v}_l)) = \mathbf{g}(\mathbf{u}_k, \mathbf{u}_l) = \mathbf{g}(\mathbf{v}_k, \mathbf{v}_l) \qquad k, l = 1, \ldots, p.$$

2. For the other vectors, non independent, $k > p$, consider the dependence relation:

$$\mathbf{v}_{k>p} = \alpha_k^i \, \mathbf{v}_i \Longrightarrow \mathbf{g}(\mathbf{v}_k, \mathbf{v}_j) = \alpha_k^i \, \mathbf{g}(\mathbf{v}_i, \mathbf{v}_j) \qquad i, j = 1, \ldots, p.$$

Since the determinant of this system of equations does not necessarily vanish (due to the linear independence of vectors), the coefficients depend only on scalar products. By replacing the scalar products of $\mathbf{v}$ by those of $\mathbf{u}$ (they have the same values by assumption), we obtain:

$$\mathbf{g}(\mathbf{u}_k, \mathbf{u}_j) = \alpha_k^i \, \mathbf{g}(\mathbf{u}_i, \mathbf{u}_j), \forall i, j = 1, \ldots, p \Longrightarrow \mathbf{u}_{k>p} = \alpha_k^i \, \mathbf{u}_i$$

to get:

$$\mathbf{Q}(\mathbf{v}_{k>p}) = q(\alpha_k^i \, \mathbf{v}_i) = \mathbf{u}_{k>p}$$

which shows that the transformation could be applied to all the vectors.

(Theorem of Cauchy continued). Let us reason through contradiction.

1. If there is no representation of the type:

$$\Phi = \tilde{\Phi}[\mathbf{g}(\mathbf{v}_k, \mathbf{v}_l), k, l = 1, \ldots, m]$$

then at least two sets of vectors $\{\mathbf{v}'_1, \ldots, \mathbf{v}'_m\}$ and $\{\mathbf{v}''_1, \ldots, \mathbf{v}''_m\}$ exist such that:

$$\mathbf{g}(\mathbf{v}'_k, \mathbf{v}'_l) = \mathbf{g}(\mathbf{v}''_k, \mathbf{v}''_l) \qquad k, l = 1, \ldots, m$$

and

$$\hat{\Phi}(\mathbf{v}'_1, \ldots, \mathbf{v}'_m) \neq \hat{\Phi}(\mathbf{v}''_1, \ldots, \mathbf{v}''_m) \tag{E.3}$$

From the previous lemma, if we have:

$$\mathbf{g}(\mathbf{v}'_k, \mathbf{v}'_l) = \mathbf{g}(\mathbf{v}''_k, \mathbf{v}''_l) \qquad k, l = 1, \ldots, m$$

then we necessarily have:

$$\exists \mathbf{Q}^* \in O \mid \mathbf{Q}^*(\mathbf{v}'_k) = \mathbf{v}''_k \qquad k = 1, \ldots, m.$$

2. It implies that, the function being isotropic by assumption, for these same m-plets:

$$\Phi = \hat{\Phi}(\mathbf{Q}^*\mathbf{v}'_1, \ldots, \mathbf{Q}^*\mathbf{v}'_m) = \hat{\Phi}(\mathbf{v}''_1, \ldots, \mathbf{v}''_m) \tag{E.4}$$

which does not fit with relation (E.3). This proves the Cauchy theorem by means of a reasoning based on contradiction.

E.2　Theorem of Cauchy–Weyl (1939)

Theorem E.2.1 (*Cauchy–Weyl, 1850–1938*) *Consider a continuum B endowed with a metric and a volume form induced from the referential body, $\mathbf{g}$ and ω_0. For any scalar-valued function Φ with vector arguments, elements of $T_M B$, the following equivalence holds, O^+ being the rotation group and p the dimension of the vector subspace generated by the m-plet:*

$$\Phi = \hat{\Phi}(\mathbf{v}_1, \ldots, \mathbf{v}_m) = \hat{\Phi}(\mathbf{Q}\mathbf{v}_1, \ldots, \mathbf{Q}\mathbf{v}_m) \qquad \forall \mathbf{Q} \in O^+$$

$$\Longleftrightarrow$$

$$\exists \tilde{\Phi} \qquad \Phi = \tilde{\Phi}[\mathbf{g}(\mathbf{v}_k, \mathbf{v}_l), \omega_0(\mathbf{v}_k, \mathbf{v}_l, \mathbf{v}_h), k, l, h = 1, \ldots, m]. \tag{E.5}$$

Proof. First, for $m < 3$ the representation admits the scalar products as only arguments. Let us give a proof of the following lemma.

Theorem E.2.2 *(Lemma) Consider two sets of ordered vector fields of $T_M B$, $\{v_1, \ldots, v_m\}$ and $\{u_1, \ldots, u_m\}$. Then, we can write:*

$$g(v_k, v_l) = g(u_k, u_l) \qquad \omega_0(v_k, v_l, v_h) = \omega_0(u_k, u_l, u_h)$$

$$k, l, h = 1, \ldots, m$$

$$\Longleftrightarrow$$

$$\exists Q \in O^+ \mid Q(v_k) = u_k \qquad k = 1, \ldots, m. \tag{E.6}$$

Indeed, let V be the vector subspace of dimension $p \leq m$ generated by the vectors $\{v_1, \ldots, v_m\}$ and U by $\{u_1, \ldots, u_m\}$. Suppose the vectors are ordered in such way that the first ones are linearly independent. Again, considering the induced volume forms ω_{0V} on V and ω_{0U} on U, we obtain:

$$\omega_{0V}(v_1, \ldots, v_p) \neq 0 \Longrightarrow \omega_{0U}(u_1, \ldots, u_p) = \omega_{0V}(v_1, \ldots, v_p) \neq 0.$$

The two subspaces U and V have the same dimension p, and therefore, there exists only one linear transformation Q from U onto V, a priori not orthogonal and not positive, such that:

$$Q\, v_k = u_k \qquad k = 1, \ldots, p.$$

This transformation is a pure rotation from U onto V since it satisfies the two following conditions:

1. It is orthogonal, first for vectors from 1 to p:

$$g(Qv_k, Qv_l) = g(u_k, u_l) = g(v_k, v_l) \qquad k, l = 1, \ldots, p$$

2. It is a pure rotation:

$$
\begin{aligned}
\omega_{0U}(Qv_1, \ldots, Qv_p) &= (\det Q)\omega_{0V}(v_1, \ldots, v_p) \\
&= \omega_{0U}(u_1, \ldots, u_p) \\
&= \omega_{0V}(v_1, \ldots, v_p)
\end{aligned}
$$

since we can observe that $\det Q = 1$.

The same transformation Q should allow us to link the other nonindependent vectors, $k > p$. For this to be possible, consider again the linear independence:

$$v_{k>p} = \alpha_k^i\, v_i \Longrightarrow g(v_k, v_j) = \alpha_k^i\, g(v_i, v_j) \qquad i, j = 1, \ldots, p.$$

The determinant of this system does not necessarily vanish (linear independence of vectors). Therefore the coefficients depend only on the scalar products. By replacing the scalar products of v with those of u (same values by assumption), one obtains:

$$g(u_k, u_l) = \alpha_k^j\, g(u_i, u_j) \qquad i, j = 1, \ldots, p \Longrightarrow u_{k>p} = \alpha_k^i\, u_i$$

to give:

$$\mathbf{Q}\,\mathbf{v}_{k>p} = \mathbf{Q}(\alpha_k^i\,\mathbf{v}_i) = \alpha_k^i\mathbf{Q}(\mathbf{v}_i) = \mathbf{u}_{k>p}.$$

(Theorem of Cauchy–Weyl continued). Let us again reason by contradiction:

1. If there is no representation:

$$\Phi = \tilde{\Phi}[\mathbf{g}(\mathbf{v}_k,\mathbf{v}_l), \omega_0(\mathbf{v}_k,\mathbf{v}_l,\mathbf{v}_h), k,l,h = 1,\ldots,m],$$

then there are at least two vector sets $\{\mathbf{v}_1',\ldots,\mathbf{v}_m'\}$ and $\{\mathbf{v}_1'',\ldots,\mathbf{v}_m''\}$ such that:

$$\mathbf{g}(\mathbf{v}_k',\mathbf{v}_l') = \mathbf{g}(\mathbf{v}_k'',\mathbf{v}_l'') \qquad \omega_0(\mathbf{v}_k',\mathbf{v}_l',\mathbf{v}_h') = \omega_0(\mathbf{v}_k',\mathbf{v}_l',\mathbf{v}_h')$$
$$k,l,h = 1,\ldots,m$$

and

$$\hat{\Phi}(\mathbf{v}_1',\ldots,\mathbf{v}_m') \neq \hat{\Phi}(\mathbf{v}_1'',\ldots,\mathbf{v}_m''). \tag{E.7}$$

Thanks to the previous lemma, if the following holds:

$$\mathbf{g}(\mathbf{v}_k',\mathbf{v}_l') = \mathbf{g}(\mathbf{v}_k'',\mathbf{v}_l'')$$
$$\omega_0(\mathbf{v}_k',\mathbf{v}_l',\mathbf{v}_h') = \omega_0(\mathbf{v}_k'',\mathbf{v}_l'',\mathbf{v}_h'')$$
$$k,l,h = 1,\ldots,m,$$

then we necessarily have:

$$\exists\mathbf{Q}^* \in O^+ \mid \mathbf{Q}^*(\mathbf{v}_k') = \mathbf{v}_k'' \qquad k = 1,\ldots,m.$$

2. The function being isotropic by assumption, we can deduce easily that:

$$\Phi = \hat{\Phi}(\mathbf{v}_1',\ldots,\mathbf{v}_m') = \hat{\Phi}(\mathbf{Q}^*\mathbf{v}_1',\ldots,\mathbf{Q}^*\mathbf{v}_m') = \hat{\Phi}(\mathbf{v}_1'',\ldots,\mathbf{v}_m'') \tag{E.8}$$

which does not fit with relation (E.7). This is the proof of the Cauchy–Weyl theorem by means of reasoning based on contradiction.

References

[1] Ashby MF., Dyson BF. Creep damage mechanics and micromechanisms. In *Advances in Fracture Research (Fracture 84)*, 1, Pergamon Press, Oxford 1986, 3–30.

[2] Audoin B. On internal variables in anisotropic damage, *Eur J Mech A/Solids*, **10**, 6 (1991), 587–606.

[3] Baldacci R., Augusti V., Capurro M. A microrelativistic dislocation theory, *Lincei Memorie Sc. Fisiche*, ecc., S. VIII, vol XV, sez II, 2, 1979, pp 23–68.

[4] Bai Y., Dodd B. *Adiabatic Shear Localization*. Pergamon Press, Oxford, 1992.

[5] Barber JR. Contact problems involving a cooled punch. *J Elasticity*, **8**, 4, 1978, 409–423.

[6] Barber JR. Some thermoelastic contact problems involving frictional heating. *Quat J Mech Appl Math*, **29**, 1976, 1–13.

[7] Barber JR. The conduction of heat from sliding solids. *Int J Heat Mass Transfer*, **13**, 1970, 857–869.

[8] Batchelor GK. *An Introduction to Fluid Dynamics*. Cambridge University Press, Cambridge, 2000.

[9] Batra RC. Thermodynamics of non-simple elastic materials. *J Elasticity*, **6**, 1976, 451–456.

[10] Batra RC. Force on a lattice defect in an elastic body. *J Elasticity*, **17**, 1987, 3–8.

[11] Betten J. Net-stress analysis in creep mechanics, *Ing Arch*, **52** (1982), 405–419.

[12] Bikerman J-J. Thermodynamics, adhesion and sliding friction. *J Lubr Technol*, **92** (1970), 198, 243–247.

[13] Bilby BA., Smith E. Continuous distributions of dislocations: a new application of the method of non Riemanian geometry, *Proc Roy Soc London*, A 231, 1955, 263–273.

[14] Bloch ED. *A First Course in Geometric Topology and Differential Geometry*, Birkhäuser, Boston, 1997.

[15] Bloom F. Modern differential geometric techniques in the theory of continuous distributions of dislocations. *Lectures Notes in Mathematics*, Dold A and Eckmann B, editors, Springer–Verlag, Berlin, 1978.

[16] Bowden FP., Thomas PM. The surface temperature of sliding solids. *Proc Roy Soc London*, Ser A, **223** (1954), 29–40.

[17] Bradley FE. Development of an Airy stress function of general applicability in one, two and three dimensions, *J Appl Phys*, **67**, (1990), 225–226.

[18] Bui HD., Ehrlacher A., Nguyen Quoc Son. Etude expérimentale de la dissipation dans les propagations de fissures par thermographie infrarouge, *C R Acad Sci Paris*, **293**, (1981), 1015–1018.

[19] Burton RA. Thermal deformation in frictionally heated contact. *Wear*, **59**, (1980), 1–20.

[20] Cameron A, Gordon AN., Symm GT. Contact temperature in rolling/sliding surfaces. *Proc Roy Soc London*, A286, (1964), 45–61.

[21] Cartan E. Sur les variétés à connexion affine et la théorie de la relativité généralisée I partie, *Ann Ec Norm*, **40**, 1923, 325.

[22] Cartan E. *On Manifolds with Affine Connection and the Theory of General Relativity*, Bibliopolis, Edizioni di Filosofia e Science, Napoli, 1986.

[23] Cartan E. *La Théorie des Groupes Finis et Continus et la Géométrie Différentielle Traitées par la Méthode du Repère Mobile*, Editions Gabay J., Paris, 1992.

[24] Cesaro E. Sulle formole del Volterra, fundamentali nello teoria della distorsioni elastiche, *Rend Accad Sci Fis Math* (Società Reale di Napoli), 1906, 311–321.

[25] Chadwick P., Powdrill B. Singular surfaces in linear thermoelasticity. *Int J Engng Sci*, **3**, (1965), 561–595.

[26] Cherepanov GP. Invariant γ-integrals and some of their applications in mechanics, *PMM* (English translation), **41**, (1977), 397–410.

[27] Cherepanov GP. Remark on the dynamic invariant or path-independent integral, *Int J Solids and Structures*, bf 25, (1989), 1267–1269.

[28] Chevray R., Mathieu J. *Topics in Fluid Mechanics*, Cambridge University Press, 1993.

[29] Choquet-Bruhat Y., DeWitt-Morette C., Dillard-Bleick M. *Analysis, Manifolds and Physics*, part I and II, North-Holland, New York, 1982.

[30] Ciarlet PG., Necas J. Unilateral problems in nonlinear, three-dimensional elasticity. *Arch Rat Mech Anal*, **87**, (1985), 319–338.

[31] Clarke FH. *Optimization and Nonsmooth Analysis*. Wiley, New York, 1983.

[32] Clifton RJ. On the equivalence of $F^e F^p$ and $F^p F^e$, *ASME Trans J Appl Mech*, **39** (1972), 287–289.

[33] Coleman BD., Noll W. The thermodynamics of elastic materials with heat conduction and viscosity, *Arch Rat Mech Anal*, **13**, (1963), 167–170.

[34] Coleman BD., Gurtin M., Thermodynamics with internal state variables, *The Journ Chem Phys*, **47**, 2, (1967), 597–613.

[35] Cooper MG., Mikic BB., Yovanovich MM. Thermal contact conductance. *Int J Heat Mass Transfer*, **12**, (1969), 279–300.

[36] Curnier A., Rakotomanana RL. Generalized strain and stress measures: Critical survey and new results, *Eng Trans Polish Academy of Sciences*, **39**, 3-4, (1991), 461–538.

[37] Curnier A., He Q-C, Telega JJ. Formulation of unilateral contact between two elastic bodies undergoing finite deformations. *C R Acad Sc Paris*, t 314, série II, 1992, 1–6.

[38] Curnier A., He Q-C, Klarbring A. Continuum mechanics modelling of large deformation contact with friction. In *Contact Mechanics*, ed. M Raous et al., Plenum Press, New York, 1995, 145–158.

[39] Coussy O. *Mechanics of Porous Continua*, Wiley, Chichester, 1994.

[40] Defrise P. Analyse géométrique de la cinématique des milieux continus. In *Publication Série B n 6 de l'Institut Royal Météorologique de Belgique*, Bruxelles, 1953, 1–63.

[41] Dienes K. On the analysis of rotation and stress rate in deforming body, *Acta Mechanica*, **32**, (1979), 217–232.

[42] Doyle TC., Ericksen JL. Non-linear elasticity. In *Advances in Applied Mechanics*, eds. H.L. Dryden and Th. von Karman, Academic Press, New York, **4**, (1956), 53–115.

[43] Duan ZP. Duality principle of conservation laws in dislocation continuum, *Int J Solids Structure*, **21**, 7, (1985), 683–697.

[44] Duan YZ., Duan ZP. Gauge field theory of a continuum with dislocations and disclinations, *Int J Eng Science*, **24**, 4, (1986), 513–527.

[45] Dunn JE., Serrin J. On the thermodynamics of interstitial working, *Arch Rat Mech Anal*, **17**, (1985), 95–133.

[46] Dunn JE. Interstitial working and a nonclassical continuum thermodynamics. In *New Perspectives in Thermodynamics*. ed. Serrin JE, Springer-Verlag, Berlin, 1986, 187–222.

[47] Edelen DGB., Lagoudas DC., *Gauge Theory and Defects in Solids*, North-Holland, New York, 1988.

[48] Ehlers J. The nature and structure of space-time. In *The Physicist's Conception of Nature*, Mehra J ed., Reidel, Dordrecht, 71–91, 1973.

[49] Emde F. Zur vektorrechnung, *Arch Math*, **3**, 24, (1915), 1–11.

[50] Epstein M. Toward a complete second-order evolution law, *Math Mech Solids*, **4**, (1999), 251–266.

[51] Epstein M., Maugin GA. On the geometrical material strucure of anelasticity, *Acta Mechanica*, **115**, (1996), 119–131.

[52] Eringen AC. A unified theory of thermomechanical materials. *Int J Eng Sci*, **4**, (1966), 179–202.

[53] Eshelby JD. The elastic energy-momentum tensor, *J Elasticity*, bf 5, (1975), 321–335.

[54] Eshelby JD. Aspects of theory of dislocations. In *Mechanics of Solids: the Rodney Hill 60th anniversary volume*, 1982, 185–225.

[55] Favard J. *Cours de Géométrie Différentielle Locale*, Gauthier–Villars, Paris, 1957.

[56] Fisher GMC., Leitman MJ. On continuum thermodynamics with surfaces. *Arch Rat Mech Anal*, 30, 225, 1968, pp 225-262.

[57] Fleck NA., Hutchinson JW. Strain gradient plasticity. In *Advances in Applied Mechanics*, Hutchinson JW, Wu TY, eds., vol. 33, Academic Press, New York, 295–361.

[58] Flugge W. *Tensor Analysis and Continuum Mechanics*, Springer-Verlag, Heidelberg, 1972.

[59] Frankel Th. *The Geometry of Physics: An Introduction*. Cambridge University Press, New York, 1997.

[60] Fried E. Thermal conduction contribution to heat transfer at contacts. In *Thermal Conductivity*, vol. 2, ed. Tye RP, Academic Press, London, 1969.

[61] Friedel J. Dislocations: An introduction, in *Dislocations in Solids*, 1, ed. FRN Nabarro, North-Holland, New York, 1979, 33–141.

[62] Gairola BKD. Nonlinear elastic problems, in *Dislocations in Solids*, 1, ed. FRN Nabarro, North-Holland, New York, 1979, 225–341.

[63] Germain P. La méthode des puissances virtuelles en mécanique des milieux continus, *J Mécanique*, 12/2, 1973, 235–274.

[64] Germain P. *Mécanique des Milieux Continus*, Editions Masson, Paris, 1973.

[65] Germain P. *Mécanique*, 2 tomes, Editions Ellipse X, Paris, 1986.

[66] Germain P., Nguyen QS., Suquet P. Continuum thermodynamics, *J Applied Mech, Trans ASME*, **50**, (1983), 1010–1020.

[67] Goldstein S. *Lectures on Fluid Mechanics*, Interscience Publishers, London, 1960.

[68] Green AE., Naghdi PM. A general theory of an elastic-plastic continuum, *Arch Rat Mech Anal*, **18**, (1965), 251–281.

[69] Green AE., Naghdi PM. A derivation of jump condition for entropy in thermomechanics, *J Elasticity*, **8**, (1978), 179–182.

[70] Green AE., Rivlin RS. Multipolar continuum mechanics, *Arch Rat Mech Anal*, **17**, (1964), 113–147.

[71] Green AE., Zerna W. *Theoretical Elasticity*, Oxford University Press, Oxford, 1975.

[72] de Groot SR., Mazur P. *Non-Equilibrium Thermodynamics*, North-Holland, Amsterdam, 1962.

[73] Gurtin ME. Thermodynamics and the possibility of spatial interaction in elastic materials. *Arch Rat Mech Anal*, **19**, (1965), 339–352.

[74] Gurtin ME., Podio-Guidugli P. Configurational forces and the basic laws for crack propagation, *J Mech Phys Solids*, **44**, 6, (1996), 905–927.

[75] Gurtin ME., Williams WO. An axiomatic foundations for continuum thermodynamics, *Arch Rat Mech Anal*, **26**, (1967), 83–117.

[76] Gurtin ME., Williams W.O. On the first law of thermodynamics. *Arch Rat Mech Anal*, **42**, (1971), 77–92.

[77] Haupt P. Thermodynamics of solids. In *Non-equilibrium Thermodynamics with Applications to Solids*, *CISM courses*, Springer-Verlag, New York, 1993, 65–138.

[78] Hasebe N., Tomida A., Nakamura T. Thermal stresses of a crack circular hole due to uniform heat flux, *J Thermal Stresses*, **11**, 4, (1988), 381–391.

[79] He QC., Curnier A. A more fundamental approach to damaged elastic stress-strain relations, *Int J Solids Structure*, **32**, 10, (1995), 1433–1457.

[80] Hoff D. Discontinuous solutions of the Navier–Stokes equations for multidimensional flows of heat-conducting fluids, *Arch Rat Mech Anal*, **139**, (1997), 303–354.

[81] Hull D., Bacon DJ. *Introduction to Dislocations*, 4th edition, Butterworth-Heineman, Boston, 2001.

[82] Ilankamban R., Krajcinovic D. A constitutive theory for progressively deteriorating brittle solids, *Int J Solids Structures*, **23**, (1987), 1521–1534.

[83] Indenbom VL., Lothe J., eds. *Elastic Strain Fields and Dislocation Mobility*, North-Holland, New York, 1992.

[84] Jaeger J.C. Moving heat sources of heat and the temperature of sliding contacts. *Proc Roy Soc NSW*, **76**, (1942), 203–224.

[85] Johansson L., Klarbring A. Thermoelastic frictional contact problems: modelling, finite element approximation and numerical realization. *Comp Meth Applied Mech Engineering*, **105**, (1993), 181–210.

[86] Kachanov M. Continuum model of medium with cracks, *J Eng Mech Div*, **106**, EM5, (1980), 1039–1051.

[87] Kachanov M. Elastic solids with many cracks and related problems. In *Advances in Applied Mechanics*, vol. 30, eds. J. Hutchinson and T. Wu, Academic Press, New York, 1994, 259–445.

[88] von Karman T., Burgers JM. General Aerodynamic Theory: Perfect Fluids, vol 2 of *Aerodynamic Theory*, Durand WF, ed., Springer-Verlag, Leipzig, 1934.

[89] Kelly JM., Gillis PP. Thermodymanics and dislocation mechanics, *J Franklin Institute*, **297**, (1974), 59–74.

[90] Kiehn RM. Retroductive determinism, *Int J Eng Science*, 14, 1976, pp. 249-254.

[91] Kiehn RM. Irreversible topological evolution in fluid mechanics. In *Some Unanswered Questions in Fluid Mechanics, ASME*, vol. 89 WA/FE-5, ed. by Trefethen LM and Panton RL, ASME, New York, 1990.

[92] Klamecki BE. A thermodynamical model of friction. *Wear*, **68**, (1980), 113–120.

[93] Klamecki BE. An entropy-based model of plastic deformation energy dissipation in sliding. *Wear*, **96**, (1984), 319–329.

[94] Klarbring A., Mikelic A., Shillor M. The rigid punch problem with friction. *Int J Engng Sci*, **29** (6), (1991), 751–768.

[95] Kleman M. *Points, Lines and Walls*. Wiley, Chichester, 1989.

[96] Kneubuhl FK. *Oscillations and Waves*, Springer-Verlag, Heidelberg, 1997.

[97] Kondo K. Non-Riemanian geometry of imperfect crystals from macroscopic viewpoint In *Memoirs of the Unifying Study of Basic Problems in Engineering Sciences by Means of Geometry*, vol. I, ed. K. Kondo, Division D: Gakujutsu Benken Fukyu-Kai, Tokyo, 1955.

[98] Korovchinski MV. Plane contact problem of thermoelasticity during quasi-stationary heat generation on the contact surface, *J Basic Engng*, **87**, (1965), 811–817.

[99] Korteweg DJ. Sur la forme que prennent les équations du mouvement des fluides si l'on tient compte des forces capillaires causées par des variations de densité considérables mais continues et sur la théorie de la capillarité dans l'hypothèse d'une variation continue de la densité, *Archiv Néerland Sci Ex Nat*, **6** (2), (1901), 1–24.

[100] Krajcinovic D. Damage mechanics, *Mech Mat*, **8**, (1989), 117–197.

[101] Kröner E. Dislocation: A new concept in the continuum theory of plasticity, *J Math Phys*, **42**, (1963), 27–37.

[102] Kröner E. Continuum theory of defects. In *Physique des Défauts, Nato Series* Ballian R et al., eds., North-Holland, New York, 1981, 219–315.

[103] Kuiken GDC. *Thermodynamics of Irreversible Processes: Application to Diffusion and Rheology*, Wiley, Chichester, 1994.

[104] Lagoudas DC., Edelen DGB. Material and spatial gauge theories of solids I: Gauge constructs, geometry and kinematics, *Int J Eng Science*, **27**, 4, (1989), 411–431.

[105] Larson RG. *Constitutive Equations for Polymeric Melts and Solutions*. Butterworths, London, 1988.

[106] Lax M. Multiple scattering of waves II. The effective field in dense systems. *Phys Rev*, **85**(4), (1951), 621–629.

[107] Le KC., Schutte H., Stumpf H. Dissipative driving force in ductile crystals and the strain localization phenomenon, *Int J Plasticity*, 1998, preprint. (To appear.)

[108] Le KC., Stumpf H. Constitutive equations for elastoplastic bodies at finite strain: Thermodynamic implementation, *Acta Mechanica*, **100**, (1993), 155–170.

[109] Le KC., Stumpf H. Nonlinear continuum theory of dislocations, *Int J Eng Science*, **34**, 3, (1996), 339–358.

[110] Le KC., Stumpf H. A model of elastoplastic bodies with continously distributed dislocations, *Int J Plasticity*, **12**, 5, (1996), 611–627.

[111] Le KC., Stumpf H. On the determination of the crystal reference in nonlinear continuum theory of dislocations, *Proc Roy Soc London A*, **452**, (1996), 359–371.

[112] Lee EH. Elastic plastic deformation at finite strain, *Trans ASME J Appl Mech*, **36**, (1969), 1–6.

[113] Lemaitre J. *A Course on Damage Mechanics*, Springer-Verlag, Berlin, 1992.

[114] de Leon M., Epstein M. The geometry of uniformity in second-grade elasticity, *Acta Mechanica*, **114**, (1996), 217–224.

[115] Levytskyi VP., Onyshkevych VM. Plane contact problem with heat generation account of friction. *Int J Engng Sci*, **34**, 1, (1996), 101–112.

[116] Lichnerowicz A. *Théorie Relativiste de la Gravitation et de l'Électromagnétisme*. Editions Masson, Paris, 1955.

[117] Lichtenstein L. *Grundlagen der Hydromechanik*, Springer-Verlag, Berlin, 1929.

[118] Ling FF. *Surface Mechanics*. Wiley, New York, 1973.

[119] Lorentz E., Andrieux S. A variational formulation for nonlocal damage models, *Int J Plasticity*, **15**, (1999), 119–138.

[120] Lovelock D., Rund H. *Tensors, Differential Forms and Variational Principles*, Wiley, New York, 1975.

[121] Madhusudana CV. *Thermal Contact Conductance*, Springer-Verlag, New York, 1995.

[122] Madhusudana CV., Fletcher LS. Contact heat transfer: The last decade, *AIAA J*, (1985), 510–523.

[123] Malliavin P. *Géométrie Différentielle Intrinsèque*. Editions Hermann, Paris, 1971.

[124] Malvern LE. Introduction to the mechanics of a continuous medium. In *Engineering of the Physical Sciences*, Englewood Cliffs, New Jersey, 1969.

[125] Mandel J. *Introduction à la Mécanique des Milieux Continus Déformables*, Editions Scientifiques de Pologne, Varsovie, 1975.

[126] Marcinkowsky MJ. The differential geometry of internal surface, *Archiwum Mechaniki Stosowanej*, **31**, 6, (1979), 763–781.

[127] Mariano PM., Augusti G. Multifield description of microcracked continua: A local model, *Math Mech Solids*, **3**, (1998), 183–200.

[128] Marigo JJ. Formulation of a damage law for an elastic material, *C R Acad Sci Paris*, série II, **292**, 19, (1981), 1309–1312.

[129] Markenscoff X., Wheeler L. On conditions at an interface between two materials in plane deformation, *J Elasticity* **45**, (1996), 33–44.

[130] Marsden JE., Hughes TJR. *Mathematical Foundations of Elasticity*, Prentice-Hall, New Jersey, 1983.

[131] Maugin GA. *The Thermomechanics of Plasticity and Fracture*, Cambridge University Press, Cambridge, 1992.

[132] Maugin G. *Material Inhomogeneities in Elasticity*. Editions Chapman and Hall, London, 1993.

[133] Maugin GA. Material forces: Concepts and applications, *Appl Mech Rev*, **48**, 5, (1995), 213–245.

[134] Michelitsch T., Wunderlin A. Fundamental solution of the incompatibility problem in three-dimensional infinite anisotropic elasticity theory, *Acta Mechanica*, **119**, (1996), 25–34.

[135] Mikic BB. Thermal contact conductance; theoretical considerations. *Int J Heat Mass Transfer*, **17**, (1974), 205–214.

[136] Minagawa S. A geometrical facet of the theory of dislocations and disclinations in a Cosserat continuum. Proceedings of the IUTAM on *Variational Methods in the Mechanics of Solids*, Nemat-Nasser S., ed., Pergamon Press, Oxford, 1978, 291–295.

[137] Minagawa S. A non-Riemanian geometrical theory of imperfections in a Cosserat continuum, *Archiwum Mechaniki Stosowanej*, **31**, 6, (1979), 783–792.

[138] Moeckel GP. Thermodynamics of an interface. *Arch Rat Mech Anal*, **57**, (1975), 255–280.

[139] Moreau J-J. Sur les lois de frottement, de plasticité et de viscosité, *C. R. Acad. Sci. Paris*, Sér. A, 271, (1970), 608–611.

[140] Moreau J-J. On unilateral constraints, friction and plasticity. In *Lecture Notes CIME*, Bressanone, Edizioni Cremonese, Rome, 1971.

[141] Mori T., Tanaka K. Average stress in matrix and average elastic energy of materials with misfitting inclusions, *Acta Metallurgica*, **21**, (1973), 571–574.

[142] Mukarami S., Ohno N. Continuum theory of creep and creep damage. In *Creep in structures*, Ponter ARS, Hayhurst DR, eds., Springer-Verlag, Berlin, 1981, 422–444.

[143] Nabarro FRN., ed., Dislocations in solids, 1: The elastic theory, North-Holland. (Dislocations: an introduction, Friedel J, 3–32; Crystal dislocations and the theory of elasticity; Kosevich AM, 33–141; Nonlinear elastic problem, Gariola BKD, 223–342), 1979.

[144] Nakahara M. *Geometry, Topology and Physics*. IOP Publishing, Bristol, UK, 1996.

[145] Nguyen QS. On the elastic plastic initial boundary value problem and its numerical integration, *Int J Num Meth Engng*, **11**, (1977), 817–832.

[146] Noll W. A mathematical theory of mechanical behavior of continuous media, *Arch Rat Mech Anal*, **2**, (1958), 197–226.

[147] Noll W. Lectures on the foundations of continuum mechanics and thermodynamics, *Arch Rat Mech Anal*, **52**, (1973), 62–93.

[148] Noll W. Materially uniform simple bodies with inhomogeneities, *Arch Rat Mech Anal*, **27**, (1967), 1–32.

[149] Ockendon H., Ockendon JR. *Viscous Flow*. Cambridge University Press, Cambridge, 1995.

[150] Olroyd JG. On the formulation of rheological equations of state, *Proc Roy Soc London*, A200, (1950), 523–541.

[151] Ottino JM. *The Kinematics of Mixing: Stretching, Chaos, and Transport*. 2nd edition, Cambridge University Press, 1997.

[152] Peach MO., Koehler JS. Forces exerted on dislocations and the stress field produced by them, *Phys Rev* II-80, (1950), 436–439.

[153] Perzyna P. Analysis of the influence of anisotropy effects on adiabatic shear band localization phenomena, *IUTAM Symposium on Anisotropy, Inhomogeneity and Nonlinearity in Solid Mechanics*, Kluwer, Dordrecht, 1995, 201–210.

[154] Peters GWM. Thermorheological modelling of viscoelastic materials. In *IUTAM Symposium on Numerical simulation of non-isothermal flow of viscoelastic liquids*, Dijksman JF and Kuiken GDC, eds., Kluwer, Dordrecht, 1995, 21–35.

[155] Petti RJ. On the local geometry of rotating matter, *General Relativity and Gravitation*, **18**, (1986), 441–460.

[156] Pietrzak G. *Continuum Mechanics Modelling and Augmented Lagrangian Formulation of Large Deformation Frictional Contact Problems*, doctoral dissertation, Department of Applied Mechanics, Lausanne, 1997.

[157] Pioletti DP., Rakotomanana LR., Benvenuti JF., Leyvraz PF. Viscoelastic constitutive law in large deformations: Application to human knee ligaments and tendons, *J Biomechanics*, **31**, (1998), 753–757.

[158] Pioletti DP., Rakotomanana LR. Non-linear viscoelastic laws for soft biological tissues *Eur J Mech A/Solids*, **19**, (2000), 749–759.

[159] Radofilao JL. *Initiation à la Géométrie Différentielle*. Editions FTM, Antananarivo, 1988.

[160] Rakotomanana RL. *Analyse Théorique et Numérique des Grandes Déformations en Mécanique des Milieux Continus*, doctoral dissertation, Department of Mechanical Engineering, EPFL, 1986.

[161] Rakotomanana RL., Curnier A., Leyvraz PF. An objective anisotropic elastic plastic model and algorithm applicable to bone mechanics, *Eur J Mech A/Solids*, **3**, (1991), 327–342.

[162] Rakotomanana RL. Contribution à la modélisation géométrique et à la thermodynamique des milieux faiblement continus, *Entropie*, **202/203**, (1997), 104–109.

[163] Rakotomanana RL. Contribution à la modélisation géométrique et thermodynamique d'une classe de milieux faiblement continus, *Arch Rat Mech Anal*, **141**, (1997), 199–236.

[164] Rakotomanana L. Heat propagation in a nonhomogeneous body. In *Proceedings of 2nd Symposium on Thermal Stresses and Related Topics '97*, 1997, 257–260.

[165] Ramaniraka A.N., *Thermo-mécanique des Contacts entre Deux Solides Déformables*. Doctoral thesis dissertation, Department of Mechanical Engineering, EPFL, 1997.

[166] Ramaniraka NA., Rakotomanana RL. Models of continuum with micro-crack distribution, *Math Mech Solids*, bf 5, (2000), 301–336.

[167] Rice JR. Path-independent integral and the approximate analysis of strain concentrations by notches and cracks, *Trans ASME J Appl Mech*, **33**, (1968), 379–385.

[168] Rocard JM. *Newton et la Relativité*. Presses Universitaires de France, Paris, 1986.

[169] Rockafellar RT., Augmented lagrangians and applications of the proximal point algorithm in convex programming. *Math Oper Res*, **1**, 2, 1976.

[170] Rockafellar RT., Wets RJB. *Variational Analysis*, Springer-Verlag, Berlin, Chern SS et al., eds., 1998.

[171] Ryhming I.L. On temperature and heat source distributions in sliding contact problems. *Acta Mechanica*, **32**, (1979), 261–274.

[172] Saanouni K., Chaboche JL., Lesne PM. On the creep crack-growth prediction by a nonlocal damage formulation, *Eur J Mech A/Solids*, **8**, 6, 1989, 437–459.

[173] Sadiki A., Bauer W., Hutter K. Thermodynamically consistent coefficient calibration in nonlinear and anisotropic closure models for turbulence. *Continuum Mech Thermodyn*, **12**, (2000), 131–149.

[174] Saffman PG. *Vortex Dynamics*. Cambridge University Press, Cambridge, 1993.

[175] Sarfarazi M. An overview of the constitutive behavior of crystalline solids, *Eng Fracture Mech*, **31**, 6, (1988), 1035–1046.

[176] Schieck B., Stumpf H. The appropriate co-rotational rate, exact formula for the plastic spin and constitutive model for finite elastoplasticity, *Int J Solids Structures*, **32**, 24, (1995), 3643–3667.

[177] Schouten JA. *Ricci Calculus*, Springer-Verlag, Berlin, 1954.

[178] Semiatin SL., Staker MR., Jonas JJ. Plastic instability and flow localization in shear at high rates of deformation, *Acta metallurgica*, **32**, (1984), 1347–1354.

[179] Serrin J. Mathematical principles of classical fluid mechanics. In *Encyclopedia of Physics VIII/1*, Flugge S, ed., Springer-Verlag, Berlin, 1959, 125–263.

[180] Signorini A. Questioni di elasticità non linearizzata, *Rend Mat*, **18**, (1959), 95–139.

[181] Simo JC., Hughes TJR. *Computational inelasticity*, Springer-Verlag, Berlin, 1998.

[182] Simo JC., Marsden J. On the rotated stress tensor and the material version of the Doyle–Ericksen formula, *Arch Rat Mech Anal*, **86**, (1984), 213–231.

[183] Simo JC., Marsden JE., Krishnaprasad PS. The hamiltonian structure of nonlinear elasticity: the material and convective representations of solids, rods and plates, *Arch Rat Mech Anal*, **104**, (1988), 125–183.

[184] Sternberg E. On the integration of the equations of motion in the classical theory of elasticity, *Arch Rat Mech Anal*, **6**, (1960), 34–50.

[185] Stout RB. Deformation and thermodynamic response during brittle fracture. In *Conference of International Union of Theoretical and Applied Mechanics*, Illinois, 1983.

[186] Suhubi ES. A generalized theory of simple thermomechanical materials, *Int J Eng Sci*, **20**, 2, (1982), 365–371.

[187] Sundararajan G. Shear bands and dynamic fracture. In *Advances in Fracture Research (Fracture 84)*, vol. 5, 1986, 3119–3126.

[188] Terrier A., Rakotomanana RL., Ramaniraka NA., Leyvraz PF. Adaptation models of anisotropic bone, *Comp Meth Biomech Biomed Eng*, **1**, (1997), 47–59.

[189] Toupin RA. Elastic materials with couple-stress, *Arch Rat Mech Anal*, **11**, (1962), 385–414.

[190] Truesdell C. *The Kinematics of Vorticity*. Indiana University Press, Indiana, 1954.

[191] Truesdell CA. *A First Course in Rational Continuum Mechanics. General Concepts*, Academic Press, New York, 1977.

[192] Truesdell C. Sketch for a history of constitutive equations. In *Rheology vol 1: Principles*, Astarita G, Marrucci G, Nicolais L, eds., Plenum Press, New York and London, 1981, 1–27.

[193] Truesdell CA., ed. *Rational Thermodynamics*, 2nd edition, Springer-Verlag, Berlin, 1984.

[194] Truesdell CA., Noll W. The nonlinear field theories of mechanics, 2nd edition. In *Encyclopedia of physics, III/3*, Flugge S, ed., Springer-Verlag, Berlin, 1965, 1993.

[195] Truesdell C., Toupin R. The classical field theory. In *Encyclopedia of Physics, vol VIII/1*, Flugge S, ed., Springer-Verlag, Berlin, 1960.

[196] Vakulenko A., Kachanov M. Continuum theory of medium with cracks, *Mechanics of Solids*, **6**, 4, (1971), 145–151.

[197] Vidal C., Dewel G., Borckmans P. *Au-delà de L'équilibre*, Hermann, Paris, 1994.

[198] Villat H. *Leçons sur la Théorie des Tourbillons*. Gauthier–Villars, Paris, 1930.

[199] Voropayev SI., Afanasyev YD. *Vortex Structures in a Stratified Fluid*, Chapman and Hall, London, 1994.

[200] Wang CC. On the geometric structure of simple bodies, or mathematical foundation for the continuous distributions of dislocations, *Arch Rat Mech Anal*, **27**, (1967), 33–94.

[201] Wang CC., Truesdell C. *Introduction to Rational Elasticity*, Noordhoff International Publishing, Leyden, 1973.

[202] Wang CC. *Mathematical principles of mechanics and electromagnetism Part A: Analytical and continuum mechanics*, Plenum Press, New York, 1979.

[203] Weitsman Y. Damage coupled with heat conduction in uni-axially reinforced composites. In *Constitutive Modeling for Nontraditional Materials, Proceedings of the Winter Annual Meeting ASME*, 1987, 161–174.

[204] Wheeler L., Luo C. On conditions at an interface between two materials in three-dimensional space, *Math Mech Solids*, **4**, (1999), 183–200.

[205] Von Westenholz C. *Differential Forms in Mathematical Physics*, North-Holland, Amsterdam, 1981.

[206] Wriggers P., Miehe C. Contact constraints within coupled thermomechanical analysis: A finite element model. *Comp Meth Applied Mech Engineering*, **94**, (1994), 301–319.

[207] Yovanovitch M.M. Recent developements in thermal contact, gap and joint conductance theories and experiment. In *Proc. 8th Internat Conf on Heat Transfer*, Tien C.L. and Carey W.P., eds., San Francisco, 1986.

[208] Zavarise G., Wriggers P., Schrefler BA. On augmented lagrangian algorithms for thermomechanical contact problems with friction. *Int J Num Meth Engng*, **38**, (1995), 2929–2949.

[209] Ziegler H., Wehrli C. The derivation of constitutive relations from the free energy and the dissipation functions. In *Advances in Applied Mechanics*, vol. 25, Wu TY., Hutchinson JW., eds., Academic Press, New York, 1987, 183–238.

[210] Zmitrowicz A. A thermodynamical model of contact, friction and wear: I Governing equations, *Wear*, **114**, 1987, 135–168.

[211] Zmitrowicz A. A thermodynamical model of contact, friction and wear: III Constitutive equations for friction, wear and frictional heat, *Wear*, **114**, (1987), 199–201.

[212] Zorawski M. *Théorie Mathématique des Dislocations*, Dunod, Paris, 1967.

[213] Zorski H. Force on a defect in nonlinear elastic medium, *Int J Eng Sci*, **19**, (1981), 1573–1579.

[214] Zubov LM. Nonlinear theory of dislocations and disclinations in elastic bodies.
     In *Lecture Notes in Physics*, Springer-Verlag, Berln, 1997.

Index

Progress in Mathematical Physics

Progress in Mathematical Physics is a book series encompassing all areas of mathematical physics. It is intended for mathematicians, physicists and other scientists, as well as graduate students in the above related areas.

This distinguished collection of books includes authored monographs and textbooks, the latter primarily at the senior undergraduate and graduate levels. Edited collections of articles on important research developments or expositions of particular subject areas may also be included.

This series is reasonably priced and is easily accessible to all channels and individuals through international distribution facilities.

Preparation of manuscripts is preferable in LATEX. The publisher will supply a macro package and examples of implementation for all types of manuscripts.

Proposals should be sent directly to the series editors:

Anne Boutet de Monvel
Mathématiques, case 7012
Université Paris VII Denis Diderot
2, place Jussieu
F-75251 Paris Cedex 05
France

Gerald Kaiser
The Virginia Center for Signals and Waves
1921 Kings Road
Glen Allen, VA 23059
U.S.A.

or to the Publisher:

Birkhäuser Boston
675 Massachusetts Avenue
Cambridge, MA 02139
U.S.A.
Attn: Ann Kostant

Birkhäuser Verlag
40-44 Viadukstrasse
CH-4010 Basel
Switzerland
Attn: Thomas Hempfling

27   WILLIAMS. Topics in Quantum Mechanics
      ISBN 0-8176-4311-7
28   OBOLASHVILI. Higher Order Partial Differential Equations in Clifford Analysis
      ISBN 0-8176-4286-2
29   CORDANI. The Kepler Problem: Group Theoretical Apects, Regularization and
      Quantization, with Applications to the Study of Perturbations
      ISBN 3-7643-6902-7
30   DUPLANTIER/RIVASSEAU. Poincaré Seminar 2002: Vacuum Energy-Renormalization
      ISBN 3-7643-0579-7
31   RAKOTOMANANA. A Geometrical Approach to Thermomechanics of Dissipating Continua
      ISBN 0-8176-4283-8
32   TORRES DEL CASTILLO. 3-D Spinors, Spin-Weighted Functions and their Applications
      ISBN 0-8176-3249-2
33   HEHL/OBUKHOV. Foundations of Classical Electrodynamics: Charge, Flux, and Metric
      ISBN 3-7643-4222-6